Life Course Research and Social Policies

Volume 11

Life course research has been developing quickly these last decades for good reasons. Life course approaches focus on essential questions about individuals' trajectories, longitudinal analyses, cross-fertilization across disciplines like life-span psychology, developmental social psychology, sociology of the life course, social demography, socio-economics, social history. Life course is also at the crossroads of several fields of specialization like family and social relationships, migration, education, professional training and employment, and health. This Series invites academic scholars to present theoretical, methodological, and empirical advances in the analysis of the life course, and to elaborate on possible implications for society and social policies applications. Topics appropriate for the series include among others: Longitudinal analyses and methods Educational and employment trajectories Occupational careers Migration Social networks and family configurations Health trajectories Non-normative events Social and individual vulnerabilities Stress over the life course Resources Accumulation of (dis) advantages and social inequalities Social mobility Personality and identity development Agency within social structures Work-family balance Social policies Volumes in this series are of interest for researchers, professionals, policy makers, and students in social sciences and related fields. Ideas and proposals for additional contributions to the Series should be sent to Laura Bernardi, Jean-Michel Bonvin, and Dario Spini, Series Editors,, Switzerland. E-mail: laura.bernardi@unil.ch. or to the publisher: Evelien Bakker, Springer Science+Business Media, P.O. Box 17, 3300 AA Dordrecht, The Netherlands. Email: Evelien.Bakker@springer.com

More information about this series at http://www.springer.com/series/10158

Julia Henke

Revisiting Economic Vulnerability in Old Age

Low Income and Subjective Experiences Among Swiss Pensioners

Springer

Julia Henke
University of Geneva
Geneva, Switzerland

ISSN 2211-7776 ISSN 2211-7784 (electronic)
Life Course Research and Social Policies
ISBN 978-3-030-36322-2 ISBN 978-3-030-36323-9 (eBook)
https://doi.org/10.1007/978-3-030-36323-9

This Springer imprint is published by the registered company Springer Nature Switzerland AG.
The registered company address is: Gewerbestrasse 11, 6330 Cham, Switzerland

Contents

Chapter 1
Introduction

The economic situation of the retired population has been the object of ongoing and lively public debates in Switzerland because of impending reforms of the old age pension scheme. Switzerland is an example that stands for many a wealthy country that has successfully driven back mass poverty in old age but that now has to tackle the flip side of longevity, the demographic aging of the population. The retirement of the numerically strong baby boomer generation and the persistently low fertility rates have raised the question about the financial sustainability of the Swiss pension system. Central to this concern is the granting of a dignified retirement for the population aged 65 and older, an objective that is laid down in the federal constitution as 'maintaining his or her previous lifestyle in an appropriate manner.'

In a time when far-reaching reforms are being forged, it is important to be clear on the kind of data that informs the public debate. In social sciences, the type of measures used importantly influences the resulting interpretation of a phenomenon. In contrast to the natural sciences, where the object of study exists prior to the observation, the creation of discrete data used in social sciences requires a number of normative decisions that are embedded in the social and legal environment of a society: there is in fact a mutual independence between the concretion of institutional practice and the reliability of societal statistics (Desrosières 2001). Rendering these normative decisions explicit and taking them into consideration in their totality is paramount in a data-driven society; in the case of the study of economic vulnerability, taking seriously the epistemological challenge that is intrinsic to all study of society is particularly pressing because the economically vulnerable are the object of social action by the welfare state.

This volume is divided into a theoretical part (I), a methodological part (II) and three parts focusing on empirical analysis (III–V): in the theoretical part we will introduce a probabilistic vulnerability framework and explain the rational for

J. Henke, *Revisiting Economic Vulnerability in Old Age*, Life Course Research and Social Policies 11, https://doi.org/10.1007/978-3-030-36323-9_1

approaching economic vulnerability from different angles of measurement according to varying degrees of subjectivity. The theoretical and conceptual contribution of this volume is to demonstrate the distinctiveness of two types of subjective social indicators, one that assesses a substance matter from a subjective perspective (self-assessed measure), and one that captures the affective response to this same observation (perceived measure). The potential of combining objective, self-assessed and perceived measures will be explained drawing from research in economics and psychology. The theoretical part concludes with the introduction of a typology that combines the three measures with the concept of vulnerability. In Part II the study design for the empirical analysis and the research questions will be introduced along with a description of the statistical techniques and the variable selection. The methods used in this volume include binary and ordinal regression models and structural equation models. The empirical part of this volume commences in Part III with a description of the prevalence of economic vulnerability – according to the three different measures – in our sample population, consisting of some 1500 Swiss residents aged 65–84 years from five cantons (Basel, Berne, Geneva, Valais, Ticino). Parts IV and V are dedicated to explaining the two subjective measures of economic vulnerability, the relationship between them and their embeddedness in the objective resource situation. This systematic analysis of the factors that cause economic strain and economic stress is filling a gap in socioeconomic and social gerontological literature as well as in Swiss policy reports on the elderly living in this country. The rational for taking an interest not only in each measurement angle separately but in the relationship between them is rooted in theoretical and empirical evidence for a potential gain of considering them jointly: first, we want to verify whether combining the measures may lead to more robust results on economic vulnerability in a wealthy country like Switzerland. Second, we posit that the divergence between the three measures sheds light on different manifestations of economic vulnerability among pensioners, allowing to better account for human agency compared to traditional measurement approaches. Thus, Part IV focuses on the analysis of the self-assessed measure of economic vulnerability, operationalized by the survey item ‘difficulties in making ends meet with current monthly income’. We find that the self-assessed measure is primarily predicted by low levels of both, income and wealth. Part V investigates more closely the perceived measure based on the survey question ‘worries about not having enough money to pay for current expenses’. Our main finding is that the influence of economic resources on the perceived measure is strongly mediated by the self-assessed measure. The final step in the pursuit of our research questions consists in operationalizing the Vulnerability Typology, which allows combining the three measures of economic vulnerability into one instrument. The result is a detailed profile of the major types of economic vulnerability that exist among pensioners in Switzerland.

Reference

Desrosières, A. (2001). Entre réalisme métrologique et conventions d'équivalence : les ambiguïtés de la sociologie quantitative. *Genèses, 43*(2), 112–127.

Part I
Theoretical Framework

Studies on the economic vulnerability of pensioners in Switzerland have mostly been conducted under the guild of poverty research. The focus has conventionally been on identifying those whose income is situated below a certain threshold, based on an absolute or a relative criterion. The limitations of monetary poverty lines – though well known in literature – are seldom discussed when interpreting the results in public reports and, more seriously, the biases this approach is likely to produce for the retired population are systematically ignored or downplayed. The economic situation of this age group is characterized by great diversity: on the resource side, the role of resources other than income is well established and on the needs side, retired adults represent some of the greatest heterogeneity of life situations compared to other age groups, including differences in health and mobility that may require more or less financial resources to attain the same average level of well-being. Thus, the multiplicity of possible constellations of resources and needs make it difficult to deduce anything meaningful about the economic quality of life from a single measure based on income.

We begin this theoretical part by introducing the concept of vulnerability and exploring what it can bring to the research on low levels of economic quality of life in old age. The concept of vulnerability lends itself well to the study of a phenomenon that is not easy to circumscribe and that is caused by the interaction of multiple factors. Moving away from a narrow focus on the risk of a drop in income after retirement, a probabilistic framework of vulnerability provides us with a heuristic tool that is furthermore able to account for individual agency. Thus uncovering those aspects that are relevant for the people categorized as 'vulnerable' will help us in constructing a meaningful measure that is apt for the analysis of cross-sectional data in the subsequent parts.

Defining the components and standards for measuring economic vulnerability is a matter of positional claims that are anchored as much in the sociocultural norms as in the objectively observable material living standard of a society. Moreover, approaching economic vulnerability from a quality of life perspective inevitably raises the question of the individual experience, a dimension that can only be assessed in a subjective manner.

Reviewing the most common measures of subjective economic vulnerability, we find that the initial disillusionment about the lack of congruence with Objective Measures has turned out to fuel fruitful interdisciplinary advances. Drawing on the work of Ruth Veenhofen and Leonard Pearlin, we suggest that there exist two different types of subjectivity that are relevant for the measurement of economic vulnerability: on the one hand, there are subjective statements of an evaluative-analytical type, referred to as *self-assessed* economic strain and on the other hand, there is an emotional response to a given situation, that is called *perceived* economic stress.

This distinction has largely gone unnoticed in social sciences though it has a great potential to enhance our understanding about the subjective experience of economic vulnerability. Based on the concept of vulnerability, we go on to demonstrate that the analysis of the combination – congruence or incongruence – of the objective, the self-assessed and the perceived measurement angle gives an indication of the risk constellation, in particular inter-individual differences in coping capabilities. Thus, we will conclude this theoretical part by postulating that the combination of the three measures into a single typology of economic vulnerability makes for an analytical instrument that allows identifying population groups that are affected by a distinct experience of economic vulnerability due to varying capabilities in terms of human agency.

Chapter 2
Vulnerability

The concept of vulnerability first emerged when natural scientists started questioning the 'naturalness' of natural disasters (O'Keefe et al. 1976). Risk research in the natural and engineering sciences has traditionally focused on the physical threat represented by extreme natural events such as tsunamis, floods and earthquakes (Burton et al. 1993). The hazard was equated with the disaster[1] it caused. This approach bore the risk of an overly technocratic approach (Hewitt 1983) and often led to generalized statements about the vulnerability of entire cities and regions, or even countries (Delor and Hubert 2000). At the beginning of the 1980s, environmental social scientists started drawing attention to the fact that not every extreme force of nature results in a catastrophe and that, in the event of a disaster, not the entire 'system' (e.g. population) was affected in the same way. The concept of vulnerability emerged as the missing link between a physical hazard and the resulting disaster (Wisner and Luce 1993). As a consequence, since the 'system' was no longer seen as passively taking in (P. Blaikie and Blaikie 1996), the focus shifted on the ability of a society to attenuate or amplify the impact through social, political and economic structures, which enable different sub-populations to respond to the hazard in different ways (Turner et al. 2003). The notion of vulnerability is now explicitly defined as a function of the external shock or stressor as well as the condition of the exposed unit. However, the focus on macro-level structures such as the national political regime and the economic system was not able to account for the diversity of experiences of the individuals at risk, as it left little room for considering their differential potential for active response (Misztal 2011). By linking the cause of vulnerability to only one or very few characteristics of the population at risk, for example certain demographic characteristics, this approach, too, was prone to causal reductionism. The challenge has henceforth been to develop models that allow for the diversity of resources (or lack thereof) that individuals and households

[1]According to Quarantelli (Quarantelli 1998) a disaster is defined as a severe disruption or rupture of the normal social order, requiring external assistance.

J. Henke, *Revisiting Economic Vulnerability in Old Age*, Life Course Research and Social Policies 11, https://doi.org/10.1007/978-3-030-36323-9_2

are able to mobilize when faced with an external risk (Delor and Hubert 2000). Different authors have adopted various ways of disaggregating the risks in order to give an account of what is generally referred to as 'human agency' and has been conceptualized as 'resilience', 'coping' or 'sensitivity', to name just the most prominent examples (Misztal 2011).

2.1 Relevance of the Concept of Vulnerability for the Study of Economic Vulnerability in Old Age

The development of the concept of vulnerability strangely echoes the evolution of the dominant view on economic difficulties in old age. In Switzerland, the time when the terms 'poverty' and 'elderly' were mentioned in one breath is not so long ago. We begin our synoptic account in the nineteenth century, when the decline of physical strength was for a large share of the elderly tantamount to economic insecurity, often characterized by a humiliating type of dependence on family, government run asylums or charity organizations (Witzig 2004). Neediness was thus seen as a 'normal' state in old age in as far as it was a direct result of the inability to provide for oneself. In Switzerland, the societal and political learning process that was addressing the structural root of the economic risk associated with old age was initiated in the 1880s (Friedli 1933) and stretched over half a century until the institution of a basic old age and survivor's pension in 1948. The role of the tax-funded pension scheme and its relationship to wage-based insurance schemes continued to divide the political spectrum until well after the overwhelming victory[2] of the wellfare state. The initially very low old age pensions were gradually raised though they never attained the level needed to grant financial security. In the meantime, occupational pensions progressively became more important, however, they were far from including the entire population.[3] In the 1960s, the understanding of social security as a mere 'basic insurance' prevailed: low levels of old age pensions were supposed to create an incentive to individual savings and wage-based provision, thus it was not in the public interest to increase pensions to a level that would provide financial security. The Federal Council supported the view that emphasized personal responsibility but also recognized the need to complement the pensions of those whose income remained insufficient. The introduction of 'supplementary benefits'[4] to old

[2] The electorate endorsed the AHV with a historic 80% (with a polling of 79%). (http://www.geschichtedersozialensicherheit.ch; 17.4.2014).

[3] Even after occupational pensions became mandatory for employees in 1985, the number of retired individuals who do not benefit from this scheme has remained high among women and people who worked in low-income jobs (Gabriel et al. 2015), as will be discussed later.

[4] Supplementary benefits are non-contributory, means tested benefits for elderly, survivors or disabled persons if they are unable to cover their basic needs in an appropriate way. These benefits are distinct from social aid benefits. Further explanations of the Swiss pension system are found on p. 28.

age pensions in 1985 was thus an attempt to do justice to individual differences in the ability to make provisions for retirement.

Providing for a safety net for those who were unable to build up the financial resources is the result of an important differentiation in our society's understanding of economic vulnerability in old age. Still, in today's official reports the discussion on economic vulnerability among pensioners continues to be focused on the structural risk induced by the loss of gainful employment at retirement age. This is reflected in the way the political discussion about poverty is framed, where the retired population is commonly referred to as an analytical category, most often in contrast to the working population. In the 'Strategy for a Swiss Aging Policy' (Conseil fédéral Suisse 2007), for example, two of four questions guiding through the part on 'The Economic Situation of Pensioners' raise the issue in terms of the relative standing of the elderly compared to the economically active population.[5] This way of posing the problem conditions the answer to be geared towards the distributional justice between the generations and less about inequality within the population aged 65 and older. This discourse is certainly not unrelated with the impending reforms of the national pension system. Switzerland is an example that stands for many a wealthy country that has successfully driven back mass poverty in old age but that now has to tackle the flip side of longevity, demographic aging. In official reports, such as the just mentioned 'Strategy for a Swiss Aging Policy', but also more specific ones as is the 'Global Strategy on Combatting Poverty in Switzerland' (Conseil fédéral Suisse 2010), experts are quick to point to summary statistics that seem to congruently speak of the relative prosperity of the population aged 65 and older: the high concentration of wealth in this population group or the relatively low number of pensioners living below the poverty-line[6] are presented as a strong evidence for the performance of the three pillar pension system. The historic improvements in the standard of living of the elderly as a result of institutional development and sustained economic growth are indeed impressive and undisputed (Gabriel et al. 2015). However, the unequal distribution of resources in this age

[5]The four guiding questions are: (Conseil fédéral Suisse 2007)

1. Ressources. Comment se situent les revenus des rentiers par rapport à ceux des actifs?
2. Perspective intergénérationnelle. Les mécanismes de solidarité mis en place dans l'AVS vont des actifs vers les rentiers. Quels sont les effets de redistribution des aînés vers les plus jeunes à travers les transferts privés comme les dons et l'héritage?
3. Egalité des chances, besoins. Quel rôle revient aux prestations du 1er pilier universel et aux prestations complémentaires pour les ménages de rentiers de condition modeste?
4. Eléments conflictuels. Niveau des rentes et financement.

[6]Both reports cite rates based on data from 1992, published in Luce (1997). The poverty line used in the 'Strategy for a Swiss Aging Policy' corresponds to the eligibility criterion for receiving supplementary benefits to old age pensions. In 1992, it was at CHF 1285.- per month and resulted in a poverty rate of 9.5% among pensioners as compared to the Swiss average of 9.8%. The 'Global Strategy on Combatting Poverty in Switzerland' cites the more conservative poverty rate at a threshold of CHF 980.- per month, which corresponds to recommendations by the Swiss Conference for Public Welfare (CSIAS). Operationalized in this way, the incidence of poverty is 3.6% for the retired population and 5.9% for the working population.

group that has been pointed out since the 1980s (Guilland et al. 1983) cannot be captured by gross categories that focus on the risk factor 'age 65+'.

The 2012 poverty report by the Federal Statistical Office represents an important step in the right direction as it takes into account further factors of exposure such as socio-demographic characteristics and sources of income. Its main findings highlight the need for a fresh look at economic vulnerability among Swiss pensioners as its figures paint a less positive picture than those presented in the two previously mentioned strategic reports[6]: according to an absolute poverty line,[7] the poverty rate among the population 65 and older was at 16.2%, which is twice the rate of the overall population (OFS 2012, p. 18). Disaggregating the retired population by household composition, the poverty rate[8] doubled for individuals living alone as compared to the average rate for the age group 65 and older (OFS 2012, p. 19). The poverty rate shoots to the same level when considering only those elderly whose primary source of income are the basic old age pensions, that is, those who do not benefit from occupational pensions nor from personal provisions.

Similar to the complex factors that interact to 'transform' an extreme natural event into a disaster, economic vulnerability among the retired emerges at the intersection between institutional structures, socioeconomic position and the life course. Because the combination of these risk factors is not random but tends to occur in an interaction between various life domains (Halleröd 2009), they need to be looked at in an integrated and systematic manner.

2.2 A Dynamic Framework of Vulnerability in Old Age

The foundational heuristic device used in this volume was specifically developed for the study of vulnerability in old age. In their conceptualization, Schröder-Butterfill and Marianti (2006) distinguish between three discrete risks, which together constitute the risk of suffering a *negative outcome*: (1) *exposure* to a threat, (2) the actual incidence of a *threat* occurring, and (3) the capacity to deal with it successfully (*coping*) by avoiding that the negative outcome actually materializes. In a mathematical language, the framework could be denoted as: $V = E + T - C$.

The 'bad outcome' we are interested in in this volume is 'exceedingly low levels of economic quality of life'; what exactly we mean by this will be explained later in this part. For the sake of this introduction to the framework, we will further define a bad outcome as an objectively measurable lack of financial resources. By 'exposure', the authors refer to individual (or contextual) characteristics of a relatively permanent nature, such as marital status or socioeconomic position, which makes

[7]The poverty line used corresponds to the recommendation by the Swiss Conference for Public Welfare (CSIAS), which was at CHF 2250.- per month for the year 2010.

[8]The concept and operationalization of the cited poverty rates will be discussed later, the important point here is the contribution of additional risk factors for understanding economic vulnerability in old age.

individuals more susceptible to encountering a certain threat. ‘Threat’ denotes the kind of negative event or process that can disrupt old-life. Some threats are rooted in social and biological processes of ageing such as bereavement[9] or illness, others arise from older people’s reduced capacity to cope with shocks and stresses that befall people at any life stage. Besides individual assets and relationships, the authors’ definition of ‘coping’ encompasses those resources represented by social networks and formal structures of social protection. By including coping as a defining component, the framework takes seriously the diversity of exposed individuals with regard to their capacity to mobilize resources when faced with a threat. However, as we will see later, the present definition of ‘coping’ is too narrow to accommodate for psychosocial mechanisms that are relevant for explaining subjective outcomes of economic vulnerability.

In real life situations, the distinction between the three types of risk factors is obviously less clean cut than presented here for the purpose of introducing the idea. A logical sequencing of the risks depends as much on the outcome in question and on the nature of the threat, for example whether it is a one-time event or a process, as on the temporal horizon under consideration. A personal characteristic such as ‘marital status’ that may at first be classified as ‘exposing’ risk factor, could subsequently show up again as ‘coping resource’. If these ambiguities render the statistical modeling task complicated, they do not diminish the usefulness of the framework for representing and thinking through the multi-causal nature of economic vulnerability. A pragmatic approach to tackling complexity is to begin with a look at real life scenarios that economically vulnerable older adults experience and from there, try to identify patterns of interaction between risk factors that are particularly common in the Swiss context.

2.3 Common Risk Patterns Among Elderly in Switzerland

In this section, we will consider what practitioners who work with economically vulnerable pensioners have identified as common risk factors among their clients and attempt to locate them within the vulnerability framework (Fig. 2.1). The data presented here stems primarily from expert interviews led in 17 cantonal Pro Senectute offices during the years 2007 and 2008. The expert interviews were transcribed and evaluated using qualitative content analysis and the main results were published in ‘Vivre avec peu de moyens. La pauvreté des personnes âgées en Suisse.’(Seifert and Pilgram 2009). We begin by a short introduction to the institutional context in which economic vulnerability unfolds. A quantitative assessment of the financial situation of pensioners in Switzerland will be presented in section 3.4.

[9]This is an example of a certain permeability of the analytical categories of risk factors: Marital status can be defined as a ‚state of exposure’ as much as a change from, for example, being married to being widowed, can constitute a ‚threat’.

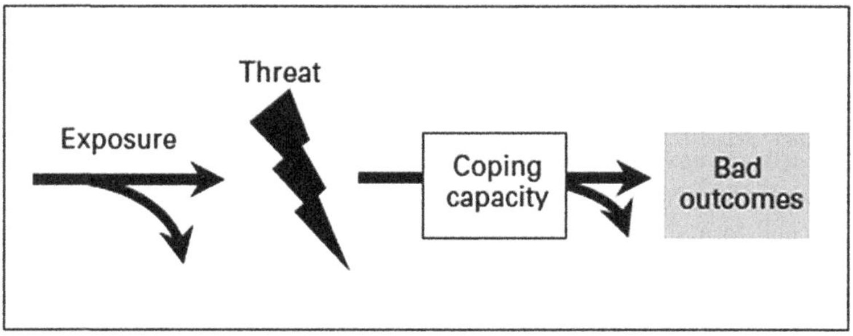

Fig. 2.1 Vulnerability framework (Schröder-Butterfill and Marianti 2006, p. 12)

The Swiss system resides on three pillars: old age, survivors and disability insurance (AHV/AVS[10]), an occupational pension plan and private investment options. Each pillar represents a different source of funding and each foresees a higher coverage level, offering the hypothetical possibility of maintaining the pre-retirement standard of living. The first pillar, AHV/AVS (henceforth referred to as 'old age pension') is expected to secure the minimum subsistence level. It is a compulsory insurance for all residents in Switzerland. The size of the pension depends on the annual average income and the number of years of contribution. In 2011, the minimum amount for a pensioner living alone was at CHF 1160.- and the maximum amount at CHF 2320.-; the maximum amount that married couples could benefit from jointly was at 3480.- (Informationsstelle AHV/IV 2010). This state-pension plan is mainly financed through the contributions paid by those insured and their employers. According to the Federal Constitution, old age pension together with supplementary benefits must be able to cover basic needs in an appropriate way. The second pillar consists of an occupational pension plan and an accident insurance, which since 1985 is compulsory for all individuals whose annual earnings are higher than CHF 20880.-. For self-employed individuals, this pension plan remains voluntary. In 2008, some 57% of pensioners were drawing on occupational pensions for their monthly income (Wanner and Gabadinho 2008).[11] When combined with benefits from old age pensions, a person is expected to earn about 60% of their final salary after retirement to help maintain their standard of living. The third pillar is made up of personal savings that are to bridge any gaps in the provision due to insufficient replacement levels by the first and second pillars. They may include life insurances or other financial assets.[12] For individuals who have reached retirement age and who do not have enough provisions stemming from the three-pillar system,

[10] For more information see http://www.ahv-iv.info

[11] This data is based on tax registers from five cantons. The actual rate of beneficiaries is expected to be higher because there exists the possibility of drawing the retirement benefits of the 2nd pillar as a lump-sum payment.

[12] A comprehensive overview can be found in Seifert and Pilgram (2009).

there exists the possibility to claim supplementary benefits. These benefits are means-tested and are only disbursed upon proof of eligibility.

Pro Senectute Switzerland is a charitable organization that is specialized in counseling retired people in precarious situations. Commissioned by the Swiss government, the organization offers individual financial assistance on a one-time off or periodic basis for pensioners who are struggling to make ends meet despite supplementary benefits to old age pensions.[13] In terms of the vulnerability framework, the 'negative outcome' we are going to focus on is the financial strain expressed in a request for help[14] at a Pro Senectute office. As the empirical evidence for the various causal influences on economic vulnerability will be discussed in more depth later, our interest here is primarily the stringing together of risk factors that have been observed in real life situations of pensioners in Switzerland, which is why only few other studies will be referenced at this stage.

Among the factors that can increase the risk of *exposure* to economic vulnerability are certain *work-life trajectories*: The Swiss pension system being financed primarily via gainful employment, a low socioeconomic position, an interrupted employment history or long spells of part-time work can lead to insufficient levels of provision for old age. Women who spent many years doing family work and migrants, who may not have paid in the complete years of contribution, are especially exposed. The same is true for people with a generally low income who experienced long-term unemployment or who were forced to retire early, be it due to health problems or the downsizing of the company. Another category of exposing factors concerns the *family situation*. Congruent with recordings in official statistics, the majority of Pro Senectute's clients live alone. Compared to couple households, single households have lesser saving capacities. Furthermore, divorce and separation can affect the financial situation on different levels: depending on the timing[15] within the life cycle, expenses related to child support, child care and the constitution of a separate household can importantly diminish the disposable income and have long-term effects on savings.

The most obvious and directly observable risk factor that can be described as *threat* is the moment of retirement: in Switzerland the transition from wage earning to pensions results in an average reduction of income of 22–30%.[16] The loss of income at retirement is especially consequential for adults that fit into one or several of the above described socioeconomic and demographic categories. The abrupt though foreseeable drop of income can require painful adjustments in life style for

[13] http://www.pro-senectute.ch/fileadmin/user_upload/_PS_Angebote/Sozialberatung/pdf/Individuelle_Finanzhilfe_-_2013.pdf (03.10.2015).

[14] In 2007, Pro Senectute counselled 30'000 older men and women, half of whom were also helped financially. (Pro Senectute Schweiz 2008).

[15] The time of occurrence of certain family events is also relevant because of the changing legal context. An example is the change in the divorce law, which regulates the splitting of occupational pensions: until the revision of the divorce law in 2000 it was not possible to benefit from the occupational pension payed in by the ex-spouse (Seifert and Pilgram 2009).

[16] Different sources arrive at different estimations, all cited in Seifert and Pilgram 2009, p. 35.

a broader share of the population, including members of the middle class. However, the ease with which this uncomfortable change can be accommodated again depends sensibly on assets and relationships as well as the ability to activate these.

Examples of strategies for *coping* with reduced income include changes in the consumption behavior, seeking support – monetary or in-kind – from family and friends or applying for supplementary benefits to old age pensions. The social and cultural capital plays an important role in these endeavors and is strongly associated with the exposure factors mentioned above.[17] Not surprisingly, among the clients counseled by Pro Senectute, people who used to work in low-income jobs and who furthermore are child-less, divorced or new in the area are clearly overrepresented. Linguistic barriers and educational deficits can be further obstacles for demanding supplementary benefits, which in the Swiss system are only disbursed upon claim (Villard 2003). Cultural capital is also relevant with regard to gender-specific roles that may hamper problem-solving skills.[18]

The vulnerability framework has now come full circle and we are currently looking at a new situation of *exposure* and possibly at a new constellation of *coping resources*. Naturally, the shrunk disposable income is most felt when faced with a sudden financial need, the occurrence of another *threat*. Especially those individuals who worked in physically demanding occupations risk to accumulate multiple types of negative outcomes – ill health and economic deprivation (Halleröd 2009), one exacerbating the other as medical treatment or adjustments in the living environment can incur unexpected and high costs. Increased costs can also be induced by changes in the family situation because of bereavement or divorce – these too represent 'threats' against which older adults are far from immune and which affect the financial situation both in a direct way as well as indirectly via the weakened coping capabilities (Höpflinger et al. 2013).

Pro Senectute's distilled experience in supporting economically challenged elderly people sheds light on the interactions between social life, health and finances that seem to be especially interrelated in the lives of older adults. While biological and social processes of aging tend to be marked by a degradation of resources in the long term, it cannot be stated enough that every experience of growing older is unique. What merits to be emphasized is therefore not so much the universality of the aging process but the idea of increased risk (Brouwer 1992). For this reasons, we argue that economic vulnerability of pensioners must be assessed in a way that accounts for the possible interactions of risks that emanate from various life domains and that can either exacerbate or attenuate the effects of low income.

[17] These overlaps point to the limitations of the framework to assign real-life situations to clearly distinguishable categories of risk.

[18] Specifically mentioned in the Pro Senectute Report is the example of widowers who are unable to perform household chores and thus need to hire someone who assists them. On the other hand, there are some widows who do not know how to manage their finances because it used to be entirely in the hands of their husbands.

2.4 Latent Versus Manifest Vulnerability

In order to make the concept of vulnerability operational, a choice needs to be made concerning the specific expression of vulnerability the researcher is going to focus on: *latent* vulnerability that is observable via the presence of risk factors or vulnerability that has materialized in a negative outcome, referred to as *manifest* vulnerability (Fig. 2.2). Schröder-Butterfill and Marianti (2006) apply their framework to both types of vulnerability, each illustrated by a case study. In a research on vulnerability to homelessness in Great Britain, the negative outcome observed at present was taken as a starting point to retrospectively identify common pathways leading to homelessness. In another example, vulnerability to a lack of care in Indonesia was examined in a probabilistic perspective: based on qualitative interviews, individuals were classified according to different risk factors constituting the overall vulnerability to a lack of care in the future. In a follow-up study, about half of the individuals that had been identified as being at risk were found to be lacking care whereas the other half had been able to activate unsuspected coping resources, obtaining access to care via their social network.

As these two case studies illustrate, it is not possible to study both sides of the equation simultaneously: individuals can either be classified according to the low quality of life they are currently suffering (manifest vulnerability) or according to their 'proximity' to future harm (latent vulnerability). Studying latent vulnerability brings with it a number of methodological challenges that are best met with a longitudinal design, ideally with some qualitative features, in order to understand the mechanisms by which risk factors – exposure, threat and coping – are combined to produce a negative outcome. For reasons that will be explained later on in this part, the starting point of this volume is the measurement of manifest vulnerability. The choice to focus this research on manifest vulnerability is also motivated by the requirement of contemporary social policy practice to identify those within the population who are currently worst off based on an observable negative outcome[19]

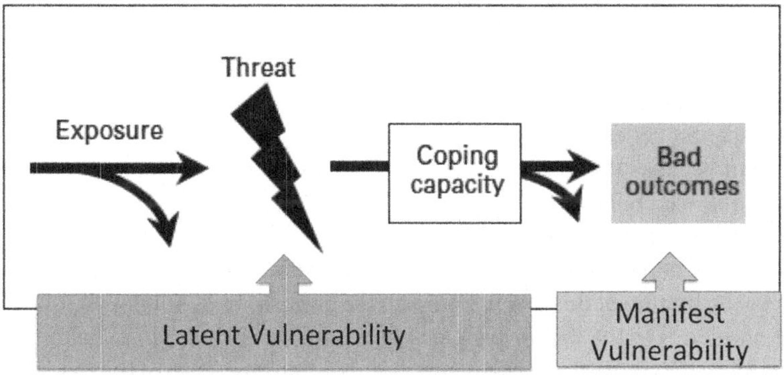

Fig. 2.2 Latent and manifest vulnerability (based on Schröder-Butterfill and Marianti 2006, p. 12)

[19] Unfortunately, it seems that the ideal of 'prevention instead of intervention' is elusive, not least because of the difficulties of identifying those in need without at the same time stigmatizing them (Phillips 2012).

and using cross-sectional data. Still, there are good reasons for starting this project by first grappling with a dynamic vulnerability framework. For one, we hope to be able to demonstrate the benefit of retrieving some information about the latent vulnerability of individuals even as we analyze the manifest outcomes. And secondly, we believe that the probabilistic features of the framework are highly relevant for the study of manifest vulnerability and should be used as a heuristic tool in order to better integrate two phases of data analysis that are frequently treated as two entirely separate steps: measurement and interpretation. The first reason will be discussed in more detail later as we will now turn to explaining why we consider an integrated approach to analysis to be particularly important when it comes to social measures.

2.5 The Convention of Equivalence Between Heterogeneous Observations

Concerning the development of an appropriate measure, understanding the interplay of risk factors is crucial because unless we understand how the negative outcome came about, we will not be able to construct a measure that registers what is most *relevant*. 'Relevance' in the sense of a systematic assessment of what is important applies to both the *components* that need to be considered in assessing economic vulnerability as an outcome, as well as the *standards* by which the economic quality of life must be considered exceedingly low. Amartya Sen put it very succinctly (italics by author): 'There are two major challenges in developing an appropriate approach to the evaluation of the standard of living. First, it must *meet the motivation that makes us interested in the concept of the living standard*, doing justice to the richness of the idea. It is an idea with far-reaching relevance, and we cannot just redefine it in some convenient but arbitrary way. Second, the approach must nevertheless be practical in the sense of being *usable for actual assessments* of the living standard. This imposes restrictions on the kinds of information that can be required and the techniques of evaluation that may be used' (Sen and Hawthorn 1987, p. 20). Depending on the 'position' or perspective adopted, different components will look relevant and thus worthy of being registered. It is precisely the normative character of any definition of economic vulnerability that requires a careful and transparent reflection on *who* may legitimately determine the domains to be considered because this seemingly 'neutral' phase of measurement demarcates the realm within which the subsequent task of analysis and interpretation takes place.

The criteria of the 'useableness' evoked by Sen has been the object of much controversy among statisticians and social scientists over the last century. Many locate the beginning of this discussion with the publishing of Stanley Smith Stevens' article entitled 'On the theory of scales of measurement' in which he outlined his famous classification of measurement levels (Stevens 1946). By extending the meaning of 'measurement' to encompass any numerical representation of an empirical structure, he laid the foundation of what has come to be called the *representa-*

tional theory of measurement (Knapp 1990). The subject matters typical of his own field of research, psychology, motivated him to push for a more liberal conception of measurement than the one defined by the classical theory, which exclusively limited measurement to ratio scaling (Michell 1986). Stevens thus introduced a definition of measurement based on correspondences or similarities (isomorphism) between objects or events and numbers. He only required that certain rules be followed in the assigning of the numbers, resulting in specific mathematical properties, scale transformation rules and the corresponding statistical operations (Stevens 1946, p. 678). He thus defined the four measurement levels – nominal, ordinal, interval and ratio – that have become standard in social sciences, though the underlying mathematical reasoning has been developed further (e.g. Luce 1997). The characteristics of these levels of measurement, shown in Table 2.1, are cumulative so that the higher order measures always include the features of lower levels measures (Stevens 1946).

The identification of categories – single or multiple/nominal or higher level – established in preparation of measurement is intrinsically linked to the analytical and interpretative task that will follow later. For the measure of manifest economic vulnerability, any of these measurement levels are a priori imaginable, but not all are 'usable' in the sense of 'meaningful' (Knapp 1990). If the study is to be relevant for policy, it must be possible to distinguish between at least two degrees of vulnerability, which points to a measurement of at least the ordinal level, though most often individuals are compared to a single benchmark below which their quality of life is considered a 'negative outcome', resulting in a dichotomous variable. The set of decisions whereby a researcher assigns individual observations to categories has pointedly been called the 'convention of equivalence between heterogeneous observations' (Desrosières 2001). At this important step of the analytical process, the dynamic vulnerability framework allows researchers to keep in mind the heterogeneity of life situations that are hidden behind the necessary but artificial categorization 'vulnerable': understanding the process that led to the observed outcome is essential for knowing how to best code the data for analysis and is indispensable for interpreting the results in a meaningful way.

Among the three types of risk under consideration in our framework, understanding the role of *coping* is especially important for the interpretation of results because it represents the locus of human agency: failure to do so would mean to override the

Table 2.1 Levels of measurement (based on N. Blaikie 2003, p. 27)

Level	Description	Types of categories
Nominal	Categories for classification	Homogeneous, mutually exclusive and exhaustive
Ordinal	As for nominal measures but ordered	Categories lie along a continuum but assumed to be equidistant
Interval	As for ordinal measures, but on a contrived measurement scale	Categories may be discrete or continuous with arbitrary intervals and zero point
Ratio	As for interval measures	Categories may be discrete or continuous but with an absolute zero point

criteria of 'relevance' that is to inform the construction of a measure of quality of life. In the next section, we will again look at risk patterns, focusing this time on the theoretically identifiable risk constellations according to the framework in order to examine the potential insight that can be gained from understanding the latent features of vulnerability, in particular the influence of differential coping resources.

2.6 Coping: Taking Human Agency Seriously

Table 2.2 schematically illustrates the possible constellation of risk factors that result in a positive or a negative outcome. For the sake of this illustration the outcome variable is assumed to have a threshold based on an objective measure of economic vulnerability, such as monthly income or consumption expenditure. In order to reduce the number of possible combinations, exposure and threat are depicted as a combined risk.[20] The arrow indicates the presence of a risk factor if it is pointing downwards (box shaded in grey), and absence, if it is pointing upwards. In the case of the risk factor 'coping', the arrow pointing downward is to be read as 'unsuccessful coping effort'.

The interpretation of constellations 1 and 2 is very straightforward: Either there was no risk of 'exposure' and 'threat' or these risks were counteracted by the activation of 'coping'. Both risk constellations result in a 'non-vulnerable' outcome.

A researcher looking at an individual with a 'vulnerable' outcome does not a priori know whether she is dealing with a member of group 3 or 4 since the latent risks are a black box at the time when the outcome is manifest. However, in this case not knowing the possible pathways (3 *or* 4) may lead to severe misinterpretation of the results. In fact, individuals concerned by risk constellation 3 could be characterized by either *absence* or *insufficiency* of assets and relationships, *ineffectiveness* of coping strategy or the *inability* to mobilize them. To put it bluntly: they either didn't have what it took in the first place or they tried but failed in the attempt to pull themselves out of the situation. What these options share in common is that there was a willingness to change something about the circumstances imposed by exposure and threat. In contrast, group 4 represents individuals that refrain from employing coping strategies because they either *do not realize* that they are vulner-

Table 2.2 Schematic overview of risk constellations

group of respondents	*latent vulnerability*		*manifest vulnerability*
risk constellation	**exposure + threat**	**coping**	**outcome (objective)**
1	↑	na	Non-vulnerable
2	↓	↑	Non-vulnerable
3	↓	↑	Vulnerable
4	↓	↓	Vulnerable

[20] For reasons of readability the matrix does not differentiate between varying degrees or accumulation of risk so that ,coping' always results in a compensation of the additive risks of exposure and threat.

Table 2.3 The role of auto-evaluation for coping

risk constellation	exposure + threat	subjective evaluation	coping	outcome (obj.)
4a	↓	neutral/positive	na	vulnerable
4b	↓	negative (adversity)	↓	vulnerable

able 4a) and therefore do not take any action, or they *do not want* to mobilize the resources (4b) (Table 2.3).

In the case of a person who does not recognize his or her economic circumstances as requiring improvement, reference to the concept of 'coping' only makes limited sense because 'coping' is by definition an active response that presupposes intentionality. Only if a situation is *recognized* as being *adverse* can we talk about an active response (Schafer et al. 2009). According to Schafer and colleagues,[21] subjectively recognized misfortune, which they call 'adversity', needs to be distinguished from mere disadvantage, which is an (objective) unfavorable position in a stratified hierarchy. In the case of economic vulnerability, we can thus say that only individuals who experience their economic quality of life as inadequate can chose to mobilize coping resources. We therefore conclude that, in addition to the diversity hidden behind the label 'vulnerable' that is represented by groups 3 and 4, there is further heterogeneity among respondents of group 4 with regard to the perception of their circumstances. Tying in with the guiding principle for the construction of an indicator put forward by Amartya Sen, this raises the question whether the label 'vulnerable' is actually *relevant* for sub-group 4a.

The second sub-group worth considering concerns those individuals who, despite recognizing their situation as being adverse, chose[22] not to use their coping resources (4b). The consideration of abstaining from coping is particularly significant for older adults and has been the object of considerable research in social gerontology.[23] The propensity to mobilize resources can be influenced by psychological attributes, culture and worldview. Also, the perception of one's time left alive may cause an old person to prioritize spending time and energy on enjoying existing relationships rather than making the effort it would take to resist a given adversity (Schafer et al. 2009). The best-known theory in this context is Baltes' model of selective optimization with compensation (Baltes and Baltes 1993). This model explains how adults confronted with age-related changes tend to become increasingly *selective* about their activities and reorder their priorities according to what they are still able to do. These things they will try to *optimize* through investing resources and adapting behavior and by doing so

[21] Schafer and colleagues utilize the term ‚resilience' defined as the ‚volitional, purposive set of responses undertaken by actors who perceive their condition to be undesirable' (Schafer et al. 2009, p. 232) which is comparable to our definition of ‚coping'.

[22] I am aware that in practice it is difficult if not impossible to clearly distinguish between a genuinely 'free' choice and a 'conditioned' one, but the point here is that if we are to take the idea of human agency seriously, we must – at least in theory – allow for the individual to opt for any situation even if others (society, scientists etc.) would consider it suboptimal or even a 'bad outcome'. The psychology underlying such choices will be discussed later.

[23] Though research has mostly focused on health-outcomes and mobility, the theoretical considerations are of general validity.

they are able to *compensate* for the loss of some other abilities. According to this theory, changes in priorities (and thus potentially also changes in relative desirability of 'outcomes' and consequently coping behavior) is part of the adjustments that people make when faced with cognitive and physical limitations.

Thinking through the risk constellations in a systematic manner and laying out the possible reasons behind coping and absence thereof promises important insights for understanding the observed outcome because it sheds light on the *heterogeneity* of respondents that are classified as 'vulnerable'. Moreover, it reveals the compromises with regard to the principle of *relevance* when relying exclusively on objective variables for assessing economic vulnerability, as exemplified by the incongruence of the label 'vulnerable' with the subjective assessment by respondents of group 4a. This incongruence, on the other hand, represents an additional source of information that can be gained when interpreting objective and subjective measures jointly: if we not only have 'matter of fact' information about the economic circumstances of a person but are able to compare this data with the 'experience' this situation produces, we are able to give some account of human agency – the very feature that distinguishes the concept of vulnerability from, for example, the concept of poverty. Thus, integrating measures based on subjective assessments opens up a window into the 'black box' of latent vulnerability even while focusing on the manifest outcome. The next section is dedicated to a discussion of what we mean by 'objectivity' and 'subjectivity' in the context of evaluative statements and some reflections on what role these angles of measurement may play in research on economic vulnerability.

References

Baltes, M. M., & Baltes, P. B. (1993). Psychological perspectives on successful aging: The model of selective optimization with compensation. In P. B. Baltes & M. M. Baltes (Eds.), *Successful aging: Perspectives from the behavioral sciences* (pp. S. 1–S. 34). Cambridge: Cambridge University Press.

Blaikie, N. (2003). *Analyzing quantitative data: From description to explanation*. London: SAGE.

Blaikie, P., & Blaikie, P. M. (1996). *Vulnerabilidad: El entorno social, político y económico de los desastres*. London: Soluciones Practicas.

Brouwer, A. (1992). The nature of ageing. In M. A. Horan & A. Brouwer (Eds.), *Gerontology: Approaches to biomedical and clinical research* (pp. S. 1–S. 10). London: Edward Arnold.

Burton, I., Kates, R. W., & White, G. F. (1993). *The environment as hazard* (2nd ed.). New York: Guilford Press.

Conseil fédéral Suisse. (2007). *Stratégie en matière de la politique de vieillesse, Rapport du Conseil fédéral donnant suite au postulat 03.3541 Leutenegger Oberholzer du 3 octobre 2003*. Bern: Conseil fédéral Suisse.

Conseil fédéral Suisse. (2010). *Stratégie globale de la Suisse en matière de lutte contre la pauvreté, Rapport du Conseil fédéral en réponse à la motion (06.3001) de la Commission de la sécurité sociale et de la santé publique du Conseil national (CSSS-N) du 13 janvier 2006*. Bern: Conseil fédéral Suisse.

Delor, F., & Hubert, M. (2000). Revisiting the concept of 'vulnerability'. *Social Science & Medicine, 50*(11), 1557–1570.

Desrosières, A. (2001). Entre réalisme métrologique et conventions d'équivalence: les ambiguïtés de la sociologie quantitative. *Genèses, 43*(2), 112–127.

Friedli, O. (1933). *Die Grundlagen einer schweizerischen Alters-, Hinterlassenen-und Invalidenversicherung*. Affoltern am Albis: Buchdruckerei Dr. J. Weiss.

Gabriel, R., Oris, M., Studer, M., & Baeriswil, M. (2015). The persistence of social stratification? A life course perspective on poverty in old-age in Switzerland. *Swiss Journal of Sociology, 41*(3), 465–487.

Guilland, P., Eichenberger, P., Lüthi, P., Baigger, G., & Guignet, J. M. (1983). *Rentiers AVS, une autre image de la Suisse: inégalités économiques et sociales*. Lausanne: Réalités Sociales.

Halleröd, B. (2009). Ill, worried or worried sick? Inter-relationships among indicators of wellbeing among older people in Sweden. *Ageing & Society, 29*(04), 563–584.

Hewitt, K. (1983). *Interpretations of calamity from the viewpoint of human ecology*. Boston: Allen & Unwin.

Höpflinger, F., Spahni, S., & Perrig-Chiello, P. (2013). Persönliche Bilanzierung der Herausforderungen einer Verwitwung im Zeit- und Geschlechtervergleich. *Zeitschrift für Familienforschung, 25*, 267–285.

Informationsstelle AHV/IV. (2010). Änderungen auf 1. Januar 2011 bei Beiträgen und Leistungen AHV – IV.

Knapp, T. R. (1990). Treating ordinal scales as interval scales: An attempt to resolve the controversy. *Nursing Research, 39*(2), 121–123.

Luce, D. R. (1997). Quantification and symmetry: Commentary on Michell, quantitative science and the definition of measurement in psychology. *British Journal of Psychology, 88*, 395–398.

Michell, J. (1986). Measurement scales and statistics: A clash of paradigms. *Psychological Bulletin, 100*(3), 398–407. https://doi.org/10.1037/0033-2909.100.3.398.

Misztal, B. A. (2011). *The challenges of vulnerability: Search of strategies for a less vulnerable social life*. Basingstoke: Palgrave Macmillan.

O'Keefe, P., Westgate, K., & Wisner, B. (1976). Taking the naturalness out of natural disasters. *Nature, 260*(5552), 566–567.

OFS. (2012). *Pauvreté en Suisse: concepts, résultats et méthodes. Résultats calculés sur la base de l'enqête SILC 2008 à 2010*. Neuchâtel: Office fédérale de la statistique.

Phillips, D. (2012). *Quality of life: Concept, policy and practice*. London: Routledge.

Pro Senectute Schweiz. (2008). *Leistungsbericht 2007*. Zürich: Pro Senectute Schweiz.

Quarantelli, E. L. (1998). *What is a disaster?: A dozen perspectives on the question*. New York: Routledge Chapman & Hall.

Schafer, M. H., Shippee, T. P., & Ferraro, K. F. (2009). When does disadvantage not accumulate? Toward a sociological conceptualization of resilience. *Swiss Journal of Sociology, 35*(2), 231–235.

Schröder-Butterfill, E., & Marianti, R. (2006). A framework for understanding old-age vulnerabilities. *Ageing & Society, 26*(01), 9–35.

Seifert, K., & Pilgram, A. (2009). *Vivre avec peu de moyens. La pauvreté des personnes âgées en Suisse*. Zürich: Edition Pro Senectute.

Sen, A., & Hawthorn, G. (1987). The standard of living. In *The standard of living*. Cambridge: Cambridge University Press.

Stevens, S. S. (1946). On the theory of scales of measurement. *Science, 103*(2684), 677–680. https://doi.org/10.1126/science.103.2684.677.

Turner, B. L., Kasperson, R. E., Matson, P. A., McCarthy, J. J., Corell, R. W., Christensen, L., et al. (2003). A framework for vulnerability analysis in sustainability science. *Proceedings of the National Academy of Sciences, 100*(14), 8074–8079.

Villard, I. (2003). Warum werden Ergänzungsleistungen nicht beansprucht? *Soziale Sicherheit, 1*, 30–32.

Wanner, P., & Gabadinho, A. (2008). Die wirtschaftliche Situation von Personen im Ruhestand: neue Daten, neue Prioritäten. *Soziale Sicherheit CHSS, 3*, 137–142.

Wisner, B., & Luce, H. R. (1993). Disaster vulnerability: Scale, power and daily life. *GeoJournal, 30*(2), 127–140.

Witzig, H. (2004). Einsamkeitserfahrungen von armen Kindern, Frauen und Männern im 19. Jahhunder. In C. Schweiz (Hrsg.), *Sozialalmanach 2005* (S. 89–98). Luzern: Caritas.

Chapter 3
Objectivity and Subjectivity

3.1 Dimensions of Measurement

In literature on subjective measures of economic vulnerability, there is usually little thought on the 'nature' of subjectivity or objectivity. This is most notably expressed in generalizing statements about the potentials and shortcomings of subjective measures. Looking carefully at the spectrum of currently existing social indicators, it becomes obvious that there exists degrees of subjectivity and objectivity and that, in fact, most survey measures of use for evaluation of economic vulnerability consist of a mixture between the two angles of measurement.

In her work on measuring subjective well-being, Ruth Veenhoven proposes that in social sciences, variables can be classified by the *substance* they measure as well as by their *mode of assessment* (Baltes and Baltes 1993), both of which can be either subjective or objective. The combination of these two dimensions yields four types of indicators at the intersection of mode and substance (Fig. 3.1). Veenhoven (2007) cites 'wealth' as example for an *objective* substance that can either be evaluated by a person's banking statement (type 1[1]) or based on the persons' own evaluation of his or her wealth (type 2). As example for a *subjective* substance, the author cites happiness, a feeling that could theoretically be assessed based on data about the incidence of suicide (type 3) or alternatively by relying on self-reports of happiness (type 4[2]) (Veenhoven 2007).

---

[1] According to a strict interpretation of this typology, most poverty indicators that are commonly referred to as 'objective' contain a subjective component if the information is elicited by asking respondents via a survey questionnaire. In the same article, Veenhoven (2007) goes on to present a more nuanced typology with various degrees of objectivity and subjectivity, which will not be discussed in this volume. Based on his argument of degrees of objectivity in the mode of assessment, we will continue to refer survey questions on income and wealth as 'objective' indicators.

[2] Veenhoven's example obviously features indicators that are measured at an individual level (type 4) with aggregated data (type 3).

J. Henke, *Revisiting Economic Vulnerability in Old Age*, Life Course Research and Social Policies 11, https://doi.org/10.1007/978-3-030-36323-9_3

Fig. 3.1 Dimensions of social indicator measures: substance and mode of assessment (based on based on Veenhoven 2007)

SUBSTANCE	MODE of Assessment	
to be assessed	Objective	Subjective
Objective	1)	2)
Subjective	3)	4)

In particular the distinction between two measures that can be classified by different degrees of subjectivity-objectivity will proof fruitful for our discussion: in this volume we will refer to type 2 in Fig. 3.1 as *Self-Assessed Measure* (only the mode of assessment is subjective) that is less subjective than type 4 (both, substance and mode of assessment are subjective), which will be called *Perceived Measure.*

In the context of measuring economic quality of life (or low levels thereof), the *objective measure* could be measured by income-based poverty lines or a deprivation index. The self-assessed measures would be operationalized by survey questions that invite respondents to make an evaluation of their financial situation based on a *qualitative criteria of evaluation* ('Given your monthly income, would you say that you are able to make ends meet?'), a *scale (*'Please indicate where you would locate your health status on a scale from 0 to 100') or by asking for a *comparison (*'Would you say that your financial situation has improved or deteriorated compared to how it was one year ago?'). In contrast, *Perceived Measures* provide information about how respondents *feel* about the matter, for example, how satisfied with or worried and stressed they are. The correlation between respondents' answers to these two types of questions is likely to be strong but they assess a different substance and for the purpose we are pursuing in this research it is important to distinguish between the two.

Though labeled differently, the theoretical distinction between the self-assessed and the perceived angle plays an important role Lazarus' theory of appraisal (Lazarus and Folkman 1984). This theory, which emphasizes the role of cognition in the evaluation process, will be addressed in more detail later, but for the sake of the present argument it is worth mentioning that Lazarus' theory is able to explain why a situation that has been evaluated as potentially stressful must not necessarily result in 'feelings' of stress. In fact, upon assessing whether a given situation is 'negative', individuals factor into their overall judgment of the situation their perceived ability to deal with it successfully (Lazarus and Folkman 1984).[3]

For the purpose of illustrating the plausibility and relevance of the distinction between self-assessed and perceived measures of economic vulnerability, we propose three stylized examples of individuals who are economically vulnerable according to one or several of the above identified measures of economic

[3] Lazarus calls the two steps ‚primary' and ‚secondary' appraisal.

vulnerability. The first two individuals are both objectively vulnerable, having less than CHF 2400 per month and disposing of no or little savings.

Imagine a young woman who has recently left her parents' home to take up her studies in a different city. She benefits from a scholarship from her home canton as well as a modest financial support from her parents. Combined with the money she makes as a waitress in a coffee shop, she has about CHF 2000 a month to live on. Though she has to watch her spending even with regard to the type of groceries she can afford, she is not worried about her modest budget. The slight difficulties she has in making ends meet every month actually fill her with a sense of pride that she manages to take care of herself and be independent.

Now, let us picture a woman, born in the late 1920s, living in a rural area in Switzerland. As a mother of four children in a traditional family constellation with her husband being the (only) bread-winner, she is used to have to manage with little money. Now, a widow of advanced age, her lifestyle of modesty bordering on self-denial is a deeply held personal value. She has a great relationship with her children, who live in the area. If we asked her the survey questions from which our indicators are derived, she would probably say that she is finding it quite easy to make ends meet. However, being aware of the tightness of her budget, she has already thought of the possibility of having to ask her children for assistance in the case of an unexpected expense. Though she knows that they would gladly help her, the idea of asking for help is displeasing to her.

At last, we are thinking of a man in his mid-sixties of a lower middle class background. Because of frequent and long-term disruptions in his employment history, he finds himself with a relatively small work pension, leaving him with a monthly income that is by some CHF 500 lower than his pre-retirement income, positioning him just above the poverty line. He has been struggling to maintain his unassuming habitual life style. Given the inherited fix costs of his previous life style, his primary strategy is to cut down on leisure and social activities. He finds it humiliating to constantly have to calculate before any minor purchase and it causes him to lose sleep at night.

These three stylized examples are plausible manifestations of economic vulnerability in Switzerland, illustrating the relevance of distinguishing between self-assessed and perceived measures for the study of a phenomenon for which the individual experience is central, as is the case for low levels of economic quality of life. The example of the student draws our attention to the fact that the timing within the life course and future prospects are crucial factors that influence the individual experience. The role of social support and the perceived efficacy and legitimacy of financial dependence are also evident in the internal processing of the elderly widow. The case of the recently retired man reminds us that psychological suffering from economic vulnerability does not start below the poverty line.

***Excerpt from Amartya Sen* 'Positional Objectivity'** in *Philosophy & Public Affairs* (Sen 1993)

It might be argued that observational statements (…) are claims 'merely' about appearance, as opposed to 'reality.' It might be tempting to take the view that the subject matter of such statements is not knowledge of the world as it is, only as it appears, so that the objectivity in question is not about the world as it is. (p. 129)

Observations are unavoidably position-based, but scientific reasoning need not, of course, be based on observational information from one specific position only. There is need for what may be called 'trans-positional' assessment - drawing on but going beyond different positional observations. The constructed 'view from nowhere' would then be based on synthesizing different views from distinct positions. The positional objectivity of the respective observations would still remain important but not in itself adequate. A trans-positional scrutiny would also demand some kind of coherence between different positional views. The 'trans-positional' assessment that we might undertake can lead to a broader understanding that makes sense of the respective (and possibly divergent) positional observations. (p. 130)

'But the scientist's ability to reason trans-positionally depends on what else she knows and on the type of reasoning she is able to use, and these, in a broad sense, are also positional features. Even the 'conceptual schemes' that mediate our understanding of the world can be fruitfully seen as general positional characteristics related to acts of observation and reflection. (p. 131)

Another line of argument for the relevance of the distinction between self-assessment and perception is found in Amartya Sen's idea of *positional* objectivity. In an article published in 'Philosophy and Public Affairs', Sen deconstructs the traditional understanding of objectivity as 'invariance with respect to individual observers and positions' (Sen 1993, p. 126). He argues that a proper concept of objectivity needs to account for the positional dependence of any observation in the sense of 'a view from a delineated somewhere' (Sen 1993, p. 127). In other words: there is no observational claim that is not made by looking at the matter from a specific position, while the term 'position' can refer to any parameter that may influence observation. Consequently, the epistemological reliability of the observational claim can theoretically be verified if other people placed in a similar position make the same observation. In the self-assessment of one's economic situation, these parameters of 'position' may include economic resources, social position and life course experiences.

Applying the same reasoning to *objective* indicators points to the fact that the perspective captured by this type of measure, too, is conditioned by a specific set of parameters. It is not a 'view from nowhere' (title of a book by Thomas Nagel 1986, cited in Sen 1993, p. 127) but from a delineated position corresponding to specific parameters that seem of relevance to the person inquiring. In the case of objective measures of economic vulnerability, experts usually determine these parameters. In the Swiss context, a private association, the conference for public welfare (SKOS/CIAS),[4] plays a central role in defining the minimum acceptable living standard. In the absence of a national law on public welfare, SKOS/CIAS coordinates the cantonal institutions in establishing guidelines and issues recommendations that take into account regional conditions with the objective of a harmonized management of

[4] Schweizerische Konferenz für Sozialhilfe (SKOS), Conférence Suisse des Institutions d'Action Sociale (CIAS).

welfare benefits. Most cantonal authorities recognize these recommendations as eligibility criteria for welfare by enshrining them in their laws and regulations.[5]

In empirical studies, the inherent 'positionality' of any measure is rarely an object of discussion. This is as true in research on 'economic' quality of life as in studies of the more general construct of 'quality of life' and 'well-being'. In the next section, we are going to confront the theoretical considerations made so far with empirical findings. We begin our review with how 'objectivity' and 'subjectivity' have been tackled in the literature about these more general concepts because – whether it is conceptualized as a constituent or as a prerequisite – *economic* quality of life is invariably connected to *overall* quality of life. As we study how objectivity and subjectivity interact within the broader stream of quality of life research, we will without fail come across the major theoretical approaches that will be relevant for understanding the relationship between the three angles of measure in the economic domain of quality of life.

3.2 Objective and Subjective Concepts of Quality of Life

In Des Gasper's review of the various conceptions of quality of life and well-being, he describes the 'fallacy of misplaced concreteness' as the temptation of making a rhetorical shortcut to presenting 'quality of life' as if it were a concrete entity (Gasper 2010). In this section, we will not be able to avoid some degree of generalization because the body of literature is very large and spans multiple fields. Our focus here is on the way that philosophical thought and empirical research on quality of life research has treated the issue of objectivity and subjectivity.

Intellectual thought on the 'good life' is as old as recorded history and it has preoccupied some of the most outstanding minds in philosophy and social sciences. What we understand by 'quality of life' is essentially a function of what we consider to be of greatest ultimate value. Our worldview – including religious and ideological orientation – determines which aspects of the human life and experience we attribute an intrinsic value to, meaning, a value that is not seen as a means to yet another greater end. Also, whether we approach the question from an individual or a societal perspective will point to different directions. Not surprisingly, no common theoretical ground has been found on which to construct a generally accepted 'neutral' concept of quality of life. This frequently ignored circumstance prompted David Phillips to provocatively state that there are strictly speaking no 'objective' – in the sense of value-independent – measures of well-being and that it would be more accurate to refer instead to 'collectively subjective' measures (Phillips 2012), that is indicators, which possess more legitimacy because they have been established in a societal process.

[5] http://skos.ch/sozialhilfe-und-praxis/haeufig-gestellte-fragen/ (19.11.2014).

Reviewing the spectrum of contemporary concepts of quality of life, several axes can be identified that shape the academic discussion surrounding it. The angle of measure, objective or subjective, is among the primary axes that have proven to be relevant (Noll 2004), transcending others such as the underlying moral philosophy or demarcations along disciplinary lines.

If we consider how these two measurement angles have been operationalized across disciplines, we find that global quality of life has most frequently been studied from an *objective angle* by the indicators of longevity and income and *subjectively* by self-reports of happiness, satisfaction and pain (Gasper 2010). Within the latter approach, focusing on individuals' own verdicts of well-being, two streams can be distinguished, eudaimonism and hedonism, both having emerged from ancient schools of thought.

Eudaimonism is central to Aristotelian ethics and has experienced a revival in the twentieth century under the term 'human flourishing' (Phillips 2012). 'Virtue' in an eudaimonic perspective is when something (or a person) performs the function well for which it was made. According to Aristotle, what is peculiar to human beings is their ability to reason; consequently, a life 'worth living' is one that is in accordance with reason (Nussbaum 1992). Assuming that only the individual person is in a legitimate position to determine the purpose for which she or he was made, the indicator used for assessing subjective well-being in an eudaimonic value system is the sense of fulfillment and the satisfaction with life (Athanassoulis 2010). Assessing the same life in terms of subjective well-being but according to a hedonistic outlook on life may come to very different conclusions: the hedonistic theories define happiness as 'pleasure' and posit that 'pleasure' and the 'avoidance of pain' – both, mental and physical – are the only things of ultimate importance (Phillips 2012).

Psychological research has brought forth considerable evidence that the understanding of subjective well-being brought forth by these two philosophical schools are congruent with the complex ways in which humans experience their own well-being. In the realm of emotions and moods, the discovery that positive and negative affect are not simply opposite poles on one and the same dimension was path-breaking (Weijers 2011) as it implied that the overall subjective well-being was a composite construct with multiple dimensions (Bradburn 1969). In the 9070s, evidence was found for a cognitive evaluation of 'life satisfaction' that is distinct from the emotion of 'happiness' (Andrews and Withey 1976). The distinctiveness of pleasant and unpleasant affect and life satisfaction was confirmed by subsequent research with more sophisticated methods and proved to be especially significant for a longer time-frame (Weidekamp-Maicher 2008). However, the divide encountered between different philosophical conceptions of 'the good life' has not been done away with in science. This is reflected in the fact that different streams of researchers dismiss parts of the components as irrelevant for global subjective well-being: for example, the validity of current mood states as indicators for subjective well-being is questioned because it disregards the vital contribution of having a meaningful purpose in life (Diener et al. 1999). Others argue just the opposite, insisting that reports of current perception of an experience are the most reliable indicator of subjective well-being (Kahneman 2003). Moreover, even among those who in principle acknowledge the validity of these diverse measures there is controversy about whether subjective well-being is to be seen as a unitary construct or whether its various components should be treated as specific constructs, given that the correlation between the different components has proven to be weak (Kim-Prieto et al. 2013).

While psychologists' main interest has been the experience of the individual, thus privileging what we have called the 'perceived' measurement angle, economists are known to focus more on the objective angle, specifically on monetary indicators. This has not always been the case. The interest of economists in matters of quality of life can be traced back to the very beginnings of economic theorizing, more precisely, to the moral philosophy of utilitarianism at the end of the eighteenth and mid nineteenth century (Frey and Benz 2002). Its founding father, Jeremy Bentham, took it for granted that the assessment of 'utility' was not only *shaped by* but essentially *consisted of* subjective appraisals of well-being. He proposed to orient and judge social policy as well as individual action according to the hedonistic maxim of 'the greatest happiness of the greatest number of people' (Alexander 2008). One of the core concerns of members of this school of thought was to make 'happiness' – defined as a mental state of pleasure – the object of a nuanced empirical analysis by differentiating the different types of utility of which there existed 'pleasure of sense, wealth, skill, amity, a good name, power, piety, benevolence, malevolence, memory, imagination, expectation, relief and the pleasures dependent on association' (Bentham 1789/1996, p. 34–35 in Frey and Benz 2002, p. 5).

In the nineteenth century, economic thought on well-being distanced itself from hedonism[6] and the psychological foundations of appraisal and embraced an increasingly positivist view of 'utility' (Robbins 1932/2007), defined by the 'satisfaction of preferences' that are observable objectively and no longer by the feeling this satisfaction evokes (Hausman 2013). The same axiom is used for deriving 'social welfare' from the consumption behavior of households (Frey and Stutzer 2002). In the 1970s, before the backdrop of a general dissatisfaction with GDP growth as sole indicator of societal progress, a surge of interest in alternative measures gave birth to what came to be known as the 'social indicator movement' (Duncan 1969, p. 1, in Land 2000, p. 2). The question of how objective and subjective measures relate to each other turned out to be a catalyst for the foundation of quality of life research as a discipline in its own right. The assumption that subjective well-being was a direct effect of observable circumstances led to an integration for psychological measures. Behind this scientific curiosity was the hope that these scientifically defined non-monetary indicators of quality could serve to orient policy. However, the lack of empirical evidence for the relationship between objective and subjective indicators disillusioned those who had envisioned using subjective indictors as a measuring stick for the overall quality and evolution of a society (Weidekamp-Maicher 2008). It was the economist Richard Easterlin who most famously expressed this perplexing

[6] Some economists in the twenty-first century continue to argue in favor of hedonism but they remain the minority.

finding: in the essay 'Does economic growth improve the human lot?' he found that increases in income only lead to temporary or small improvements in subjective well-being and general life satisfaction (Easterlin 1974). Since both measurement angles clearly represent different aspects of economic quality of life, something might be gained by looking at them in combination and paying attention to what the discrepancy could mean.

3.3 Incongruence Between Objective and Subjective Measures

Since Easterlin's poignant statement of the income-happiness paradox, the weak anchoring of subjective appraisals in the actual circumstances has spurred much interest among economists. Some of the most promising explanations for the weak relationship between objective and subjective measures have come from attempts to expand the narrow behavioral model of 'homo economicus' according to which higher income should lead to increased levels of subjective well-being because more goods and services can be consumed (Frey and Stutzer 2002). The satisfaction of preferences theory is able to give some account for the incongruence by referring to the law of diminishing marginal returns to income. This law states that an increase in one factor of production (income) – all other factors remaining constant – will eventually yield a lower per-unit return. For this reason, the same increase in income does not necessarily boost every person's well-being in the same manner, it actually depends on the wealth level he or she has already got: an increase in wealth leads to a greater increase of subjective well-being among less wealthy people compared to wealthier people. Several theoretical models have sprung up at the intersection between social psychology and economics, contributing to enlarging the concept of 'preferences' to take into account alternative types of individual utility with the objective of finding a more convincing model for explaining the relationship between subjective and objective quality of life (Frey and Stutzer 2002).

The concept of 'reference utility' is probably the most encompassing theoretical framework, accommodating several strands of explanation for the objective-subjective incongruence while drawing on different disciplinary traditions. At the heart of the theory, we come across the positional concerns we reflected on earlier: individuals value their well-being by comparing their actual position with regard to different reference points, which influence their aspirations. The earliest sign of a scientific interest for the role of aspirations is usually attributed to Torstein Veblen. In his work, Veblen describes the phenomenon of 'conspicuous consumption', corresponding to the modern motivation for purchasing goods and services as a demonstration of wealth, which had become the basis for social status in industrialized societies (Veblen 1899/2005). First formalized in economics as a hypothesis of relative income (Duesenberry 1949) aspirations framed by social comparison were also adduced to explain the afore mentioned 'Easterlin paradox' (Corazzini et al. 2011): though wealthy people are on average happier, a general raise in income of *all* does not increase the happiness of all, because what is more important for subjective well-being is to have a higher income than others (McBride 2001). The idea that experience of economic quality of life is a result of mechanisms of social comparisons has influenced the way economic vulnerability has been conceptualized, most notably in the development of a relative deprivation approach (Townsend 1979; Boarini and d' Ercole 2006) but also in advances of subjective poverty lines (Flik and Praag 1991), both of which will be discussed later.

Individual aspirations are also affected by past consumption levels (Frey and Benz 2002). Theories that explain comparisons a person makes in reference to their own biography are usually summed up under the term 'theory of adaptation'. Traditionally, these theories were developed by psychologists but they have increasingly drawn the attention of economists interested in happiness and welfare. To be mentioned first is the theory of the 'hedonic treadmill' because of its explanatory power in accounting for the divergence of self-assessed measures from observable circumstances. The model dates back to the work of Brickman and Campbell (1971) and states that individuals have set points of happiness that are generally only altered – towards more or less happiness – on a temporary basis before returning to their original level. The theoretical underpinning is found in the automatic habituation processes in which psychological systems return to current levels of well-being by permitting that constant stimuli fades into the background (Diener et al. 2006; Frederick and Loewenstein 1999). This theory has gained considerable empirical support, the most famous studies including hedonic adaptations among lottery winners and paraplegia patients (Brickman et al. 1978). Though the claims of the hedonic treadmill model have been revised and nuanced, the main hypothesis that happiness and satisfaction primarily reflect 'changes in circumstances rather than the overall desirability of the circumstances themselves' (Diener et al. 2006, p. 305) remains relevant for the analysis of subjective economic vulnerability.

The intellectual wrestling with the disjunction of objective and subjective measures has provided the basis for cross-fertilization between psychology and economics. Painting with a big brush, we have now become acquainted with the theoretical paradigms that are potent for advancing our discussion on the measure of economic vulnerability and the integration of the three measurement angles. While each of them falls short of being able to tell the whole story, allowing them to 'talk with one another' promises to take us closer to a measurement approach of 'economic vulnerability' that is able to identify those with low 'capabilities' of living a good life (Wells 2012).

References

Alexander, J. M. (2008). *Capabilities and social justice: The political philosophy of Amartya Sen and Martha Nussbaum*. Burlington: Ashgate Publishing Limited.

Andrews, F. M., & Withey, S. B. (1976). *Social indicators of well-being: Americans' perceptions of life quality*. New York: Plenum Press.

Athanassoulis, N. (2010). Virtue ethics. In *Internet encyclopedia of philosophy*. http://www.iep.utm.edu/virtue(20.02.2014)

Baltes, M. M., & Baltes, P. B. (1993). Psychological perspectives on successful aging: The model of selective optimization with compensation. In P. B. Baltes & M. M. Baltes (Eds.), *Successful aging: Perspectives from the behavioral sciences* (pp. S. 1–S. 34). Cambridge University Press.

Boarini, R., & d'Ercole, M. M. (2006). Measures of material deprivation in OECD countries. In *OECD social, employment and migration working papers 37, OECD publishing*.

Bradburn, N. M. (1969). *The structure of psychological well-being*. Chicago: Adeline.

Brickman, P., & Campbell, D. T. (1971). Hedonic relativism and planning the good society. In M. H. Appley (Ed.), *Adaptation level theory: A symposium* (pp. S. 287–S. 302). New York: Academic.

Brickman, P., Coates, D., & Janoff-Bulman, R. (1978). Lottery winners and accident victims: Is happiness relative? *Journal of Personality and Social Psychology, 36*(8), 917–927.

Corazzini, L., Esposito, L., & Majorano, F. (2011). Exploring the absolutist vs relativist perception of poverty using a cross-country questionnaire survey. *Journal of Economic Psychology, 32*(2), 273–283.

Diener, E., Suh, E. M., Lucas, R. E., & Smith, H. L. (1999). Subjective well-being: Three decades of progress. *Psychological Bulletin, 125*(2), 276–302.

Diener, E., Lucas, R. E., & Scollon, C. N. (2006). Beyond the hedonic treadmill: Revising the adaptation theory of well-being. *The American Psychologist, 61*(4), 305–314.

Duesenberry, J. S. (1949). *Income, saving, and the theory of consumer behavior*. Harvard: Harvard University Press.

Easterlin, R. (1974). Does economic growth improve the human lot? Some empirical evidence. In M. Abramovitz, P. A. David, & M. W. Reder (Eds.), *Nations and households in economic growth: Essays in honor of Moses Abramovitz*. New York: Academic.

Flik, R. J., & Praag, B. M. S. v. (1991). Subective poverty line definitions. *De Economist, 139*(3), 311–330.

Frederick, S., & Loewenstein, G. (1999). Hedonic adaptation. In D. Kahneman, E. Diener, & N. Schwarz (Eds.), *Well-being: The foundations of hedonic psychology* (pp. S. 302–S. 329). New York, NY, US: Russell Sage Foundation.

Frey, B., & Benz, M. (2002). Ökonomie und Psychologie: Eine Übersicht. In D. Frey & L. von Rosenstiel (Eds.), *Enzyklopädie der Witschaftspsychologie*. Hogrefe Verlag: Göttingen.

Frey, B. S., & Stutzer, A. (2002). What can economists learn from happiness research? *Journal of Economic Literature, 40*(2), 402–435.

Gasper, D. (2010). Understanding the diversity of conceptions of well-being and quality of life. *The Journal of Socio-Economics, 39*(3), 351–360.

Hausman, D. M. (2013, Winter). Philosophy of economics. In E. N. Zalta (Hrsg.), *The Stanford encyclopedia of philosophy*. Stanford: Stanford University.

Kahneman, D. (2003). Objective happiness. In E. Diener, N. Schwarz, & D. Kahnemann (Eds.), *Well-being: Foundations of hedonic psychology* (pp. S. 3–S. 25). New York: Russell Sage Foundation Publications.

Kim-Prieto, C., Diener, E., Tamir, M., Scollon, C., & Diener, M. (2013). Integrating the diverse definitions of happiness: A time-sequential framework of subjective well-being. In A. D. Fave (Ed.), *The exploration of happiness* (pp. S. 47–S. 75). Dordrecht: Springer.

Land, K. (2000). Social indicators. In E. F. Borgotta & R. V. Montgomery (Eds.), *Encyclopedia of sociology*. New York: Macmillan.

Lazarus, R. S., & Folkman, S. (1984). *Stress, appraisal and coping*. New York: Springer.

McBride, M. (2001). Relative-income effects on subjective well-being in the cross-section. *Journal of Economic Behavior & Organization, 45*(3), 251–278.

Monroe, S. M., & Kelley, J. M. (1997). Measurement of stress appraisal. In S. Cohen, R. C. Kessler, & L. U. Gordon (Eds.), *Measuring stress: A guide for health and social scientists* (pp. S. 122–S. 147). New York: Oxford University Press.

Nagel, T. (1986). *The view from nowhere*. New York: Oxford university press.

Noll, H.-H. (2004). Social indicators and quality of life research: Background, achievements and current trends. In *Advances in sociological knowledge* (pp. S. 151–S. 181). In N. Genov (Ed.): VS Verlag für Sozialwissenschaften.

Nussbaum, M. C. (1992). Human functioning and social justice in defense of Aristotelian essentialism. *Political Theory, 20*(2), 202–246. https://doi.org/10.1177/0090591792020002002.

Phillips, D. (2012). *Quality of life: Concept, policy and practice*. London: Routledge.

Robbins, L. (2007). *An essay on the nature and significance of economic science*. Auburn: Ludwig von Mises Institute.

Seligman, M. E. P. (2002). *Authentic happiness: Using the new positive psychology to realize your potential for lasting fulfillment*. New York: Simon and Schuster.

Sen, A. (1993). Positional objectivity. *Philosophy & Public Affairs, 22*(2), 126–145.

Townsend, P. (1979). *Poverty in the United Kingdom: A survey of household resources and standards of living*. Berkeley: University of California Press.

Veblen, T. (2005). *The theory of the leisure class; an economic study of institutions*. Delhi: Aakar Books.

Veenhoven, R. (2007). Subjective measures of well-being. In M. McGillivray (Ed.), *Human Well-being: Concept and measurement* (pp. 214–239). New York: Palgrave Macmillan.

Weidekamp-Maicher, M. (2008). *Materielles Wohlbefinden im späten Erwachsenenalter und Alter* (Text).

Weijers, D. (2011). Hedonism. In *Internet encyclopedia of philosophy*. http://www.iep.utm.edu/hedonism/(27.08.2014)

Wells, T. (2012). Sen's capability approach. In *Internet encyclopedia of philosophy*. http://www.iep.utm.edu/sen-cap(20.01.2014)

Chapter 4
Measures of Economic Vulnerability

In this chapter, we continue laying some more theoretical ground for the construction of the measures that will be used in the empirical part of this volume. At the beginning, there is a methodological question that needs to be answered in order to be able to more selectively focus on the most promising theoretical concepts: is economic vulnerability, defined as 'exceedingly low levels of economic quality of life', to be seen as the lower end of a *continuum of states,* or is it possible and necessary to distinguish between categories in the sense of *distinct levels* of economic quality of life? While theoretically, either continuous or categorical variables may be used for constructing measures of economic vulnerability, the interpretation will be easier if the variable is categorical. In following Desrosières's idea of a 'convention of equivalence between heterogeneous observations' (Desrosières 2001), we find that not all methods for constructing thresholds have equal legitimacy if judged by what is most relevant to the respondent's experience of quality of life.

4.1 The Absolute Core of Manifest Vulnerability

Thinking of economic vulnerability as a categorical concept implies the idea of a threshold, which in turn has implications for the choice of appropriate variables for operationalizing the concept. In view of the focus of our research on manifest outcomes, three arguments – a theoretical, a conceptual and an empirical one – unanimously speak in favor of conceptualizing economic vulnerability as categorical rather than as a continuous construct.

The theoretical argument comes from Sen's philosophical considerations on the 'absolute core' of poverty. In a response to the increasing popularity of relative indicators for assessing economic vulnerability in wealthy countries, he developed the argument that the escape from poverty is necessarily associated with an absolute requirement of capabilities, to be distinguished from the commodities that may be necessary to meet this requirement. He summed up his argument as follows: 'There

J. Henke, *Revisiting Economic Vulnerability in Old Age*, Life Course Research and Social Policies 11, https://doi.org/10.1007/978-3-030-36323-9_4

is a difference between achieving relatively less than others, and achieving absolutely less because of falling behind others' (Sen 1983, p. 155). Before we can further develop this particular argument, we need to briefly introduce the Capability Approach, the implications of which will be important for this volume on a more general ground.

At the core of Sen's Capability Approach, we find a focus on 'opportunities'. According to the Indian economist and philosopher, the right focus for assessing quality of life is an individual's capability of living the kind of life she or he has reason to value (Sen 1999). 'Capabilities' refer to the range of possible 'doings and beings' a person can achieve, for example, nutrition, health, education, and social recognition. Those 'capabilities' that the person ends up choosing are called 'functionings'. The originality of Sen's approach is the focus on people's options. The Aristotelian heritage is clearly recognizable in the criteria for evaluating the good life, which is neither the subjective experience (e.g. not every state of pleasure is ethically satisfactory) nor goods and services (not everyone has the ability of converting goods into outcomes) but the ability to pursue worthwhile activities and states of being. In contrast to theories where the desired result is equality of outcome for all, the capability approach focuses on equality of opportunity that can be seized by exercising individual agency (Alexander 2008). Whether people make use of it is important but more significant still is that they are given the option. Consequently, evaluation of economic quality of life should take into account both, actual achievements (so called 'functionings') as well as the substantial freedom of choice ('capabilities') (Wells 2012). An intuitive example that illustrates the importance of this distinction is the case of the nutritional status of people who are fasting: if the only indicator is the caloric intake, one would be led to conclude that their well-being is comparable to people who are starving (Sen 1999). However, 'capabilities' are more difficult to evaluate than 'functionings' because they are not directly observable (Krishnakumar 2007). The story behind how Sen got on to the track of the importance of freedom in matters of economic vulnerability echoes the origin and discovery of the concept of vulnerability: his famous analysis entitled 'Poverty and Famines' (Sen 1981) showed that famines are not primarily the result of a shortage of food supply as used to be thought but come about because of socio-political structures that keep people from purchasing food or from a lack of safety nets that prevents them from going hungry.

With regard to the discussion on the 'absolute core' of poverty, Sen is building on Adam Smith's concept of 'necessaries' by which he referred to basic commodities, including those items that 'the custom of the country renders it indecent for creditable people, even the lowest order, to be without' (Smith 1776, pp. 351–352, in Sen 1983, p. 159). Smith explains this type of commodity using the example of leather shoes: they were a necessary part of the outfit without which people would feel ashamed in public in eighteen-century England. Expounding on this illustration, Sen demonstrates that the capability of 'avoiding shame in public' could only be granted if a person had access to leather shoes. What matters for quality of life is 'not so much having equal shame as others, but just not being ashamed, absolutely' (Sen 1983, p. 161). While the commodity space is relative to the standard of living

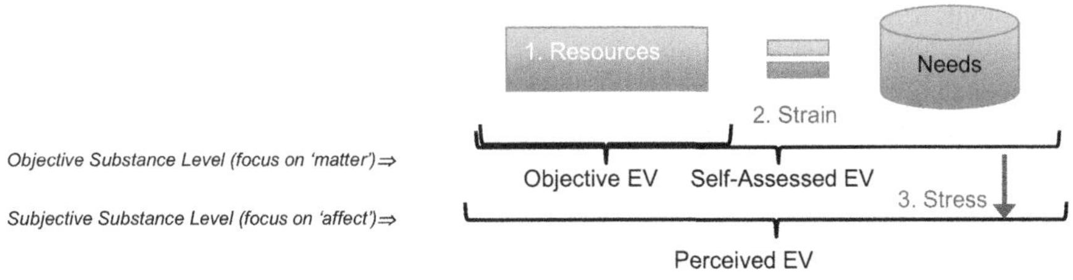

Fig. 4.1 The substance level of three measurement angles of economic vulnerability (EV)

at a given time in a specific society (i.e. the need for leather shoes arose because *everyone* wore them, implying that here, the situation of others is highly significant), the granting of a specific capability (e.g. not being ashamed) is associated with an absolute requirement of certain commodities. Following this line of argument, we conclude that there exists for each individual in each society an absolute threshold below which she or he is deprived of a capability.

The second argument in favor of economic vulnerability being a categorical concept is based on how 'coping' is conceptualized within our vulnerability framework: there is an underlying assumption that economic vulnerability needs to be qualitatively distinct and especially so from a subjective perspective because there exists, for each individual, an absolute threshold below which she or he experiences deprivation, or, as discussed on page 18, where 'disadvantage' turns into 'adversity'. Analogous to Sen's argumentation, we find that, while the experience of economic vulnerability is conditioned by different reference points (and is therefore relative), the threshold relevant for quality of life at the individual level is an absolute one.

Figure 4.1 depicts the spectrum of what is taken into account for the evaluation of quality of life by each of the three angles of measure: as the measure moves towards increasing subjectivity, the range of factors that influence the measure is expanded. As a direct consequence of the varying components that each angle encompasses, the threshold – in the sense of a 'radical' change which occurs after surpassing a limit – looks different for each measurement angle: objective thresholds of economic vulnerability such as a monetary poverty line tend to focus on the amount of *resources* that are sufficient to provide for a previously defined set of basic needs. The self-assessed measure of economic vulnerability entails a comparison of the available economic resources with the perceived needs: the needs are greater than the available resources, the economic situation will be described as inadequate, resulting in an experience of economic *strain*. The perceived measure of economic vulnerability expresses whether this situation (strain) generates a sense of *stress*.[1]

As we move towards increasing subjectivity, we expect the inter-individual variability given the same amount of resources to increase. Granted the elasticity of this

[1] The mechanisms leading to the experience of stress will be discussed later.

threshold, it nonetheless represents – as Adam Smith's examples of English shoes showed – an absolute requirement for each individual for avoiding economic strain or stress.

The details of operationalization will be addressed in a later section but for the purpose of rendering this argument more concrete, we will illustrate the implication of this line of reasoning based on examples of real measures, among them, the measures we are going to use for the empirical analyses.

The construction of a monetary threshold is usually done by dividing observations into a category 'above' and a category 'below' a specific level of income that has been identified as sufficient for covering basic needs. The two groups will most likely exhibit different population-level statistics for most socio-demographic characteristics that are of interest to social policy research. The arbitrary nature of this way of establishing a threshold is revealed when comparing individuals located just above or just below the threshold: they are most likely going to be close to identical with regard to these same socio-demographic characteristics.

When economic vulnerability is measured in a self-assessed manner, resources and needs are compared to one another in order to determine their 'adequacy', defined as 'a perception that one thing, event, or state of affairs is equal to another thing, event, or state of affairs' (Hazelrigg and Hardy 1997, p. 2). Among the survey questions mentioned earlier as examples for operationalizing the self-assessed measurement angle with regard to economics, not all are equally suited for the study of low levels of economic quality of life: if we view economic quality of life as range of states with 'equal distance' between them, any survey question with an in-built ranking is appropriate for constructing an indicator. However, if we think of economic vulnerability as *exceedingly low* levels of economic quality of life, we can expect more relevant results if the survey items offer a natural cut-point, indicating a qualitative drop in economic quality of life that breaks the continuum of states. Such a qualitative criterion of evaluation is provided by the question 'Do you have difficulties in making ends meet given your household's monthly income' where the inbuilt threshold is located between response items 2 and 3: (1)'with difficulties, (2)'with some difficulties', (3)'rather easily', (4) 'very easily'. In contrast, response items that are set up in the form of a neutral ranking – be it with word labels or numbers – tend to suggest equal distances between the items. They are more difficult to interpret, resulting in a greater risk of attributing dissimilar experiences to one and the same category.

The same difficulties arise with survey questions about the individual's perception of his or her economic vulnerability. This angle of measurement tells us about the emotional reaction in view of the economic situation. Operationalizing this measure as a continuum, for example, a scale from 1 to 10 in response to the question 'how satisfied with your financial situation are you?' leaves us with the difficult task of classifying the numbers in a relevant, non-arbitrary way. Using this type of question will hamper the interpretation of results because we cannot assume that the actually experienced levels of welfare are consistent among respondents (Stanovnik 1992). While a wording that implies a qualitative rupture does not resolve the methodological challenges surrounding the interpersonal comparability of subjective answers, we argue that the probability that two respondents who give the same answer have a similar emotional experience is higher, if the survey question has an inbuilt threshold than if both had checked the same rank in an undefined monotonous scale.

Empirical findings provide the third argument for the absolute core of the concept of economic vulnerability by affirming the discontinuous nature of low levels of economic quality of life. Indeed, the correlation of subjective measures with objective measures varies according to whether it is an average objective situation or whether it is an extremely negative objective situation. To mention just one example, Cummins investigated the correlations between a variety of objective and subjective measures of quality of life and found evidence for a homeostatic process enabling individuals to maintain their subjective well-being within a narrow range when faced with a variety of more or less desirable objective conditions (i.e. low correlation). However, the disassociation between objective circumstances and subjective experience was only true if the objective conditions were not exceedingly bad: 'very poor objective conditions can defeat homeostasis and, once this occurs, the objective and subjective measures display stronger covariation' (Cummins 2000, p. 1). This statistical evidence lends supports to a categorical understanding of the concept of economic vulnerability. In the subsequent discussion of indicators, we will therefore focus on those that define economic vulnerability as a qualitatively distinct category in the sense of an exceedingly low level of economic quality of life.

4.2 Measuring Economic Vulnerability Objectively

Objective measures are based on the assumption that it is possible to identify a certain number of 'basic needs', the satisfaction of which grants the same (minimum) level of economic quality of life to everyone. The most ancient way of conceptualizing low levels of economic quality of life is the classical definition of 'absolute poverty' based on a strictly material understanding of the concept: it is defined as a state of extreme deprivation, the threshold being a minimum standard of material resources required to meet basic needs such as food, clothes and shelter etc. – just enough to keep the person alive.[2] From a sociological perspective, the only way 'economic quality of life' can be conceptualized in a meaningful way is by considering the interdependence of the material situation with other areas of life as well as with the overall standard of living of the society in question (Klocke 2000). This multidimensionality and 'relativity' in terms of embeddedness of the definition of economic quality of life in a given societal standard of living has become widely accepted among academia and policy makers. It is reflected in the definitions used

[2]A prominent example is B.S. Rowntree's 'Poverty: A Study of Town Life', from 1901, republished by The Policy Press in 2000.

in contemporary policy documents, such as the official Swiss strategy for fighting poverty: 'As poor are considered those individuals, families or groups who dispose of so little means (material, cultural and social) that they are excluded from the life style that is considered an acceptable minimum in the member State where they live'[3] (Conseil fédéral Suisse 2010, p. 15). This phrasing echoes Adam Smith's far-reaching definition of 'necessities' as it acknowledges that financial resources are but a means to a less tangible end that includes opportunities of social participation. The sociocultural definition of poverty also shapes the different measures that are used in contemporary poverty research in wealthy countries: as different as they may be, they invariably allow for a minimum level of social participation. As definitions of economic vulnerability move further away from mere subsistence, they are bound to involve more and more normative decisions with regard to the threshold below which a person is considered unable to participate in society in an acceptable manner. With the rise of consumerism, the line between 'basic needs' and 'luxuries' has become increasingly blurred: while it is true that even the definition of adequacy of basic needs such as of food and shelter is not without room for discussion (e.g. does it have to be a building out of brick stones or is paper board construction enough to provide shelter?), the spectrum of possible levels of standard of living that can be considered 'minimal' in a *sociocultural* sense has not ceased to widen as our societies have grown wealthier.

In view of the cultural diversity of today's societies, the idea that everyone basically ought to have the same needs is a strong assumption and its validity is stretched even more when applying it to an age group as diverse as Swiss residents aged 65 and older. Having lived through a time of major socioeconomic transformations, the cohorts who are at retirement age today have very diverse expectations in terms of living standard and sociocultural activities (Spini 2009). The biographies of members of this population group comprise experiences that range from those who struggled through the financial crisis of the 1930's and the hardship caused by World War II to those who have only known the seemingly never ending expansion of economic and social opportunities. The trend towards increasingly diverse family and work patterns is undisputed, though the degree of destandardization and pluralisation of the traditional life course is object of scientific debate (Kohli 2007; Widmer and Ritschard 2009). The diversity of today's elderly is reflected in both the resource side and the needs or expectations side of the adequacy equation, which has important implications for the conceptualization of economic vulnerability. Differences in health status represent another source of heterogeneity among older adults that is also rooted in the life course and that tends to widen with advancing age. Naturally, persons with mental and physical disabilities need more financial resources to achieve comparable levels of well-being, which raises serious questions about the ethics behind using resource-based methods for interpersonal comparison of eco-

[3] The original French version reads: 'Sont réputés pauvres les personnes, les familles ou les groupes qui disposent de si peu de moyens (matériels, culturels et sociaux) qu'ils sont exclus du mode de vie considéré comme un minimum acceptable dans l'Etat membre où ils vivent.'

nomic quality of life.[4] For the purpose of studying long-term trends, however, objective poverty measures undoubtedly offer important benefits that allow comparing the standard of living across time and space. If interpreted with the necessary care, their use may be a legitimate approach to capturing something as elusive as a minimum level of economic quality of life.

Granted that it would theoretically be possible to define a minimal living standard, one then has to decide on the appropriate indicator for measuring it. Two main approaches can be distinguished: the 'resource approach', which uses the *potential* supply in form of finances as a proxy and the 'life condition approach',[5] which attempts to capture the *actual* supply situation and therefore takes into account multiple dimensions of life (Leu et al. 1997).

The resource approach is most frequently operationalized as an *absolute monetary poverty line*: a set of basic needs and services is determined, which is then converted into a monetary value at average price levels. The individual economic welfare is considered 'poor' if the disposable income falls below the thus established absolute[6] poverty line. There is no particular scientific reason for the use of monetary poverty lines (Klocke 2000) but the evidence for the value of money as proxy for less tangible aspects of poverty (Glatzer and Hübinger 1990) makes absolute poverty lines particularly appealing for the monitoring of social policy interventions. In this volume, we will incorporate an absolute monetary poverty line into the empirical analysis not least because it remains one of the most widely methods used.

One of the first difficulties one is confronted with when assessing the economic resources of elderly people in monetary terms is to unite all relevant data (Grundy and Holt 2001). Many studies rely on survey data, in which case the probability of error for self-reports is relatively higher for pensioners, whose income stems from multiple sources, than for the economically active, who for the most part can simply state the amount of their monthly salary. In general, self-reports of income (and wealth) are known to frequently be incorrect or missing (Heady et al. 2009), whether for reasons of privacy or lack of accurate knowledge.

Another challenge concerns the comparability of various types of resources and needs. Equivalent scales have been developed for the purpose of comparing the living costs of individuals living in differently sized households, or in regions that differ with regard to important items of expenditures such as taxes, health insurance primes and house rents, etc. With regard to differences in resources, an important decision concerns the integration of wealth in form of savings, financial and tangible assets. While monthly income flows are a reasonable proxy for assessing the economic situation of the working population, it is well established that wealth

[4]This shortcomings of the resource approach was one of the main concerns in Sen's development of the Capability Approach (Sen 2009).

[5]In German: Lebenslagen-Ansatz.

[6]Despite an explicit relation to a given society at a given time, it continues to be considered 'absolute' as opposed to a 'relative' definition of poverty that takes the overall distribution of income as basis of calculation.

plays an important role in the disposable income of Swiss pensioners: Firstly, the optional individual provident measures, the so called 'third pillar' of the Swiss pension scheme,[7] explicitly foresees personal wealth as an integral part of one's provision for old age. Furthermore, there is an important number of pensioners[8] who opt for a lump-sum payment of their occupational benefits, which de facto increases wealth at the expense of future monthly income. Secondly, there is evidence for a concentration of wealth in the age group 50 and older due to the dynamic of wealth accumulation over time (Moser 2006), on the one hand, and the cumulative effect of inheritances in this age group on the other hand (Stutz et al. 2007).[9] For these reasons, the decision whether and how wealth is to be included in the assessment of economic resources of pensioners has important implications on the poverty rates obtained. The spectrum of possibilities reaches from excluding wealth altogether to factoring the totality of wealth into the disposable income (Leu et al. 1997). If a single indicator had to be chosen, the disposable income is thought to best represent the economic situation, not least because of its proven high correlation with wealth. This pragmatic argument of the strong correlation prevails in many studies that use income as a proxy to depict the overall economic resources of pensioners (Weidekamp-Maicher and Naegele 2007).

If wealth is to be taken into account, the string of ensuing decisions includes the question which type of wealth and what proportion of it should be included. Not all forms of wealth are suited for providing protection against the risk of income poverty. Real estate, for example, cannot easily be converted into liquidity. Social welfare institutions in Switzerland that allocate means-tested benefits usually do take into account a share of individual wealth for determining eligibility. The supplementary benefits to old age pensions, for instance, foresees different amounts of allowances,[10] depending on whether the person is single or married and according to the home ownership situation. One tenth of any wealth exceeding the so determined threshold is counted as 'income'; for individuals living in an institution up to one fifth of personal wealth is thus taken into account (Centre d'information AVS/AI 2014). A report on the economic situation of the active and the retired population in Switzerland, mandated by the Federal Office for Social Insurances, recommends to factor in 5% of the capital benefits[11] stemming from occupational benefits or individual provident measures into disposable income (Wanner et al. 2008). The authors acknowledge, however, that this information is difficult to obtain.

---

[7] The Swiss pension system will be discussed in more detail on p. 28.

[8] For the year 2005, as many as one third of all individuals entering into retirement chose this option (Bundesrat 2011).

[9] Based on tax data from five cantons, Wanner and Gabadinho (2008) found the median gross wealth of pensioners aged 65–69 to be at 375,000 CHF and at 229,000 CHF for the 85–59 year-olds, compared to just under 100,000 CHF for the overall population.

[10] In 2011, for example, the allowance for single households was 37,000 and 60,000 for couples (Informationsstelle AHV/IV 2010).

[11] Occupational pensions can either be drawn as monthly payments or as a lump-sum payment which is called capital benefit.

The methodological difficulties surrounding measures of adequacy not only hamper the comparison of poverty statistics cross-nationally. As a population grows increasingly heterogeneous the complexity of the interpretative task does, too. The diversity of resources and needs being particularly great among the elderly, the use of summary statistics based on monetary indicators is especially misleading among this age group (Quinn 1987). Whichever side of the adequacy equation is used as a proxy – resources or needs – there is a temptation to confuse the means with the end: the money or the goods and services are merely contributing factors to enable people 'to be and to do' those things that are of value to them, to put it in Sen's words (Sen 1999, p. 75). The realization that conventional poverty measures based on income and consumption only partly capture the experienced reality has played an important role in the serach for alternative subjective indicators (Ravallion 2012).

4.3 Measuring Economic Vulnerability Subjectively

The interest in subjective measures of economic vulnerability goes back to the 1960s and spans multiple fields. While in economics the emphasis has been on alternative concepts of poverty (Hazelrigg and Hardy 1997), psychologists have mainly been interested in financial strain as a predictor of various (mental) health-related outcomes (Alley and Kahn 2012). Quality of life research has embraced self-assessments of economic resources as independent variable because there is evidence that it has more predictive power for health-related outcomes than objective measures of socioeconomic status (Rios and Zautra 2011).

As discussed earlier with regard to the incongruence between objective and subjective measures of quality of life, the dominant theory in the context of subjective measures is based on the concepts of reference utility and aspirations. There are many possible reference points that could be relevant for economic vulnerability, but most can be classified as either social comparison or intra-biographical referencing. Individuals are known to judge their own well-being by comparing their lot with that of important others who are similar to themselves. Satisfaction with one's income or living standard, for example, is thus influenced by the endowment of a reference group of similar age, sex, profession and educational attainments (Frey and Stutzer 2002). With regard to intra-biographical referencing, subjective appraisal is influenced by past experiences but also by expectations[12] people used to have in the past or what they anticipate to happen in the future. There has been multiple evidence for the influence of past experiences of poverty on current evaluation of economic well-being (Castilla 2010; Crettaz and Suter 2013). This phenomena, called adaptive preferences, is one expression of the mechanisms of the more

---

[12] The most comprehensive theory with regard to the influence of expectations on subjective appraisals is Michalos's (1985), Multiple Discrepancy Theory', which has been classified among the *relative* theories of quality of life because it states that life satisfaction is higher, the closer the actual life situation is, relative to the reference points, of which he states seven.

general theory of adaptation that was discussed earlier: the underlying assumption is that due to the homeostatic processes influencing appraisal, individuals with prior experience of economic vulnerability tend to lower their expectations regarding their standard of living, for example, by opting for those goods and services they can afford while mistakenly believing that they have freely 'chosen' this modest life style.[13] In a sense, every individual can be said have an implicit theory of 'expected returns' that is unique to his or her biography and cohort history (Hazelrigg and Hardy 1997). The process of internal bargaining leads to trade-offs between current well-being and anticipated future well-being (Kim-Prieto et al. 2013): people may subconsciously justify or rationalize current disadvantage in order to avoid a sense of disappointment or dissatisfaction in the long-term. Such a trade-off may also happen consciously, as discussed previously (p. 18), in the case where individuals do recognize their circumstances as being adverse but choose not to invest resources to change it.

In most research designs on subjective economic outcomes, the distinctive nature of the self-assessed and the perceived measurement angle is not recognized. While economists prefer self-assessed measures, we have found that psychologists tend to use both measurement angles indifferently in their operationalization of, for example, financial satisfaction or financial strain. Remembering Veenhoven's matrix, financial satisfaction has a subjective substance matter, whereas financial (in)adequacy has an objective substance matter. As the following definition of financial satisfaction in the context of well-being research shows, the distinctiveness of these substance matters is often ignored: 'financial satisfaction is the subjective evaluation of the degree to which one's financial resources are adequate versus inadequate or bring satisfaction versus dissatisfaction' (George 1992, p. 72, in Weidekamp-Maicher and Naegele 2007, p. 74). An example from stress research confirms the casual approach to this nuance in the field of psychology: Alley and Kahn's study (Alley and Kahn 2012) examines subjective economic vulnerability within a reserve-capacity framework in order to test the effect of psychosocial variables on financial strain among older adults. They constructed a dependent variable based on four items: (1) satisfaction with the current financial situation, (2) difficulties meeting monthly expenses, (3) distress caused by ongoing financial strain, and (4) degree of control over their financial situation. Items 1 and 3 would qualify as perceived and items 2 and 4 are self-assessed measures. In this research design, too, the underlying assumption seems to be that 'If one does not have enough money to meet monthly expenses (…) we assume that the individual finds such a situation stressful' (Angel et al. 2003, pp. 538–539). Thus, economic strain is often equated with stress.

[13]This hypothesis has been tested by Crettaz and Suter (2013) on data from the Swiss household panel for a self-assessed measure (difficulties in making ends meet) and a perceived measure (satisfaction with income/financial situation). Controlling for current income level and various sociodemographic characteristics, they found that an additional year spent in poverty significantly increased the odds of being more satisfied and reduced the odds of having more difficulties in making ends meet. This study will be discussed in the conclusion of this volume.

In literature on subjective poverty, self-assessed indicators of economic vulnerability dominate. Two main families have been distinguished by economists: an indirect money-metric and a direct categorical approach (Ravallion 2012). The first one asks respondents to quantify different levels of economic welfare in monetary terms. The most well-known measures in this family are survey questions that ask respondents to indicate the amount needed to make ends meet, the so called 'Minimum income Question' (Kapteyn et al. 1988; Stanovnik 1992; Milanovic and Jovanovic 1999) or what income level they would consider as bad/sufficient/good in their current circumstances, the so called 'income Evaluation Question' (van Praag 1968). The objective of the money-metric approach is to assess the level of utility experienced at a given income level. It is a model-based approach, since the thus generated subjective poverty-line is not directly based on the survey question but needs to be estimated by a statistical model that explains inter-individual variation (Kapteyn et al. 1988). The underlying assumption is that respondents have a comparable welfare level in mind when they respond to the question, making interpersonal comparisons possible. Kapteyn and collegues are well aware of the implications when they explain (Kapteyn et al. 1985, p. 39): 'the approach is subjective in the sense that a poverty line is not defined in terms of some pre-specified commodity bundle that a household should be able to afford. It is only the respondent's own opinion of what is minimally needed that is the basis for this definition. This puts a heavy burden on the wording of the question and involves the assumption that somehow make ends meet has the same meaning to everyone, at least approximately' (Kapteyn et al. 1985, p. 39). This bold assumption is difficult to test empirically (den Bosch et al. 1993). Moreover, the utilitarian underpinning of the money-metric approach causes it to qualify for some of the same criticisms as the absolute monetary poverty line, in particular, the reproach that Quinn called 'the parsimony of the mean' (Quinn 1987): even if the model accounts for a few characteristics that may be relevant for assessing income adequacy, it must fall short of doing justice to the diversity of economic needs and resources in this age group.

The second stream of subjective poverty measures called 'qualitative categorical approach' asks respondents to situate their economic utility level within a given welfare space, usually based on verbal labels. It then takes this assessment at face value by classifying respondents directly as 'poor' or 'non poor'. Examples are the economic-ladder-question with a range from 'poor' to 'rich' (Riffault 1991; Howe et al. 2011) and the Deleeck question[14] about how well the household is making ends meet based on a likert-type scale starting from 'with great difficulties' to 'very easily' (Flik and Praag 1991; Castilla 2010).

The criticism voiced against subjective indicators of economic vulnerability has mainly to do with the very mechanisms of referencing that have been described earlier: any self-assessment of vulnerability is bound to be influenced by cohort and cultural differences of what is commonly considered 'acceptable' suffering (Hughes

[14]A version of the Deleeck question will be used as Self-Assessed Measure in the empirical part of this volume.

1990). Furthermore, the subjective mode of measurement may reflect an embellished picture of the actual situation because individual expectations are tuned down as a result of downward social comparison or adaptation to suboptimal states of well-being (Heidrich and Ryff 1993). The pitfall of trusting such frames of reference that are acquired unconsciously throughout the life course, shaped by institutions, legal frameworks, and socialization, has been pointed out a long time ago by Marxists, who called these frames a 'false consciousness'. Also, they are one of the main arguments Sen evokes against using subjective measures in the satisfaction of preferences theory (Alexander 2008). Another difficulties that has been broadly recognized pertains to the previously mentioned issue of interpersonal comparability. This problem may be worsened when the survey is passed in different languages where biases may be introduced because of slight though inevitable differences in the connotations of words and expressions.

Self-assessed measures of economic vulnerability have mainly been used as explanatory variables. The following two studies are among the few examples of research on older adults that have focused on income adequacy as an outcome variable. Litwin and Sapir (2009) set out to test the validity of self-assessed income adequacy as measure for economic status among older adults (50+) from 12 countries using data from the Survey of Health, Ageing and Retirement in Europe (SHARE). For the logistic regression, subjective economic hardship was operationalized by a binary coding of the Deleeck question 'difficulties in making ends meet'. Independent variables included socio-demographic variables, economic resources (standardized monthly household income, net wealth and relative income), health status measures, and expectations regarding one's financial future. Their findings confirm the overriding importance of economic resources – primarily wealth, followed by employment status and income – though the effect of objective factors grew weaker with advancing age of pensioners. Both physical and mental health predicted poor subjective economic status. Stoller and Stoller (2003) focused their research specifically on what has been called the 'income paradox', namely that elderly people generally find their incomes to be adequate, even if it is relatively low. They developed three indexes to tap into 'perceived income adequacy'. They were specifically interested in the influences of household income and self-assessed health: The first indicator spanned six items from 'money takes care of needs poorly' to 'money takes care of needs very well, and respondent can buy what he or she wants'. The second indicator elicited whether respondents had encountered a series of financial limitations during the past 12 months (unable to pay bills etc.). At last, respondents were asked how they would spend an extra $100 (on rent, food, medical care etc.). The results confirm that old age is associated with a more positive evaluation of financial adequacy. Objective measures of income and number of persons supported by that income only explained 18% of the variation in subjective financial adequacy. Congruent with Litwin's findings, older people who assess their health more negatively or who had experienced recent declines in health status had a higher probability of assessing their financial resources as inadequate when controlling for income levels.

Before looking at the potential of the three measurement angles combined, we return to a brief overview of what is known about the prevalence and manifestation of economic vulnerability among old adults in Switzerland. This review of the main conclusions drawn in a national poverty report shows flawed interpretation that we think is the result of, firstly, a failed integration of the steps of measurement construction and subsequent analysis, and secondly, an isolated analysis of objective and subjective measures of economic vulnerability.

4.4 Economic Vulnerability Among Swiss Pensioners

The most recent data on economic vulnerability among pensioners in Switzerland stems from a poverty report by the Federal Office for Statistics (OFS). The main results and conclusions put forward in the report provide an illustration of some of the theoretical and conceptual challenges of measurement discussed in this Part, highlighting the need for an integrated perspective of objective and subjective measures.

The poverty line used in the OFS report is based on the recommendations by the Swiss Conference for Public Welfare (CSIAS). The CSIAS defines the sociocultural existence level based on the living costs for maintenance, housing and a lump sum for other inevitable expenses and estimates the necessary minimum income accruing from gross income[15] minus direct tax, social security contribution and health insurance premium. In 2010, the thus defined minimum existence level was set at CHF 2250 disposable equivalent income. Based on this absolute poverty line, the OFS found the poverty rate to be twice as high for the population aged 65 and older compared to the overall population (16.2% as compared to 7.9%).

In the ensuing discussion, the report systematically relativizes the meaningfulness of the thus established poverty rates for the older population: It is suggested that there is good reason to believe that the economic situation of the population 65 and older is cushioned by additional financial resources other than income, specifically by savings. This is said to be confirmed by the above average share of older respondents who use their wealth for current expenses (18.4%, compared to 7.8% for the overall population) (OFS 2012, p. 18). As a consequence, and in the absence of any data on wealth,[16] the report concludes that the poverty rates for the older population is most likely overestimated (ibid, p. 14). The same allusion is made with regard to adults living alone, who constitute a high-risk group: 'Unmarried individuals, too, stand out from among other groups: while they record a particularly high rate of poverty according to the absolute concept at any age, individuals

[15] The income sources taken into account as pertaining to pensioners are: old age pensions, occupational benefits (without capital benefits), individual provident measures (without capital benefits), salaries, social benefits (supplementary benefits to old age pensions, welfare benefits), revenues derived from property and other financial assets.

[16] No explanation is given as to why this variable was not included in the survey questionnaire.

who are older than 65 years of age seem particularly exposed to the risk[17] of poverty (29.5%).' (ibid, p. 23–24). In a footnote to this statement it further says: 'It is albeit advisable to interpret with prudence the situation of older adults since the possible utilization of their wealth has not been taken into account.' As further indicators of the positive financial situation of pensioners, the report mentions the high average rate of satisfaction with their financial situation: 67.9% marked 8 or higher on a scale from 0 to 10. Another variable that is used to fill the interpretative gap is the comparably low share of pensioners who have arrears with payments (3% versus 8.8% for the overall population).

The conclusions put forward by the OFS illustrate the difficulties that arise when the measurement angles are interpreted separately, without linking them on at individual level. The less frequent arrears in payment among the age group 65 and older, which is interpreted as an indicator for a relatively better economic standing of pensioners, could just as well reflect a lesser need for purchasing consumer goods in this phase of life. Furthermore, there might be cohort differences in mentality towards money handling in particular and a sense of duty in general (George 1992). A more important line of argumentation is based on omissions of data on wealth and the ensuing ambiguities. It is true that there exists as of this writing no data on the distribution of wealth at the household level (Eidgenössische Steuerverwaltung 2014) and it seems that it was not among the priorities of the poverty research agenda to remedy this lack of public data. Still, the reasoning of the OFS report is surprising, as it seems to ignore the commonplace that income and wealth are strongly correlated since both are the result of one and the same work life trajectory. Furthermore, the worldwide phenomena that wealth tends to be more unequally distributed than income also applies to the Swiss context, as evidence based on tax registers from the canton of Zurich shows (Moser 2006). Thus, contrary to what is suggested in the report, the observed inequality of income situations would have been exacerbated if it had been possible to take into account wealth.

This example illustrates that relying exclusively on income-based measures represents an ill-adapted approach to capturing economic vulnerability among the elderly, firstly, because of the methodological difficulties of taking into account all relevant income sources of pensioners. Neglecting the diversity of resources among the elderly leads to biases in comparison of the resulting poverty rate with the poverty rate of the working population.[18] Furthermore, such an approach is based on the assumption that the same level of quality of life can be achieved with the same

[17] This result is based on a relative poverty line, which is set at 50% of the median income.

[18] Or the age group ‚65 plus' is entirely omitted, as in the report on the distribution of wealth in Switzerland: 'Pensioners finance part of their livelihood by the drawdown of wealth, as various results show (for example in the chapter 'Saving'). Unfortunately there exists as of this writing no data on the distribution of wealth on the level of private households. Thus an important component for comparing households prior to and after retirement with regard to their financial situation is lacking and especially so for pensioners. For that reason this report limits its detailed analysis on the overall population and on the working population.' (Eidgenössische Steuerverwaltung 2014, p. 3).

basket of goods and services, an assumption that disregards the heterogeneity of needs that is characteristic for this age group. Forgetting about these important factors may be the result of a disjunction between the two steps of data analysis, the construction of the measuring instrument and the interpretation of the results. It may also reflect ignorance and a lack of care about what is really relevant for the economic quality of life of older adults in this country.

The question of 'relevance' propels us to return to the Pro Senectute report. The following section reviews the main findings of the qualitative study described earlier (p. 12) based on what practitioners have observed as being the most important psychosocial consequences and symptoms in individuals who are struggling economically.

Pro Senectutes' distilled impression from innumerable consulting sessions shows that even among those who ask for financial assistance, there exist important differences in how clients assess and experience their economic vulnerability: 'Some people feel poor even though they benefit from a small occupational pension. And then there are those who really have very little at their disposition but experience their situation as very different.' (Seifert and Pilgram 2009, p. 73) Old adults suffer especially from their lack of resources if they are *unable to meet social conventions or fulfill social roles*. It could be that they don't have enough money for dressing properly, getting a regular hair cut or that they cannot afford the socially expected financial obligations towards their children and grandchildren. People living in more traditional areas are often stressed at the thought of not having money to put aside as provision for a dignified burial. For many, not being able to pay their bills on time can result in stress, not least because going into debts is less acceptable among these generations and the collection by claim would in most cases be considered a shame. The inability to fulfill these expectations – by society, family and self – tends to increase the perception of one's own economic vulnerability and can generate anxiety. A second source of suffering for economically vulnerable older adults is the *experience of discrimination and the fear of social isolation*. The former is especially marked in the area of housing where the quality of living arrangements and the shame that is associated with not being able to afford a certain standard has been found to exacerbate the feeling of vulnerability. Also, many of Pro Senectute's clients struggle with not being able to fully participate in social life. Building and cultivating relationships costs money but the emotional 'price' of social isolation and loneliness is often assessed as more costly than material deprivation.

The possibility that these factors generate psychological stress is further increased by the permanent nature of the economic situation in old age. At the lower end of the income spectrum, where there is no or very little wealth and no additional income from continued professional activity, the monthly budget of pensioners is very likely to remain stable. Thus, the feeling of being trapped in a situation that cannot be changed can further provoke feelings of helplessness and powerlessness. This aspect is important from a quality of life perspective but is often ignored when comparing at-risk-groups across the same population.

References

Alexander, J. M. (2008). *Capabilities and social justice: The political philosophy of Amartya Sen and Martha Nussbaum*. London: Ashgate Publishing Limited.
Alley, D., & Kahn, J. R. (2012). *Demographic and psychosocial predictors of financial strain in older adults*. In Paper submitted at the 2012 annual meeting of the Population Association of America.
Angel, R. J., Frisco, M., Angel, J. L., & Chiriboga, D. A. (2003). Financial strain and health among elderly Mexican-origin individuals. *Journal of Health and Social Behavior, 44*(4), 536–551. https://doi.org/10.2307/1519798.
Bundesrat. (2011). *Bericht des Bundesrates zuhanden der Bundesversammlung über die Zukunft der 2. Säule*. Bern: BSV.
Castilla, C. (2010). *Subjective poverty and reference-dependence. Income over time, aspirations and reference groups*. Helsinki: World Institute for Development Economics Reserach.
Centre d'information AVS/AI. (2014). *Prestations complémentaires à l'AVS et à l'AI*.
Conseil fédéral Suisse. (2010). *Stratégie globale de la Suisse en matière de lutte contre la pauvreté, Rapport du Conseil fédéral en réponse à la motion (06.3001) de la Commission de la sécurité sociale et de la santé publique du Conseil national (CSSS-N) du 13 janvier 2006*. Bern: Conseil fédéral Suisse.
Crettaz, E., & Suter, C. (2013). The impact of adaptive preferences on subjective indicators: An analysis of poverty indicators. *Social Indicators Research, 114*(1), 139–152.
Cummins, R. A. (2000). Objective and subjective auality of life: An interactive model. *Social Indicators Research, 52*(1), 55–72. https://doi.org/10.1023/A:1007027822521.
den Bosch, K. V., Callan, T., Estivill, J., Hausman, P., Jeandidier, B., Muffels, R., & Yfantopoulos, J. (1993). A comparison of poverty in seven European countries and regions using subjective and relative measures. *Journal of Population Economics, 6*(3), 235–259.
Desrosières, A. (2001). Entre réalisme métrologique et conventions d'équivalence : les ambiguïtés de la sociologie quantitative. *Genèses, 43*(2), 112–127.
Eidgenössische Steuerverwaltung. (2014). *Verteilung des Wohlstands in der Schweiz. Bericht des Bundesrates in Erfüllung des Postu-lats 10.4046 von Jacqueline Fehr vom 07.12.2010*.
Flik, R. J., & Praag, B. M. S. v. (1991). Subective poverty line definitions. *De Economist, 139*(3), 311–330.
Frey, B. S., & Stutzer, A. (2002). What can economists learn from happiness research? *Journal of Economic Literature, 40*(2), 402–435.
George, L. K. (1992). Economic status and subjective well-being: A review of the literature and an agenda for future research. In N. E. Cutler, D. W. Gregg, & M. P. Lawton (Hrsg.) (Eds.), *Aging, money, and life satisfaction: Aspects of financial gerontology* (pp. S. 69–S. 99). New York: Springer Publishing.
Glatzer, W., & Hübinger, W. (1990). Lebenslagen und Armut. In D. Döring, W. Hanesch, & E.-U. Huster (Hrsg.) (Eds.), *Armut in Wohlstand* (pp. S. 31–S. 55). Frankfurt: Suhrkamp.
Grundy, E., & Holt, G. (2001). The socioeconomic status of older adults: How should we measure it in studies of health inequalities? *Journal of Epidemiology and Community Health, 55*(12), 895–904.
Hazelrigg, L. E., & Hardy, M. A. (1997). Perceived income adequacy among older adults issues of conceptualization and measurement, with an analysis of data. *Research on Aging, 19*(1), 69–107.
Heady, B., Krause, P., & Wagner, G. (2009). Poverty redifined as low consumption and low wealth, not just low income: Psychological consequences in Australia and Germany. Gehalten auf der Joint OECD-University of Maryland Conference, "*Measuring Poverty, Inequality and Social Exclusion: Lessons from Europe*". Paris: OECD.
Heidrich, S. M., & Ryff, C. D. (1993). The role of social comparisons processes in the psychological adaptation of elderly adults. *Journal of Gerontology, 48*(3), P127–P136.

Howe, L. D., Hargreaves, J. R., Ploubidis, G. B., De Stavola, B. L., & Huttly, S. R. A. (2011, May). Subjective measures of socio-economic position and the wealth index: A comparative analysis. *Health Policy and Planning, 26*(3), 223–232. https://doi.org/10.1093/heapol/czq043.

Hughes, B. (1990). Quality of life. In S. M. Peace (Hrsg.) (Ed.), *Researching social gerontology: Concepts, methods and issues* (pp. S. 46–S. 58). London [etc.]: Sage/British Society of Gerontology.

Informationsstelle AHV/IV. (2010). *Änderungen auf 1. Januar 2011 bei Beiträgen und Leistungen AHV – IV.*

Kapteyn, A., van de Geer, S., & van de Stadt, H. (1985). *The impact of changes in income and family composition on subjective measures of well-being* (NBER Chapters) (S. 35–S. 68). Cambridge, MA: National Bureau of Economic Research, Inc.

Kapteyn, A., Kooreman, P., & Willemse, R. (1988). Some methodological issues in the implementation of subjective poverty definitions. *The Journal of Human Resources, 23*(2), 222–242.

Kim-Prieto, C., Diener, E., Tamir, M., Scollon, C., & Diener, M. (2013). Integrating the diverse definitions of happiness: A time-sequential framework of subjective well-being. In A. D. Fave (Hrsg.) (Ed.), *The exploration of happiness* (pp. S. 47–S. 75). Dordrecht: Springer.

Klocke, A. (2000). Methoden der Armutsmessung. Einkommens-, Unterversorgungs-, Deprivations- und Sozialhilfekonzept im Vergleich. *Zeitschrift für Soziologie, 29*(4), 313–329.

Kohli, M. (2007). The institutionalization of the life course: Looking back to look ahead. *Research in Human Development, 4*(3–4), 253–271.

Krishnakumar, J. (2007). Going beyond functionings to capabilities: An econometric model to explain and estimate capabilities. *Journal of Human Development, 8*(1), 39–63.

Leu, R. E., Burri, S., & Priester, T. (1997). *Lebensqualität und Armut in der Schweiz*. Bern: Haupt.

Litwin, H., & Sapir, E. V. (2009). Perceived income adequacy among older adults in 12 countries: Findings from the survey of health, ageing, and retirement in Europe. *The Gerontologist, 49*(3), 397–406.

Michalos, A. C. (1985). Multiple discrepancies theory (MDT). *Social Indicators Research, 16*(4), 347–413.

Milanovic, B., & Jovanovic, B. (1999, September 30). Changes in the perception of the poverty line during the depression in Russia, 1993–96. *World Bank Economic Review, 13*(3), 539–559.

Moser, P. (2002). *Alter, Einkommen und Vermögen. Eine Analyse der Zürcher Staatssteuerstatistik 1999*. Zürich: Statistisches Amt des Kantons.

Moser, P. (2006). *Einkommen und Vermögen der Generationen im Lebenszyklus. Eine Querschnitts-Kohortenanalyse der Zürcher Stuerdaten 1992–2003*. Zürich: Statistisches Amt des Kantons.

OFS. (2012). *Pauvreté en Suisse: concepts, résultats et méthodes. Résultats calculés sur la base de l'enqête SILC 2008 à 2010*. Neuchâtel: Office fédéral de la statistique.

Quinn, J. F. (1987). The economic status of the elderly: Beware of the mean∗. *Review of Income and Wealth, 33*(1), 63–82.

Ravallion, M. (2012). *Poor, or just feeling poor? On using subjective data in measuring poverty* (SSRN Scholarly Paper No. ID 2004930). Rochester: Social Science Research Network.

Riffault, H. (1991). How poverty is perceived. In K. Reif & J.-R. Rabier (Hrsg.) (Eds.), *Eurobarometer: The dynamics of European public opinion: Essays in honour of Jacques-René Rabier*. New York: St. Martin's Press.

Rios, R., & Zautra, A. J. (2011). Socioeconomic disparities in pain: The role of economic hardship and daily financial worry. *Health Psychology, 30*(1), 58–66.

Seifert, K., & Pilgram, A. (2009). *Vivre avec peu de moyens. La pauvreté des personnes âgées en Suisse*. Zürich: Edition Pro Senectute.

Sen, A. K. (1981). *Poverty and famines: An essay on entitlement and deprivation*. Oxford: Clarendon Press.

Sen, A. (1983). Poor, relatively speaking. *Oxford Economic Papers, 35*(4), 153–169.

Sen, A. K. (1999). *Development as freedom*. Oxford: Oxford University Press.

Sen, A. (2009). *The idea of justice*. Cambridge, MA: Belknap Press of Harvard University Press.

Smith, A. (1776). *An inquiry into the nature and causes of the wealth of nations*. London: Home University Library.
Spini, D. C. L. (2009). *Styles de vie et activités de formation chez les seniors du Canton de*. Vaud: Enquête. LIVES Workingpaper. https://doi.org/10.12682/lives.2296-1658.2012.10.
Stanovnik, T. (1992). Perception of poverty and income satisfaction: An empirical analysis of Slovene households. *Journal of Economic Psychology, 13*(1), 57–69.
Stoller, M. A., & Stoller, E. P. (2003). Perceived income adequacy among elderly retirees. *Journal of Applied Gerontology, 22*(2), 230–251.
Stutz, H., Bauer, T., & Schmugge, S. (2007). *Erben in der Schweiz. Eine Familiensache mit Volkswirtschaftlichen Folgen*. Zürich/Chur: Rüegger.
van Praag, B. M. S. (1968). *Individual welfare functions and consumer behavior: A theory of rational irrationality*. Amsterdam: North Holland Publishing Company.
Wanner, P., & Gabadinho, A. (2008). Die wirtschaftliche Situation von Personen im Ruhestand: neue Daten, neue Prioritäten. *Soziale Sicherheit CHSS, 3*, 137–142.
Wanner, P., Gabadinho, A., & Pecoraro, M. (2008). *La situation économique des actifs et des rentiers. Rapport technique et tableaux commentés*. Berne: OFAS.
Weidekamp-Maicher, M., & Naegele, G. (2007). Economic resources and subjective well-being in old age. H. Mollenkopf & A. Walker (Hrsg.), *Quality of life in old age: International and multidisciplinary perspectives* (Softcover reprint of hardcover 1st ed. 2007, S. 65–S. 84). Springer.
Wells, T. (2012). Sen's capability approach. In *Internet Encyclopedia of Philosophy*. http://www.iep.utm.edu/sen-cap (20.01.2014)
Widmer, E., & Ritschard, G. (2009). The de-standardization of the life course: Are men and women equal? *Advances in Life Course Research, 14*(1), 28–39.

Chapter 5
Integrating Objective and Subjective Measures

A quality of life approach to the question of economic vulnerability obliges us to take seriously the experience of individuals that are confronted with insufficiency of resources. As Pro Senectute's portrayal of clients reveals, lack of financial means is associated with a wide range of subjective experiences.

The foundational theoretical model for this volume is the vulnerability framework by Schröder-Butterfill and Marianti, which proposes a definition of coping focused on the mobilization of resources such as social networks and formal structures of social protection. However, if we want to grasp the processes that lead to the subjective vulnerability outcomes, we need to open the black box of 'coping' by borrowing from socio-psychological models. In fact, the means by which individuals can 'deal with the threat successfully by avoiding that the negative outcome materializes' need to be expanded when the outcome under consideration is economic strain or stress. The most accurate and at the same time parsimonious model for integrating the three measurement angles – objective, self-assessed and perceived – into one model was found to be Leonard Pearlin's stress process model. We will use it as a heuristic instrument to outline our understanding of what happens between the incident of the threat and the bad outcome. It will provide important components for our final analytical model by conceptually explaining why some individuals with a similar endowment of resources experience different levels of economic strain and why not all those who are experiencing strain are equally stressed about it.

5.1 Pearlin's Stress Process Model

The stress process model was first published by Pearlin and his colleagues in 1981 and has since shaped sociological research on this question: its focus on the importance of social structure introduced a change in paradigm in the study of mental health, stirring away from an exclusive focus on pathology (Avison et al. 2010).

J. Henke, *Revisiting Economic Vulnerability in Old Age*, Life Course Research and Social Policies 11, https://doi.org/10.1007/978-3-030-36323-9_5

Over the last three decades Pearlin's sustained interest in the concept has led him to continue writing about it at several instances. In a book chapter entitled 'The stress process revisited', he humbly states that no single pathway of stress has been identified to guide all research in social sciences but rather, 'the stress process should be regarded as an orienting framework that can guide the thinking of researchers about potentially stressful circumstances, and suggest to them fruitful lines of analysis and interpretation of their effects. It is intended to be of special use to sociologists who seek to incorporate and emphasize features of social and economic life into accounts of (...) well-being of people' (Pearlin 2006, p. 396). The explicit purpose of the stress process model ascribed to it by its originator is very well aligned with our research questions, which center on measures that are capable of capturing low levels of economic quality of life. We will draw on Pearlin's model because of its heuristic qualities not because of its etiological potential. In contrast to the explicit purpose of the stress process model, our objective is not primarily the analysis of the economic stress process but the theoretical and empirical inquiry of the added value of combining objective, self-assessed, and perceived measures to capture multiple expressions of economic vulnerability. As outlined earlier, we assume that the causal pathways through the interaction between multiple risk factors happened ex-ante the measurement via our three measures of economic vulnerability and can therefore only be deduced indirectly. However, the model can gainfully be applied to explain divergence between the three measures of vulnerability in the same individual.

In his original model, Pearlin distinguishes between three main conceptual domains that constitute the stress process: the *source* of stress, the *mediators,* and the *manifestation* of stress. Among the *sources of stress,* Pearlin differentiates between the direct impact of life events and the indirect influence via changes in the self-concept or through challenges to the social role. The literature we have reviewed on the theory of reference utility suggests that both of these types of influences emerge in an appraisal process to which the individual is not exogenous but the degree of financial strain is evaluated relative to an individual frame of reference. Moreover, people who report income inadequacy may have very different emotional responses: when confronted with an objectively similar situation and even when self-assessing it in a similar manner, some individuals experience more stress than others. According to Pearlin's stress model, if people able to preserve their sense of self, they are likely to be less worried about income inadequacy. In contrast, if the economic vulnerability results in role strains, and/or the individual experiences a diminishment of self, there is a higher probability of experiencing stress and more intense levels thereof. Concerning the global concept of *diminishment of self,* he identifies two subordinate concepts that are particularly relevant: mastery and self-esteem. Mastery refers to the sense of being in control of one's life. By self-esteem Pearlin means a person's judgment of his or her own self-worth. The erosion of self-esteem and mastery are seen as the final step in the process leading to stress because 'hardships that are an enduring testimony to one's lack of success or to the inadequacy of one's efforts to avoid problems would seem to pose the most sustained affront to one's conceptions of self-worth and of being in control over personal destiny' (Pearlin et al. 1981). With regard to *role strains,*[1] Pearlin's own research showed that life events provoke an intensification of preexisting strains or the emergence of new strains[2] which have their own direct effect on stress (Pearlin and Lieberman 1979). In that way, role strains represent the missing link that combines with life events to create chronic stressors. In this context, living in or close to poverty is mentioned as a chronic stressor that cuts across different social roles and affects people in all of their relationships (Pearlin 1989).

According to Pearlin's stress model,[3] whether a person is able to avoid stress when faced with economic hardship depends critically on counteracting role strains and the diminishment of self and this, in turn, depends on the quality of support and coping resources a person has at his/her disposition.[4] These 'mediators' as Pearlin calls them are, on the one hand, a kind of *social support* that is characterized by high quality relationships, more so than by the mere size of the social network. On the other hand, Pearlin differentiates between two types of *coping* that differ by their function, one that is oriented towards the problem and the other that is of a cognitive order, modifying the interpretation of the situation or managing the negative emotions it causes.

At this point it becomes obvious that in Pearlin's model some aspects of coping are impossible to separate from the actual activity of appraisal (Alley and Kahn 2012), which is at the core of our research question. Indeed, appraisal represents an iterative process over time, taking into account past and current situations, expected possibilities in the future as well as personality and values (Monroe and Kelley 1997). Richard Lazarus, one of the most widely acknowledged experts of the measurement of appraisal, describes it as a cognitive process by which people constantly evaluate the realities of their experience with respect to its significance to personal well-being: appraisal provides a means for protecting themselves from

---

[1] Examples for role strains include interpersonal conflicts, role captivity and role restructuring (Pearlin 1989, p. 245)

[2] In later articles, he modifies the original model by adding so called 'secondary stressors', that is stressors that are the direct consequence of 'primary stressors' and are therefore second in sequence but that, in terms of their impact, may be just as strong as primary stressors (Pearlin 1989). In a more restricted sense they are present in the original model in form of 'role strains'. Since the vulnerability framework takes into considerations this type of downward spiral by assuming that any manifest outcome potentially becomes a new starting point (exposure) we refrain from going into a very detailed account of all possible components that could interfere in the stress process and focus on those that bring an added value to our final theoretical model.

[3] Pearlin uses the example of job loss to illustrate the direct and indirect factors that lead to low levels of psychological well-being. Estimating several path models he showed that economic strain functions as a mediator of the effect of job loss on depression. Economic strain was operationalized by the self-assessed difficulty in acquiring necessities of life and items that reveal a high standard of economic quality of life. He went on to demonstrate that part of the influence of economic strain on depression is indirect as it is mediated through the respondent's self-concept, operationalized by a scale of mastery and a measure of self-esteem.

[4] While in the original stress process model, self-concept is mentioned among 'sources of stress' Pearlin categorizes it later as a mediating resource alongside with coping and social support. There are arguably pertinent reasons for both conceptualizations.

threats but can also result in a change in perspective so that a threat may begin to appear more like a challenge that can be tackled (Lazarus and Folkman 1984, p. 22, p. 34). Despite the closeness of the two mechanisms in everyday experience, it is important to theoretically distinguish between appraisal and coping when studying the stress process (Lazarus and Folkman 1984). For the sake of disentangling the two in our own research, we deem it necessary to expound briefly on Lazarus theory of appraisal before returning to the outline of our theoretical framework based on the stress process model.

5.2 Lazarus's Appraisal Theory

Central to Lazarus theory is the distinction between a 'primary' and a 'secondary' appraisal that occur in interaction when people categorize an experience. The former refers to the assessment of 'what is at stake',[5] the latter takes this thought a step further by inquiring 'what can be done about it' (Lazarus and Folkman 1984, p. 31). The primary appraisal yields three possible results, depending on the bearing it has (or is thought to have in the future) on the person's well-being: firstly, the encounter can be seen as benign or even positive; secondly, it can be judged as irrelevant, or it can be perceived as stressful. In the case of its appraisal as stressful, an encounter can be further classified through the secondary appraisal by taking into consideration the availability and effectiveness of coping strategies. Thus, a stressful encounter is appraised as *harm/loss* if the person has suffered damage, as *threat* if damage is anticipated but coping seems possible, or as *challenge*, if the appraisal focuses on the potential gain or growth that might result from the situation after successfully coping with it. On a cognitive level, a situation can often appear as a threat and as a challenge at the same time, but the emotional response differs in that a challenge evokes a sense of excitement and exhilaration, whereas a threat is more likely to be accompanied by feelings of fear, anxiety, and anger (Lazarus and Folkman 1984, p. 33). Though the names of the two aspects of appraisal may suggest differently, 'primary' does not mean an antecedent timing in the process nor a greater importance for the outcome. In fact, similar to the concept of risk chains, the two constitute 'discrete and complementary sources of knowledge on which evaluation of the personal significance of an encounter rests' (Lazarus and Folkman 1984, p. 133 in Monroe and Kelley 1997, p. 125).

We find that appraisal theory lends further support to the theoretical distinction between two types of subjective measures (p. 25) and the validity of one of our primary research objectives: to arrive at a meaningful interpretation of the contrasting vulnerability profiles that emerge at the intersection of the three measurement angles and specifically, to exploit the potential insights that the 'inside-out'

[5] Lazarus mentions the following key questions that are typical for the primary appraisal: „Am I in trouble or being benefitted, now or in the future, and in what way?' (Lazarus and Folkman 1984, p. 31).

perspectives of self-assessed and perceived measures can shed on individual agency. Indeed, Lazarus' explanation of the complex interactions between the two appraisals in the production of stress sounds familiar: '(...) other things being equal, if the person is helpless to deal with a demand, stress will be relatively great because the harm/loss cannot be overcome or prevented. If the person has a high stake in the outcome, meaning that it touches a strong commitment, helplessness is potentially devastating. Even when people believe they have considerable power to control the outcome of an encounter, if the stakes are high any doubt can produce considerable stress' (Lazarus and Folkman 1984, p. 35). According to Lazarus, cognitive as well as affective mechanisms are at work in the process of appraisal: while the secondary appraisal is primarily of a cognitive order, both intervene in the primary appraisal: a stressful encounter that is appraised as a *challenge* would be the anticipation of gain or growth (cognitive component) and create a feeling of excitement (affective component) (Lazarus and Folkman 1984, p. 33).

These 'echoes' of familiarity we may hear at this point require careful verification because though the two concepts intersect with regard to the phenomena we are researching, they are interested in it for different reasons: Our research question focuses on the measurement level and considers the process as a black box. In our concept, we propose to disentangle the affective from the cognitive: the self-assessed measurement angle is appealing to the analytical faculty of the individual to evaluate the (economic) situation with respect to a certain criteria (e.g. the difficulties in making ends meet). The perceived measure, on the other hand, is thought to primarily tap into the affective dimension. Though it is impossible to separate the emotional experience entirely from the cognitive evaluation of the objective substance level, or vice versa, we believe that the converging perspectives of several theoretical models make a strong case in favor of this approximate distinction not just as a heuristic instrument but as the basis on which to construct complementary indicators for measuring the complex phenomena of low levels of economic quality of life in a more holistic manner.

5.3 A Typology of Economic Vulnerability

The considerations we have outlined so far about the nature of objective and subjective measure of economic vulnerability and the theoretical models that explain their mutual interdependence have prepared us to take a fresh look at the table of risk constellations we presented in the section 'Coping: taking human agency seriously' (p. 22) where the basic idea of the concept was illustrated using objective outcomes. We will now attempt to also derive the manifest outcomes for the self-assessed and the perceived measure by thinking through the risk constellations, paying special attention to the role of coping. Firstly, our goal is to extend our survey of the major theoretical situations of subjective economic vulnerability in order to tap into the heterogeneity that is concealed behind the binary categorization 'vulnerable' versus 'non vulnerable'. Related to the first concern, we expect to gain understanding about the criteria for relevance in labeling certain population groups 'vulnerable'. Thirdly, thinking through the risk chain in view of the probable manifest outcome for each measurement angle provides us with a systematic catalog, linking all theoretically possible combinations of risk factors with the most probable resulting juxtapositions of the three measurement angles. This index will be important for the empirical part of this volume, where the order of analysis will be reversed: data analysis will mainly focus on manifest outcome variables and only address risk factors in as far as they can be expected to be of a more permanent nature.

From the reviewed literature and theoretical considerations surrounding the vulnerability concept, we can expect that self-assessed and perceived outcomes will frequently diverge from objective outcomes, for each given hypothetical risk scenario. In order to move as closely as possible to the type of questions investigated in the empirical part of this volume, the following discussion is based on the same subjective indicators that will be employed in the statistical analysis. The *Self-Assessed Measure* of economic vulnerability will be operationalized based on the survey question 'Do you have difficulties in making ends meet given your household's monthly income?' with the vulnerable outcome (= 1) corresponding to the response options 'very difficult' and 'quite difficult', and the non-vulnerable outcome (= 0) to the responses 'very easy' and 'easy'. The *Perceived Measure* of economic vulnerability is based on the question 'Does the following situation worry you? – Not having enough money for current expenses.' Here, the vulnerable outcome (= 1) is identified when respondents said 'very worried' and 'quite worried', whereas the non-vulnerable outcome (= 0) is attributed to the response items 'not worried', 'not worried at all' and 'does not concern me'.

Table 5.1 shows all theoretically possible risk constellations leading to different combinations of objective and self-assessed outcomes. Like before, the schema was simplified by depicting exposure and threat as one single risk and by comparing just two measurement angles – the comparison between the self-assessed and the perceived measure will be shown below. The arrow indicates the presence of a risk factor if it is pointing downwards (box shaded in grey), and absence of risk if the arrow is pointing upwards.

The interpretation of risk constellation 1 does not change when moving from the objective to the self-assessed measure, representing economic *strain*. Those groups

Table 5.1 Risk constellations for self-assessed as compared to objective outcomes of economic vulnerability

Group of respondents	*latent vulnerability – risks*		*manifest vulnerability – outcomes*	
Risk constellation	**exposure + threat**	**coping**	**objective**	**self-assessed**
1	↑	not applicable	non-vulnerable	non-vulnerable
2	↓	↑ success	non-vulnerable	Vulnerable
3	↓	↑ absent, insufficient	vulnerable	Vulnerable
4 a)	↓	↓ do not realize	vulnerable	non-vulnerable
4 b)	↓	↓ do not want to mobilize coping	vulnerable	Vulnerable

of respondents who are exposed and who are facing a threat on the other hand, are likely to self-assess themselves as 'having difficulties in making ends meet', even if they are successfully coping with their income inadequacy (group 2) but certainly if their coping resources are lacking or insufficient, resulting in the experience of economic strain (group 3). Members of group 4a will not consider themselves as vulnerable because they do not recognize the adversity of their situation. There are various possible reasons for this type of incongruence, where the self-assessed measure indicates adequacy of resources relative to needs while the objective measure shows a lack of resources, and some have already been mentioned in earlier sections of this Part: For one, objective measures often do not capture all relevant resources or, according to reference theory, individuals have gotten used to living with limited means (adaptive preferences) respectively, they do not find that they have less than important others (social comparison). Similarly but distinct in the ethical implications, it is possible that individuals value a modest life style and that their limited material resources reflect a conscious choice. As for those individuals who choose not to mobilize their resources (group 4b), it is more difficult to predict whether they are going to consider themselves as vulnerable or not. While they do recognize the adversity of their situation, it is possible that they adjust their evaluation at this point by lowering their expectations and comparing themselves to others with less. However, according to the assumptions of this theoretical model, they would affirm that they are having difficulties in making ends meet.

As far as the heterogeneity of life situations between these theoretical groups is concerned, we find that only risk constellations 3 and 4b are classified in an identical way, namely as 'vulnerable according to both the self-assessed and the perceived measure. All other groups can be distinguished by their unique combination of the objective and the self-assessed measure. With regard to the discrete risks of the two identically labeled groups we see that they only distinguish themselves by their willingness or lack thereof to activate coping resources; other than this difference in attitude, there is nothing that allows distinguishing a different degree of vulnerability on theoretical grounds. Therefore, the label 'vulnerable' is relevant for describing both indiscriminatingly.

Moving on to the comparison between *perceived* and *self-assessed* outcome measures, we are now looking at the *stress* that is caused by the economic strain and exacerbated by a diminished sense of self or the feeling of lacking control over the situation (Table 5.2). Since constellation 1 is not possible in theory, we move directly to risk constellation 2a: in the presence of economic strain, respondents are likely to say that they are not worried about their financial situation if the coping strategy is working well and is sustainable (Krause 1997). However, if it causes them to experience a diminishment in the self-concept or a loss in mastery, the financial situation is likely to generate stress (2b). For risk constellation 3 and 4a), there is a high probability for congruence between the two subjective angles: scenario 3 describes the outcome of a failed attempt at coping, where economic strain is not relieved and feelings of stress are likely to be the result. 4a refers to a situation where low levels of resources are not experienced as deprivation, thus leading to neither strain nor stress. Risk constellation 4b) evokes the same question as for the

Table 5.2 Risk constellations for perceived as compared to self-assessed outcomes of economic vulnerability

Group of respondents	*latent vulnerability – risks*		*manifest vulnerability – outcomes*	
Risk constellation	**exposure + threat**	**coping**	**self-assessed**	**perceived**
1	↑	not applicable	non-vuln.	non-vuln.
2 a)	↓	↑ success, sustainable	vulnerable	non-vuln.
2 b)	↓	↑ success, role-strain	vulnerable	vulnerable
3	↓	↑ absent, insufficient	vulnerable	vulnerable
4 a)	↓	↓ do not realize	non-vuln.	non-vuln.
4 b)	↓	↓ do not want to mobilize coping	vulnerable	non-vuln.

self-assessed angle: in this scenario it is likely that individuals are not emotionally bothered by their experience of economic strain because if the concern was very strong, they would most likely make it a priority to resolve the problem, placing them in group 2 or 3.

Revisiting the question of concealed heterogeneity, following the logic of vulnerability framework yields a number of concordant outcome-pairs between the self-assessed and the perceived measure, resulting from very distinct risk constellations. It needs to be assessed whether the mixing up of these individuals jeopardizes the relevance of the observed outcome. Groups 1 and 4a are both 'non-vulnerable' according to both subjective measures. However, while group 1 has been spared from any risk factors, group 4 is highly vulnerable because in addition to being exposed and threatened, no coping resources are mobilized since these respondents fail to recognize their situation as adverse. In order to be able to distinguish the two groups, the Objective Measure needs to be taken into account.

The second concordant pair of outcomes is group 2a and group 4b, both of which are 'vulnerable' by the Self-Assessment Measure and 'non-vulnerable' by the Perceived Measure. The two groups are, however, distinct in terms of their level of actualized quality of life. Cognitive awareness of the adversity has led to different responses: while group 2 has managed to find a lasting solution to remedy the dire economic situation, group 4b has merely adjusted to low levels of quality of life, maybe because other things are felt to be more important at this point in life. Again, the Objective Measure will allow telling one scenario from the other. The third pair of subjective outcomes, groups 2b and 3, shares in common that they are 'vulnerable' according to both measures. Here, too, we have to reckon with substantial differences in the actual experience of quality of life: group 2b has successfully moved out of the immediate precarious state, though the performance of getting out and staying there causes them to affirm that they are facing 'difficulties in making ends meet' and that they are 'worried about not having enough money for current expenses'. While responding in the exact same pattern to these survey questions, members of group 3 have not been able to change their objective situation. This is yet another example of the added-value of looking at the three measures jointly.

As the comparison of one-by-one measurement angle demonstrated, combining the three outcomes – objective, self-assessed and perceived – provides a typology that can serve as a shortened index offering information on latent vulnerability derived from the underlying risk chain constellations (Table 5.3). For better readability we use the letters 'A' to denote absence of vulnerability and 'B' for presence of vulnerability; the order of the letter indicates the measurement angle from objective to subjective so that, for example, Type ABA denotes a manifest outcome that is non-vulnerable by the objective, vulnerable by the self-assessed, and non-vulnerable by the perceived measure. Two vulnerability types are not naturally derived from the risk constellations: Types AAB and types BAB. The first one reflects the situation of respondents who are worried about their finances even though they objectively have sufficient economic resources and they themselves do not consider their situation as being inadequate. For this group of respondents, the vulnerability framework does not account for, though we did find some individuals with this response pattern in empirical data (shown later). It is likely that among this group, there is an increased probability of displaying neurotic personality traits. Type BAB appears to be a more unlikely combination of vulnerability outcomes because of the way the pathways leading to economic stress is thought to be mediated by the experience of economic strain.

This classification of vulnerability has to be located at the nominal level of measurement even though a ranking order can be identified between some of the types. It is very clear, for example, that type BBB is more vulnerable than type AAA. However, the ranking between types BAA and types ABA is less clear as it would require establishing an order of importance between measurement angles (in this example, between the objective and the self-assessed angle). An important objective of our research will therefore be to empirically establish which of these types are more vulnerable in terms of their life circumstances and coping resources.

Table 5.3 Typology based on manifest outcomes of objective, self-assessed, and perceived vulnerability

Group of respondents	*latent vulnerability – risks*		*manifest vulnerability – outcomes*			*Vulnerability Typology*
Risk constellation	**exposure + threat**	**coping**	**objective**	**self-assessed**	**perceived**	
1	↑	not applicable	non-vuln.	non-vuln.	non-vuln.	**Type AAA**
2 a)	↓	↑ success, sustainable	non-vuln.	vuln.	non-vuln.	**Type ABA**
2 b)	↓	↑ success, role-strain	non-vuln.	vuln.	vuln.	**Type ABB**
3	↓	↑ absent, insufficient	vuln.	vuln.	vuln.	**Type BBB**
4 a)	↓	↓ do not realize	vuln.	non-vuln.	non-vuln.	**Type BAA**
4 b)	↓	↓ do not want to mobilize coping resources	vuln.	vuln.	non-vuln.	**Type BBA**

5.4 Conclusion

The objective of this first part has been to lay the theoretical foundation for approaching the measurement of economic vulnerability among pensioners in Switzerland from a quality of life perspective, integrating both objective and subjective measures.

Before reviewing the main propositions presented in this theoretical part, it seems appropriate to briefly explain our insistence on the concept of vulnerability to study a topic that could just as well have been analyzed through the lens of more widely used concepts such as poverty or precarity. While the heuristic benefits and distinguishing features of the probabilistic concept of (latent) vulnerability have been amply discussed, the question of why we cling to the concept of vulnerability even though our empirical analysis of cross-sectional data will focus on *manifest* vulnerability is legitimate: why this apparent detour instead of using the more specific concepts of income poverty, economic strain or stress as they have traditionally been handled within their respective fields of economics, social-gerontology and psychology? The reason is that using the more general concept of 'manifest economic vulnerability' for denoting the outcomes of the risk cycle provides a better basis for the conceptual contribution of this volume. In combining the concepts "vulnerability" and "economic quality of life" we obtain a construct – vulnerability to low levels of economic quality of life – that lends itself well to an interdisciplinary approach. While it would have been an option to simply label the manifest outcomes of vulnerability according to their specific manifestation of low economic quality of life (income-poverty, economic strain or stress) and thus emphasizing the disciplinary origins of the individual concepts, we find that the term 'economic vulnerability' in combination with the respective measurement angle (objective, self-assessed and perceived) enhances conceptual clarity by emphasizing the interconnectedness of the three dimensions. In other words, we decided to employ a concept and a terminology that underscores our hypothesis that income-poverty, economic strain and economic stress are all aspects of the *same* phenomenon (though circumscribed more or less narrowly), the study of which benefits from a triangular approach. Instead of approaching the concept of low levels of economic quality of life from a disciplinary perspective, we approached it from a measurement perspective. We therefore started out with a definition of the 'smallest common denominator' of the phenomena, a lack of financial resources (objective measure) in order to then turn our attention to the operationalization of the subjective evaluation. Thus, in our approach, arriving at a clearly circumscribed concept of *what it is* that we are measuring when we speak of 'low levels of subjective economic quality of life' does not represent the starting point but the goal of what we attempt to find out empirically. For this reason our drawing on various disciplinary perspectives was focused on the measurement aspect. Similarly, our research objective of assessing low levels of *economic* quality of life made the more general definitions of quality of life research of limited use to our purpose.[6]

[6] It was only for the sake of our theoretical discussion on objective and subjective measures that we looked at quality of life research in more global terms in Chap. 3, p. 29. If the measures of eco-

We have argued that a relevant definition of 'economic vulnerability' is a matter of positional claims that must be anchored as much in the sociocultural norms and material living standard of a society as in the life experience of the persons identified as 'vulnerable'. The vulnerability framework by Butterfill-Schröder and Marianti helped expand our understanding for the interaction of a series of risk factors that together constitute the vulnerability to suffering from exceedingly low levels of economic quality of life. Though this volume concentrates on the manifest negative outcome ex-post, we have shown that understanding the probabilistic nature of vulnerability is crucial for the construction of measures that are capable of capturing what is meaningful to the individuals concerned as well as for informing social policy. Moreover, a theoretical peek into the black box of 'coping' has shed light on the great heterogeneity of respondents that are classified as 'vulnerable' by an objective standard but who do not, when asked personally, identify with this diagnosis.

In a brief excursus on 'objectivity' and 'subjectivity' in the context of social indicators we introduced the theoretical distinction between two types of subjective measurement, self-assessment and perception, which in the context of economic vulnerability can be denoted as self-assessed economic *strain* and perceived economic *stress*. Exploring measurement theory on objective and subjective concepts of quality of life prepared the ground for a closer look at the most important measures that have been used in economics and psychology.

One of our main conclusions regarding objective measures of economic vulnerability concerns the shortcomings of the absolute poverty line: well-known reservations about the adaptability of monthly income as an indicator for low levels of economic quality of life are all the more valid for the study of the retired population because of the great heterogeneity in this age group: on the one hand, the hypothesis that everyone has the same financial needs is a problematic assumption about an age group that is among the most diverse sub-population, especially in terms of health and disability. On the other hand, retired people are known to draw on resources other than income for their daily living, in particular wealth and home ownership: these resources play a more important role in their budget than for the working population. Despite the challenges related to measuring economic vulnerability based on an absolute threshold, the advantages of an indicator that facilitates comparisons of poverty rates in different places and at different times makes monetary poverty lines an indispensable instrument for research and policy.

nomic vulnerability used in this volume had to be located within the broader streams of quality of life research, the income-based measure would correspond to the Scandinavian tradition, that defines welfare as objective living conditions, assessed in a way that explicitly avoids any influence by the individual's own evaluation of the situation (Erikson 1993). The subjective measures, on the other hand, clearly resonate with a more recent rediscovery of a very old utilitarian tradition that states that 'if men define situations as real, they are real in their consequences' (Thomas and Thomas 1928: 571–572, cited in Noll 2004, p. 159).

The potential alternative of using subjective measures to assess economic vulnerability has been experimented with for half a century and across disciplinary lines. The initial disillusionment about the lack of congruence between subjective and objective measures has turned out to fuel fruitful interdisciplinary advances, most notably in the fields of happiness economics and positive psychology. Here, we pointed out that measuring economic vulnerability using self-reports is not without pitfalls: the very mechanisms based on comparisons to an internal frame of reference make them susceptible to variations that cannot be accounted for and that may hamper inter-individual comparisons.

Drawing on Pearlin's and Lazarus' work, we postulate that the differentiation between the two types of subjective measures that has thus far gone unnoticed in social sciences has the potential to yield important information about human agency. In order to harness the potential of each of the three measurement angles and to systematically extract the insights offered by the congruence or incongruence between them, we proposed to combine them into a typology. We furthermore integrated the three measures as manifest outcomes into the dynamic vulnerability framework, which revealed that each outcome constellation reflects a different theoretical risk chain. Thus, we postulate that a triangular typology of economic vulnerability provides an analytical instrument for distinguishing among subsets of distinct groups among the elderly population who are affected by different manifestations of economic vulnerability.

The empirical part of this volume can now be summarized by two overarching research objectives while the specific hypotheses will be introduced in Part II 7): the first general aim is to describe the prevalence of economic vulnerability in our sample population according to each of the three measures in terms of their demographic and socioeconomic characteristics. Based on the presented theoretical and conceptual foundation, we expect being able to empirically show that the Self-Assessed Measure (economic strain) outperforms the Objective Measure (low income) in terms of its ability to capture those individuals who are most economically vulnerable. The second overall aim of our volume is to assess the relationships between these two measures of economic vulnerability and integrate a third indicator, the Perceived Measure (economic stress). By relying on a tri-angular measure we expect to obtain more nuanced and robust results about the prevalence of economic vulnerability in our sample population (Bradshaw and Finch 2003). Moreover, we posit that the divergence between the three measures reveals relevant information on different manifestations of economic vulnerability including inter-individual differences in coping capabilities. For the purpose of verifying this hypothesis, we are going to operationalize the Vulnerability Typology that combines Objective, Self-Assessed, and Perceived Measures.

References

Alley, D., & Kahn, J. R. (2012). *Demographic and psychosocial predictors of financial strain in older adults*. Paper submitted at the 2012 annual meeting of the Population Association of America.

Avison, W., Aneshensel, C. S., Schieman, S., & Wheaton, B. (Hrsg.). (2010). *Advances in the conceptualization of the stress process – Essays in honor of Leonard I. Pearlin*. New York: Springer. Abgerufen von.

Erikson, R. (1993). Descriptions of inequality: The Swedish approach to welfare research. In M. Nussbaum & A. Sen (Eds.), *The quality of life* (pp. 67–87). USA: Oxford University Press.

Bradshaw, J., & Finch, N. (2003). Overlaps in dimensions of poverty. *Journal of Social Policy, 32*(04), 513–525.

Krause, N. (1997). Anticipated support, received support, and economical stress among older adults. *Journal of Gerontology: Psychological Sciences, 52B*, P284–P293.

Lazarus, R. S., & Folkman, S. (1984). *Stress, appraisal and coping*. New York: Springer.

Monroe, S. M., & Kelley, J. M. (1997). Measurement of stress appraisal. In S. Cohen, R. C. Kessler, & L. U. Gordon (Eds.), *Measuring stress: A guide for health and social scientists* (pp. S. 122–S. 147). New York: Oxford University Press.

Noll, H.-H. (2004). Social indicators and quality of life research: Background, achievements and current trends. In N. Genov (Ed.), *Advances in sociological knowledge* (pp. 151–181). VS Verlagfür Sozialwissenschaften.

Pearlin, L. I. (1989). The sociological study of stress. *Journal of Health and Social Behavior, 30*(3), 241–256.

Pearlin, L. I. (2006). The stress process revisited. In C. S. Aneshensel & J. C. Phelan (Hrsg.), *Handbook of the sociology of mental health* (S. 395–415). New York: Springer Science & Business Media.

Pearlin, L. I., & Lieberman, M. A. (1979). Social source of emotional stress. In *Research in community and mental health* (pp. 217–248). Greenwich: JAI Press.

Pearlin, L. I., Menaghan, E. G., Lieberman, M. A., & Mullan, J. T. (1981). The stress process. *Journal of Health and Social Behavior, 22*(4), 337–356.

Thomas, W. I., & Thomas, D. S. T. (1928). *The child in America; behavior problems and programs*. New York: A. A. Knopf.

Part II
Research Design

This part introduces the technical aspects of this volume. We begin by introducing the data set and the survey design, before moving on to posing the research questions and hypothesis and a discussion of the corresponding analytical methods that will be used in Parts III, IV and V. We proceed with a presentation of the variables used and a discussion on the treatment of missing values, which will take us to the description of the final sample.

Chapter 6
Data Set

The data used in this volume stems from an interdisciplinary survey that was collected in 2011–12 among community-dwelling and institutionalized pensioners living in Switzerland. «Vivre/Leben/Vivere» (VLV) is a cross-sectional survey that represents the cultural diversity of the retired population living in the three linguistic regions of the country.

6.1 Background of the Survey «Vivre/Leben/Vivere »

VLV is the third wave of a study that was originally launched in 1979[1] in French-speaking Switzerland (Geneva and Central Valais) and repeated 15 years later[2] in the same regions (Lalive d'Epinay et al. 2000). The title of the latest replication, which means 'to live' in French, German and Italian, reflects the overarching purpose of the study, namely the assessment of living and health conditions of older adults. All three waves of the survey were conducted by the Centre for the Interdisciplinary Study of Gerontology and Vulnerability (CIGEV) of the University of Geneva, Switzerland. The VLV survey's principal source of funding[3] came from the Swiss National Science Foundation via the Sinergia Grant n° CRSII1_129922

[1] The extended title of the 1979 survey was 'Withdrawal and dependency of older adults' and was part of the National Research Program N°3 «Problems of Social Integration in Switzerland».

[2] The extended title of the survey (1994) was 'Autonomy and sociocultural environment of older adults'; it was part of the National Research Program N°23 «Aging».

[3] The VLV survey also received financial support from Pro Senectute Schweiz, and logistic support from the following institutions: Département de la solidarité et de l'emploi du Canton de Geneva; Département des affaires régionales, de l'économie et de la santé du Canton de Geneva; Département de la sécurité, des affaires sociales et de l'intégration de l'Etat du Valais; Département des finances, des institutions et de la santé de l'Etat du Valais; Pro Senectute beider Basel, Dipartimento della sanità e della socialità del Canton Ticino.

J. Henke, *Revisiting Economic Vulnerability in Old Age*, Life Course Research and Social Policies 11, https://doi.org/10.1007/978-3-030-36323-9_6

and by affiliation with the National Centre of Competence in Research LIVES of which VLV represents individual project 13.

From its conception, the study was designed for promoting a holistic approach to the research on aging, understood as a life-spanning development that encompasses social and cultural dimensions, side by side with biological processes. One of the main findings of the original survey – which has since become state of the art in social gerontology – was the great diversity of aging experiences (Lalive d'Epinay and Christe 1983). The follow-up study in 1994 revealed an unspectacular but profound trend towards general improvements in both health and living conditions. These two tendencies, enhancement of quality of life and heterogeneity of aging experiences, constitute the base-line on which the VLV survey has built the objective of verifying the sustainability of these developments. The overarching research question has poignantly been phrased as 'democratization of old age?',[4] hinting at the historic novelty of 'old age' as a life stage in its own right, but also at the fragility of this social achievement and the risk that it may not continue to remain accessible to everyone.

6.2 Questionnaires and Procedures

In line with the overall purpose of the study, the survey questions tap into all life domains, assessing availability and use of resources in the areas of physical, cognitive and psychological health, social and family relations, leisure and occupation, living conditions and finances as well as opinions, values and beliefs. For the purpose of comparability, one third of the questions were maintained from the previous waves (Tholomier 2011). This criteria and compromises that come with working in an interdisciplinary team inevitably influenced the choice of questions included (Ludwig et al. 2014). Still, the broad and interdisciplinary scope of the resulting data represents an important strength for the empirical analysis: it includes detailed information on objective economic status, including wealth and housing tenure in addition to current income and past socio-professional status and also a wide variety of psychosocial measures.

Three types of questionnaires were used: a paper-pencil questionnaire, a computer assisted personal interview (CAPI) and a life history calendar, though the latter was not used in the analysis presented in this volume. The CIGEV decided to refrain from mandating a social research institute to conduct the study in order to be able to keep a close eye on the details of the surveying procedures. This concern was inspired by the historic heritage of the present study and the foreseeable challenge to include the most vulnerable segments of the surveyed population (Oris and Nicolet 2016). Interviewers were hired locally and received a 1-week training. The

[4] 'Behind the Democratization of Old Age: Inequalities within Progress' is the subtitle of the VLV survey in the project proposal submitted to and accepted by the Swiss National Science Foundation.

contact information of potential respondents was obtained from cantonal and national population registers. A personalized letter was sent to potential respondents, informing about the survey and announcing the subsequent phone call by an interviewer. If the person agreed to participate, the paper-pencil questionnaire and the life history calendar were sent by mail. The CAPI usually took place in the respondent's home, which also provided the possibility to briefly check the quality (i.e. completeness and coherence) of the self-administered questionnaires.

Throughout the survey design and data collection process, attention was given to ensure the greatest possible representativeness of the population group 65 and older, and especially those considered most vulnerable. Both, community-dwelling and institutionalized individuals were included and, in the case of cognitive problems or a physical condition preventing a personal response, a shortened *proxy*-questionnaire was filled out with a family member or a staff of the institution. If interviewers were unable to reach potential participants by phone, they were instructed to visit them at their home. The response rate was at 35%,[5] which is comparable to similar studies, such as the Survey on Health, Aging and Retirement in Europe (SHARE) (Oris and Nicolet 2016).

6.3 Population Surveyed

The VLV survey represents the diversity of the country by including all three linguistic regions (and by ensuring a mixture of areas that are predominantly catholic or protestant, and taking into account different degrees of urbanization. The cantons[6] included are: French speaking Geneva (GE) and Valais (VS), German speaking Berne (BE) and Basel (BS and BL) and Italian speaking Ticino (TI).

The sample was drawn at random and is stratified by sex, age (ranging from *65–69, 70–74* till *90 plus)* and canton, corresponding to 60 strata of each 60 individuals. The targeted sample size of 3600 individuals was met for most strata though some data sets are partial due to the above described proxy-procedure.

Of an overall target of 3600 interviews, a total of 3080 complete interviews were conducted (Appendix).

Splitting up the distribution by the stratifying variables sex and age reveals that men responded more frequently than women in all cantons. For the age groups 65–69, 70–74 and 75–79 the strata are largely complete (more than 90%). Among the octogenarians we reached an average of 80% of the targeted strata size. For the age group 90 and older we had to resort more frequently to proxy-interviews, resulting in an average of only 35 full data sets instead of the 60 that were targeted for one

[5] These figures were calculated for Geneva and Central Valais.

[6] In the canton of Valais, only the region ‚Central Valais' was covered; in the canton of Berne, the survey was limited to the regions Oberland, Seeland and Mittelland. For reasons of simplicity we will refer to these cantonal regions as ‚cantons' even though Valais and Berne were only partly sampled whereas ‚Basel' represents both, Basel-Landschaft and Basel-Stadt.

stratum. All cantons and both sexes taken together the subsample 90+ is still among the largest data sets of the 'oldest old' that have been collected in Europe.

Appendix: Sample Distribution of the VLV Survey, Without Proxy Data

This table does not include proxy-data that were excluded from our analysis since the shortened questionnaire (filled out by a third party) obviously did not feature any of the subjective questions relevant for this research.

	Geneva		Valais		Berne		Basel		Ticino		
Age groups	Women	Men	Women	Men	Women	Men	Women	Men	Women	Men	Total
65 – 69	58	58	58	59	62	70	67	67	58	56	613
70 – 74	56	54	58	59	65	62	57	59	57	67	594
75 – 79	58	55	50	60	63	65	57	57	55	52	572
80 – 84	43	49	45	51	54	60	56	49	43	49	499
85 – 89	44	45	35	43	49	58	40	44	44	49	451
90 +	23	35	28	30	37	39	31	52	34	42	351
Total	282	296	274	302	330	354	308	328	291	315	**3080**

References

Lalive d'Epinay, C., & Christe, E. (1983). *Vieillesses: situations, itinéraires et modes de vie des personnes âgées aujourd'hui*. Saint-Saphorin: Ed. Georgi.

Lalive d'Epinay, C., Bickel, J.-F., Maystre, C., & Vollenwyder, N. (2000). *Vieillesses au fil du temps 1979–1994: une révolution tranquille: santé, situations de vie, formes de participation et visions du monde des personnes âgées en Suisse*. Lausanne: Réalités sociales.

Ludwig, C., Cavalli, S., & Oris, M. (2014). "Vivre/Leben/Vivere": An interdisciplinary survey addressing progress and inequalities of aging over the past 30 years in Switzerland. *Archives of Gerontology and Geriatrics, 59*(2), 240–248.

Oris, M., & Nicolet, M. (2016). Mesures et capture de la vulnérabilité dans une enquête sur les conditions de vie et de santé des personnes âgées. L'expérience de VLV (Vivre-Leben-Vivere) en Suisse. In Oris, Michel & Cordazzo, Philippe & Bellis, Gil & Brown, Elizabeth & Parant, Alain. Les populations vulnérables. Actes du XVIe colloque national de démographie (207–224). Aix-en-Provence - 28-31 mai 2013 - Bordeaux: CUDEP (Conférence Universitaire de Démographie et d'Étude des Populations).

Tholomier, A. (2011). *De l'huile dans les rouages de VLV. Les questionnaires d'une enquête interdisciplinaire sur les personnes âgées en Suisse*. Master thesis. Geneva: University of Geneva.

Chapter 7
Theoretical Model and Research Questions

The overarching objectives of this volume is to assess the contribution of a triangular approach to measuring low levels of economic quality of life in view of a more nuanced profiling of economic vulnerability in wealthy countries like Switzerland, and as part of this endeavor, to evaluate the added value of distinguishing between two types of subjective measures. Figure 7.1 shows the theoretical model on which the empirical analysis is based. In line with the vulnerability framework presented earlier, the three indicators are conceptualized as simultaneously measurable manifestations of economic vulnerability: low income, economic strain and economic stress. Though they are not considered to be different points in a process, it helps to represent them in the order of increasing subjectivity (left to right), moving from the objective to the self-assessed to the perceived measure. This way of linking the three measures moreover visualizes the increasing complexity of factors that are captured by each outcome measure as we move from objective to subjective, as well as the relative closeness between objective and self-assessed measures (shared substance level), and self-assessed and perceived measures (shared mode of evaluation). So just as shown in Fig. 4.1, where the relation between the objective and the self-assessed measure is depicted as an extension of the measure of one and the same phenomenon, a lack of financial resources (measured by the objective indicator), the same overlap is hypothesized for the perceived measurement angle: having its basis in the cognitive appraisal of the economic situation (self-assessed measure), it is moreover influenced by a series of psycho-social factors. As discussed in Part I, the analysis of how the objective and the self-assessed measure are linked has already received considerable attention across disciplinary lines; what is original in the approach presented here is the attempt to further nuance between the self-assessed and the perceived measure, while preserving the embeddedness in the objective resource situation.

Let it be said clearly, however, that our objective here is not the analysis of the two processes of appraisal that lead from low income to economic strain and from economic strain to economic stress. Rather, we are proposing a series of analyses that take each of the measures of manifest economic vulnerability as outcome

J. Henke, *Revisiting Economic Vulnerability in Old Age*, Life Course Research and Social Policies 11, https://doi.org/10.1007/978-3-030-36323-9_7

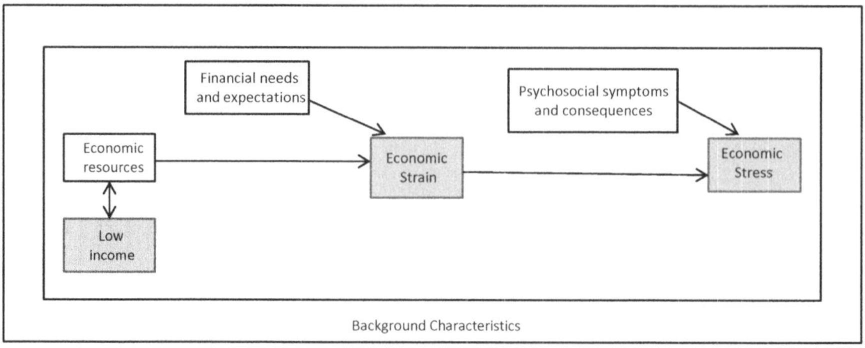

Fig. 7.1 Model for measuring economic vulnerability in a triangular approach

variables in regression models, or simultaneously (self-assessed and perceived measures) in structural equation models (SEM). The three outcomes of low levels of economic quality of life therefore represent a snapshot *ex-post* the risk cycle, when the mediating model components (Financial Needs and Expectations and Psychosocial Symptoms and Consequences) are thought to have reached a set-point as a result of an individual homeostatic process. The arrows in the theoretical model (and in the SEMs) therefore only represent pathways of causality in a very restricted sense: the relationship between the three measures and the covariates is grounded in our theoretical argument and cannot be ascertained.

The fact that we consider the measurement of all of these variables to be at 'the end' of the risk cycle (latent vulnerability) means that we are looking at a chronic type of economic vulnerability, where the primary stressor of a diminished income at retirement[1] has been met with (if available and wanted) coping strategies, both, those aimed at changing the situation and those that merely modify the subjective meaning of the situation. Therefore, variables pertaining to self-concept and mastery, too, reflect the success or – put negatively – the personal 'cost' these strategies have incurred for the affected individual in terms of psychological and social consequences. The path-model is embedded in what we refer to as *Background Characteristics*, encompassing contextual factors and demographic characteristics. These also include social statuses that may potentially influence any of the other model components because these characteristics tend to reflect the position of people within the larger systems of society (Pearlin 2006, p. 397, 398) and regulate access to social and cultural resources (Mayer and Wagner 1996).

Let us consider each of the model components. L*ow income* (objective measure) is connected to the more general model component of *economic resources*, which includes home ownership and wealth, with the potential to jointly result in *economic strain* (self-assessed measure). The relationship between the model

[1]The decision to focus on, low income' rather than, economic resources' as the objective indicator reflects the desire to make the empirical analysis pertinent for the social policy debate where income-based poverty measures are prevalent.

component economic resources and economic strain is mediated by (higher) *expectations* of a certain living standard and (increased) *financial needs*. It is not always easy to distinguish between the two. Tough we will attempt to differentiate them in the interpretation of our statistical models we conceptualize them as one single source of influence on economic strain. The added value of our model lies exactly in the possibility of taking the analysis a step further: two people who report the same level of economic strain, that is they experience some degree of inadequacy of resources, do not suffer of it in same way psychologically. We therefore posit that, in addition to the direct effect of economic strain on *economic stress* (perceived measure), the influence of economic strain is mediated by *psychosocial symptoms and consequences*, for example (higher) social isolation, (lower) sense of self and a (lower) sense of mastery.

Before we move on to the specific research questions, we need to briefly expound on the theoretical distinction between the sense of mastery and the self-concept on the one hand (as part of model component Psychosocial Symptoms and Consequences) and the perceived measure of economic vulnerability. In a discussion of the relationship between the above-mentioned psychological concepts and their predictive power for depression, Pearlin raises the legitimate question whether 'lowered self-esteem and mastery are inherent and inseparable symptoms of depression itself, not independent conditions whose loss contributes to the exacerbation of depression' (Pearlin 1989, p. 346). The same question must be asked about the distinctiveness of these concepts from economic stress as outcome variable. If it is true that, as we posit in our model, a lower self-esteem and a decreased sense of mastery are independent conditions that contribute to economic stress, we should be able to observe situations were economic stress exists but independently from lower levels of mastery and self-esteem. Indeed, we argue that that the presence of these psychosocial conditions reflects a perceived ineffectiveness or inadequacy of coping capacities resulting in an increased probability that resource inadequacy (economic strain) provokes stress. Inversely, elders possessing a strong sense of mastery may be more efficient in working out strategies to modify a potentially stressful environment. This problem-oriented coping can enable older adults to reduce financial difficulties in the future and to effectively contain economic hardship so that it does not intrude into other aspects of life and threaten their role and relationship functioning. Therefore, successful problem solving can prevent distress associated with difficult financial circumstances. Additionally, elders with high mastery may be more likely to appraise financial difficulties as controllable, focusing on those aspects of their financial situation that are amenable to change (Lazarus and Folkman 1984; Taylor and Aspinwall 1996). Analogous to Pearlin's logic, we may test this possibility by assessing the variations in mastery and self-esteem, on the one hand, and economic stress, on the other hand. If they vary independently from each other, we have found evidence that they are not merely measures of the same underlying condition of economic vulnerability.

From the general aims of this volume, stated at the end of Part I, arise the following research questions and hypotheses that will guide our empirical research:

I. The objective and the self-assessed angle of economic vulnerability

Is a self-assessed measure of economic vulnerability (difficulties in making ends meet) able to capture inadequacy of economic resources and is it more sensitive to the lack of economic resources as compared to an objective measure (absolute monetary poverty line)?

(i) Is there a significant difference between the two (overlapping) population groups that are vulnerable (a) according to the self-assessed measure and (b) according to the objective measure with regard to *background characteristics*?
The two groups are expected to be similar in terms of their socio-demographic background as both groups are at the lower side of the socio-economic gradient. Therefore, this set of variables, representing social strata, contextual factors and personality traits, is hypothesized to have a small but significant influence but only after controlling for economic resources (ii).

(ii) Is there a significant difference between the two (overlapping) population groups that are vulnerable (a) according to the self-assessed measure and (b) according to the objective measure with regard to types and level of *economic resources*?
Significant differences regarding type and level of economic resources are expected because of the differential scope the two measures are assumed to cover: the self-assessed measure is thought to assess adequacy of financial resources by taking into account resources other than monthly-income; moreover, the heterogeneity of levels of income among the population group that is vulnerable according to self-assessment is – by definition of the absolute monetary poverty line – greater than among the population group that is vulnerable by the objective measure.

(iii) Is there a statistically significant relationship between economic resources and the self-assessed measure?
The self-assessed measure is expected to be strongly and positively correlated with low levels of income and wealth, and negatively correlated with home ownership.

(iv) Is there a significant difference between the two (overlapping) population groups that are vulnerable (a) according to the self-assessed measure and (b) according to the objective measure with regard to their *financial needs and expectations*?
The self-assessed measure represents an evaluation of adequacy of all available resources in view of actual financial requirements. For this reason it is expected to be more sensitive than the objective measure to vary as a result of factors that may increase the (perceived or actual) need for resources such as a poor health status and social participation, which is expected to be less frequent among individuals who are having difficulties in making ends meet, all other things constant.

(v) Is there evidence that financial needs and expectations play a mediating role between economic resources and the self-assessed measure?
Controlling for socio-demographic characteristics and economic resources, part of the observed incongruence between the objective and the self-assessed at the individual level is thought to be explained by differences in financial needs and expectations; for example, among two individuals with very similar profiles in terms of background characteristics and economic resources, one may report finding it easy to make ends meet while the other one finds it difficult because he or she has a poor level of health, requiring additional resources to attain (maintain) the same level of economic quality of life.

II. The self-assessed angle and the perceived angle of economic vulnerability

Is it possible and relevant to distinguish between two types of subjective measures of economic vulnerability, a self-assessed measure (economic strain: difficulties in making ends meet) and a perceived measure (economic stress: worries about not having enough money for current expenditures)?

(i) Is there a significant difference between the two (overlapping) population groups that are vulnerable (a) according to the self-assessed measure and (b) according to the perceived measure with regard to *background characteristics*?
The two groups are expected to be similar in terms of background characteristics as members of both groups are thought to come from similar social contexts.

(ii) Is there a significant difference between the two (overlapping) population groups that are vulnerable (a) according to the self-assessed measure and (b) according to the perceived measure with regard to types and levels of economic resources?
The association between the perceived measure and economic resources is expected to be less strong than with the self-assessed measure because, while both indicators are subjective in nature, the perceived measure (economic stress) is moreover thought to be comparatively less determined by the cognitive aspects of appraisal.

(iii) Is there a statistically significant relationship between *symptoms and consequences of economic vulnerability* and the perceived measure?
We hypothesized that the perceived measure (economic stress) is the result of an emotional response to economic strain that is influenced by questions of identity and the ability to maintain social roles, both of which could be negatively affected by low levels of economic quality of life.

(iv) Is there a significant difference between the two population groups that are vulnerable (a) according to the self-assessed measure and (b) according to the perceived measure with respect to symptoms and consequences of economic vulnerability, as measured by psychological and social conditions that are characteristic of a stressful economic situation and of failing or insufficient coping mechanisms?

For reasons mentioned above, the perceived measure is supposed to be associated more strongly with symptoms of symptoms and consequences of economic vulnerability as compared to the self-assessed measure.

(v) Is there evidence that psychosocial symptoms and consequences play a mediating role between the self-assessed and the perceived measure?
All other things being equal, part of the observed incongruence between the self-assessed and the perceived measure at the individual level may be attributed to differences in how economic strain is experienced psychologically; for example, among two individuals who are similar in terms of background characteristics, economic resources and economic strain, one may not worry at all about his or her financial situation while the other one is very stressed because the financial difficulties are negatively affecting the self-concept and/or the ability to fulfill social roles.

(vi) Is there a statistically significant mediation effect of the self-assessed measure between income and the perceived measure?
We posit that the relation between the objective, the self-assessed measure and the perceived measure is rooted in one and the same situation, a lack of financial resources (measured by the objective indicator) but that the affective response (perceived measure) happens in direct response to how this situation is assessed by the individual.

The idea of examining the overlap between economically vulnerable population groups that have been identified based on different measures has been the focus of several studies. To mention just two examples, from a Western country and from contexts where wide-spread poverty dominates, that both come to a similar conclusion that other studies have, namely that there exists surprisingly little overlap between different poverty measures: in a study based on the Poverty and Social Exclusion Survey of Britain, three measures were used to identify the poor – lacking socially perceived necessities; being subjectively poor and having a relatively low income (Bradshaw and Finch 2003). The second article takes as a starting point the diverging definitions of poverty and the implications this has for targeting the poor and designing poverty reduction strategies, in particular in Peru and India. Comparing four concepts of poverty (monetary, capability, social exclusion and participatory approaches) the authors found that the lack of theoretical congruence is also strongly reflected in empirical data with little overlap between the groups identified as poor according to each of the four indicators (Laderchi 2000). Our research objective of systematically comparing measures of economic vulnerability with regard to their ability to target the most vulnerable is thus a well-established practice in the field of poverty research. What distinguishes our theoretical model is that the focus is on the idea of 'measurement angles' as opposed to different concepts of economic vulnerability/poverty. While it would be legitimate to argue that each measurement angle represents a different concept, especially when addressing it from various disciplinary perspectives, we are suggesting that these measures represent dimensions of the same phenomenon (see conclusion of Part I) that encompasses material as well as psychological aspects that are relevant to the experience of low levels of economic quality of life. This triangular approach has, to our knowledge, not yet been applied to the study of poverty nor economic vulnerability, neither has the integration of measurement angles into one single indicator, as we will do by means of the Vulnerability Typology.

References

Bradshaw, J., & Finch, N. (2003). Overlaps in dimensions of poverty. *Journal of Social Policy, 32*(04), 513–525.

Laderchi, C. R. (2000). *The monetary approach to poverty: A survey of concepts and methods*. Queen Elizabeth House: University of Oxford.

Lazarus, R., & Folkman, S. (1984). *Stress, appraisal, and coping*. Springer Publishing Company.

Mayer, K. U., & Wagner, M. (1996). Lebenslagen und soziale Ungleichheit im hohen Alter. In *Die Berliner Altersstudie Ein Projekt Der Berlin- Brandenburgischen Akademie Der* (pp. S.251–S.276). Berlin: Akademie Verlag.

Pearlin, L. I. (1989). The sociological study of stress. *Journal of Health and Social Behavior, 30*(3), 241–256.

Pearlin, L. I. (2006). The stress process revisited. In C. S. Aneshensel & J. C. Phelan (Hrsg.), *Handbook of the sociology of mental health* (S. 395–415). Dordrecht: Springer.

Taylor, S. E., & Aspinwall, L. G. (1996). Mediating and moderating processes in psychosocial stress: Appraisal, coping, resistance, and vulnerability. In *Psychosocial stress: Perspectives on structure, theory, life-course, and methods* (pp. 71–110). San Diego, CA, US: Academic.

Chapter 8
Analytical Methods

In this chapter we are going to introduce the methods that we will use to empirically answer our research questions. The objective is to provide a brief overview of our methodological choices by expounding on possible alternatives, and to provide the background information necessary for understanding the results presented in the subsequent chapters. We begin by determining the measurement level of ordinal-level variables and the resulting choices for bivariate statistics before specifically addressing the two main methods used, regression models and structural equation models. All statistical analysis presented in this volume was performed using the software Stata 12.

8.1 Determining Measurement Levels

In social sciences, carefully determining the level of measurement of variables is an especially salient question for ordinal-level data because of the debated option of applying statistical methods designed for interval-level data: being parametric in nature, interval-level variables fulfill the mathematical assumptions required for more powerful inferential techniques. In this chapter we will outline our guiding principles in the choice of treating ordinal variables as interval data, while in Chap. 9, where the variables are introduced, we will specifically consider the question of measurement level for each ordinal-level variable.

The general approach to determining the measurement level of our variables is as follows: we first distinguish between likert scale and likert-type items. Likert scales are treated as interval data if they follow a close to normal distribution. Likert-type items and other ordinal variables, too, are examined as to their distributional characteristics and if the distribution is approximately symmetric, summary statistics for parametric variables are provided as an overview. The strength of their association with our (binary) indicators of economic vulnerability will be measured using statistics for ordinal variables. As far as their use in regression analysis is concerned,

J. Henke, *Revisiting Economic Vulnerability in Old Age*, Life Course Research and Social Policies 11, https://doi.org/10.1007/978-3-030-36323-9_8

ordinal variables are treated as interval variables if the significance level remains invariant (Cliff 2014). The use of ordinal variables in Structural Equation Models is more complex and will be discussed in the corresponding section.

8.1.1 Ordinal and Interval Data

As mentioned in the theoretical part, the measurement perspective underlying this volume is informed by representative theory. According to this theory of measurement, 'numbers are used to represent empirical relations between objects' (Michell 1986, p. 398). The application of certain inferential procedures to ordinal-level variables is considered legitimate as long as the conclusions reached are implied by the empirical data itself, and could have theoretically been reached independently of the numbers assigned (Michell 1986). In the case of ordinal-level data, the number assigned to a category is therefore a label, indicating a specific order in a rank or the assumption of an underlying continuous variable, with the objective of taking advantage of the quantitative nature of the variable (Agresti 2013).

The differentiation of levels of quantification introduced by Stevens (1946) represents a major contribution to the philosophy of science because it outlines a classification of scales that is based on admissible types of data transformations (Cliff 2014). Stevens argued that the difference between ordinal-level and interval-level data can be summarized by the kind of transformations that are acceptable for each type of data: for ordinal-level data this is the *monotonic increasing* transformation whereas for interval data it must furthermore be a *linear* transformation (Stevens 1946). Though there are other ways of explicating the difference (Michell 1986), these mathematical arguments share in common that they point out the following: At the most basic level, interval-level data distinguishes itself from ordinal-level data in that the differences between different points of the scale are all equal, which makes the differences additive (Cliff 2014). This characteristic of interval-level data is the premise for the use of any inferential statistics that are based on the comparison of distances on a measurement scale. Consequently, the difficulties with treating ordinal-level data as intervals is that the categories cannot be assumed to be equidistant: the numerical difference only symbolically represents an ordered scale, but it cannot be inferred that the change from one point on the scale to the next (for example in a measure of attitude) is always of the same intensity or degree. Applying the arithmetical manipulations required to calculate the mean (and standard deviation) to ordinal data therefore means adopting assumptions that pertain to interval-level data and this bears the risk of wrongly interpreting the statistical results (Jamieson 2004). In research practice, however, many studies have come to the conclusion that parametric results remain robust even if these assumptions are violated (Norman 2010). For this reason, the approach followed in this volume is to make an informed choice based on theoretical considerations, while at the same time examining in practice the potential costs of losing statistical power.

Among various types of ordinal levels, Likert scales deserve special consideration because this type of survey questions were originally designed to be added up for building a measuring scale that is to be interpreted at the interval level.

8.1.2 Likert Scales and Likert-Type Items

Likert-scales are frequently used in psychometric studies for eliciting opinions and as an indirect measure of behavior. They were invented by Rensis Likert (1932) who originally developed this procedure for measuring attitudes. He asked respondents a series of questions with five response options and attributed to each one a number: strongly approve (1), approve (2), undecided (3), disapprove (4), and strongly disapprove (5). He then added the numbers from the series of questions to a score, thus creating a measurement scale.

Though these types of questions fall into the ordinal-level of measurement, they are almost always analyzed as interval data (Blaikie 2003). In order to avoid losing explanatory power by analyzing these variables at the lower ordinal-level of measurement, in this volume, too, Likert scales will be analyzed as interval-level variables if it is verified that they follow an approximately normal distribution.

In contrast to the original Likert scales, *Likert-type items* are not designed to be combined into a composite scale though they do use similar response alternatives and are often used as a series of questions in a survey (Clason and Dormody 1995). For Likert-type items and other ordinal variables such as questions on frequencies, the crucial criterion for deciding whether the variable can be treated as interval-level data is the number of categories and the symmetry of the distribution among the categories. For determining symmetry and asymmetry we used the following criteria, inspired from Rhemtulla et al. (2012): for five categories, which is frequent in our data set, perfect symmetry condition corresponds to the peak of the distribution lying in the middle category, moderate asymmetry (= approximate symmetry) is identified if the peak is found in the category left or right of the center, and the extreme asymmetry condition is fulfilled when the categories to the extreme left or right contains most cases. In the case of extreme asymmetry, we refrained from analyzing the variable with interval-level methods. Previous research has shown that for variables with less than five categories, a continuous methodology is generally not recommended (Rhemtulla et al. 2012), which is a principle that was heeded in this volume. Exceptions, such as treating ordinal variables with only four or three items at the interval level for estimating a regression model, will be mentioned explicitly.

8.2 Bivariate Analysis

The empirical research presented in this volume is based on four main dependent variables – the three measures of economic vulnerability and a composite measure, the Vulnerability Typology, based on the combination of the three measures. A first and important step in testing some of our hypotheses is describing the relationship between the various independent variables and each of the three measures by determining the statistical significance of the association and the effect size.

The appropriate measure of association is determined by the 'lowest measurement level' of the independent and dependent variable. Our three indicators of economic vulnerability being of a binary type, they can statistically be considered as either ordinal or nominal (Jann 2005). The Vulnerability Typology is a nominal variable with eight categories.

For assessing nominal relationships we use Cramér's V or Phi. Cramér's V is a chi-square-based measure of association that can be used for both nominal- and ordinal-level data but it is particularly appropriate when at least one level is nominal (Blaikie 2003). It is based on a contingency table, comparing the difference between observed and expected frequencies in each cell. Phi is also based on Pearson's correlations but is used to measure the strength of the relationship between two dichotomous variables. Both Cramér's V and Phi produce a coefficient that can take values between 0 and 1. For verbally describing the effect size, we use the benchmark labels recommended by Healey (2009), which are based on Cohen's seminal work (Cohen 1988). A coefficient of less than 0.1 indicates the absence of or a very weak association, 0.1 to 0.3 indicates a moderately strong and higher than 0.3 indicates a strong relationship (Healey 2009, p. 293).

Relationships at the ordinal level are assessed using the symmetric coefficients gamma and tau-b. Goodmanns and Kruskals' gamma (γ) is a measure of association based on the difference of proportions of concordant and discordant observations (Toutenburg and Heumann 2008). γ is -1 when there are no concordant observations and 1 when all observations are concordant. For the interpretation of the coefficient in terms of effect size we follow Acock's recommendations that presume a weak relationship for a coefficient below 0.2, a moderate relationship for a value between 0.2 and 0.49, and a strong relationship for any value above 0.49 (Acock 2008, p. 122). Gamma does not take into account ties nor the size of the contingency table (Cramer 1998). Consequently, the coefficient is larger when there are many observations that take the same score or position for both variables and when there are few categories. Kendall's Tau-b is similar to gamma but is less sensitive to the choice and frequency of categories as it takes into account tied pairs (Agresti 2012): Tau-b is based on the difference between the probability that the observed data are in the same order for both variables versus the probability that the observed data are in a different order for the two variables. In the case of many categories and few tied observations, gamma and tau-b yield similar effect sizes. The literature does not discuss the question when to use gamma and when tau-b with unanimity: gamma's characteristics can be seen as an advantage for ordinal variables with few categories, granted that it is statistically well founded[1] to code the variable with few categories and especially if there is reason to believe that the correlation is not of a diagonal nature but stronger between the corners of the contingency tables (Galtung 1970, p. 223; in Benninghaus 2007, p. 164). On the other hand, Agresti recommends using tau-b because it is less affected by a change in categories; he moroever argues that the quality of a measure of association is also defined by the stability of the results it yields (Agresti 2012, p. 191). For a more complete description of the relationship between two ordinal variables we will report both measures and decide from case to case, which seems more appropriate for interpreting the results. At last, a Point Biserial Correlation will be used for examining the association between binary indicators and interval-level variables (Weinberg and Abramowitz 2002). These correlations are interpreted analogous to Pearson's correlation coefficients.

8.3 Logistic Regression Models

Regression techniques are widely used statistical methods for quantitatively assessing the relationship between a dependent variable and two or more predictor variables. The underlying mathematical principle referred to as 'least squares method' was first published at the beginning of the nineteenth century by A.M. Legendre and C.F. Gauss in the field of astronomy (Chabert 1989). The general purpose of multiple regression was formalized by E.S. Pearson (Pearson 1967) and continues to be a burgeoning field of statistical inquiry. The following section is largely based on the work of Ulrich Kohler and Frauke Kreuter (2008) and J. Scott Long and Jeremy Freese (2006).

The basic objective of regression models is to find the equation that best predicts the values of the dependent variable, based on the values of one or more explanatory variables. Mathematically, the relationship between the independent variable x and the dependent variable y can be annotated as

$$y = f(x) + \epsilon$$

or as follows, if there are several independent variables

$$y = f(x_1, x_2, \ldots, x_n) + \epsilon$$

where ϵ is the random error.

The specific functions f we are going to use in this volume are a nonlinear *binary logit model*, usually referred to as binary regression model (BRM), and an *ordinal logit regression model* (ORM). There are various ways to derive the mathematical

[1] Some researchers may be tempted to collapse categories in order to artificially inflate the gamma coefficient.

model for BRM and ORM; here, we will briefly introduce the latent-variable approach, based on Long and Freese (2006). A thorough treatment can be found in Long (1997).

Regression models for binary outcomes are well suited to help us answer those research questions aimed at exploring how each explanatory variable affects the probability of an individual to be economically vulnerable. According to our theoretical considerations, economic vulnerability is a latent construct. Though we said it is preferable to operationalize it based on survey questions with an inbuilt cut-point,[2] the continuous nature of all of our dependent variables can mathematically be described as latent variable γ^*ranging from $-\infty$ to ∞. For a model with only one independent variable, the annotation of the structural component of BRM would be

$$\gamma_i^* = \alpha + \beta x_i + \epsilon_i$$

The measurement equation expressing the relationship between the observed binary variable γ and the latent variable γ^* for economic vulnerability would then be

$$\gamma_i = \begin{cases} 1 \text{if } \gamma_i^* \leq 0 \\ 0 \text{if } \gamma_i^* > 0 \end{cases}$$

In the concrete case of a regression model that takes one of our three measures of economic vulnerability as dependent variable, respondents with γ^* equal 1 are those who answered that their monthly income was CHF 2′400 or less (objective measure), they said they were finding it 'difficult' or 'very difficult' to make ends meet with their current income (Self-Assessed Measure) or they responded that their financial situation caused them to 'worry somewhat' or 'worry strongly' (Perceived Measure). Not all respondents classified as 'vulnerable' by, for example, the Self-Assessed Measure, are vulnerable to the same degree or with the same certainty since the coding includes two response categories ('difficult' and 'very difficult'). Furthermore, each of these categories encompasses a wide range of actual life situations and experiences.

The binary logistic model assumes that errors ϵ are distributed logistically with Var (ϵ)= $\pi^2/3$ leading to the following equation (Long and Freese 2006, p. 134)[3]:

$$\Pr(\gamma = 1 \mid x) = \frac{\exp(\alpha + \beta x)}{1 + \exp(\alpha + \beta x)}$$

The BRM is often interpreted in terms of changes in odds, i.e. the relation of the probability $\Pr(\gamma_i = 1)$ to the inverse probability of $1 - \Pr(\gamma_i = 1)$, or $\Pr(\gamma_i = 0)$ respectively, because this is easier to interpret than changes in β coefficients.

[2]The Self-Assessed Measure fulfills this criterion, but not the Perceived and the Objective Measures.

[3]exp(b) = odds ratios, subsequently abbreviated as OR.

It follows that

$$\text{Odds}\left(\gamma_{\frac{1}{0}}\right) = \frac{\Pr(\gamma_i = 1)}{1 - \Pr(\gamma_i = 1)} = \frac{\Pr(\gamma_i = 1)}{\Pr(\gamma_i = 0)}$$

The regression parameters estimated in BRM are fitted by maximum likelihood: Using a start value, a numerical method continues to iterate until the values of the parameters that have the greatest likelihood of generating the observed data (given the model assumptions) is found. The thus estimated coefficients can be tested using Wald test and likelihood-ratio test. Both tests presume a null hypothesis H_0, which implies constraints on the model's parameters, though for the likelihood test it is necessary to fit two (nested) models. The overall goodness of fit statistic is based on an analogy to the $\mathcal{R}^2$ for linear regressions of explained variance (Long and Freese 2006, p. 99).

The technique of ordinal regression models (ORM) will be used to test those hypotheses related to the Vulnerability Typology in parts IV and V. As the name implies, the outcome variables in these models are measured at an ordinal level, coded as consecutive integers from 1 to the highest number of categories. ORM are nonlinear models and from among the different models that have been developed, we are going to apply a version – the proportional odds model – that assumes an underlying latent variable (McCullagh 1980).

The structural component of the ordinal regression model is the same as for the binary regression model, defining γ^*as a latent variable, that can be thought of as the propensity of a respondent to be in one of the ordinal categories. The measurement model defines how γ^* is divided into J ordinal categories:

$$\gamma_i = m \text{if} \tau_{m-1} \leq \gamma_i^* < \tau_m \quad \text{for} m = 1 \text{to} J$$

where the τ is the cutpoint that is estimated and that determines where the continuous latent variable γ^*crossess a threshold and the observed category changes. For given values of the independent variables x, the probability of observing $\gamma = m$ is

$$\Pr(\gamma = m \mid x) = F(\tau_m - x\beta) - F(\tau_{m-1} - x\beta)$$

where F is the cumulative distribution function for the random error ϵ; it is logistic with Var $(\epsilon) = \pi^2/3$.

As for the BRM, the magnitude of the effect of a change in the independent variable on the outcome variable depends on the levels of all the independent variables. Thus, the challenge of interpreting the results consists in distilling those effects that are substantive from among a great number of possible dynamics conveyed in the output. While it is possible to simply look at standardized coefficients just like coefficients for the linear regression models or to consider the above described odds ratios, in this volume we follow the recommendation to use different methods of interpretation for describing the relationship between the explanatory variables and the outcome probabilities, more specifically, we will focus on changes in probability outcomes.

8.4 Structural Equation Models

Structural equation modeling (SEM) is a statistical technique that is widespread in social sciences because of its ability to combine latent variables with qualitative causal assumptions. In terms of other methods it can best be described as a combination of confirmatory factor analysis[4], representing the measurement part, and path analysis, making for the structural part of SEM (Acock 2013). Structural equation modeling represents a convergence of two traditions, bringing together the psychometric focus on unobserved (latent) constructs and the econometric emphasis on causal prediction (Chin 1998).

The first use of structural equations is likely to go back to Sewall Wright, who in the 1920s addressed the problem of simultaneity in the measure of demand (for a discussion, see Goldberger 1972). Several decades passed before SEM started to receive widespread attention, and only at the end of the 1980s as research questions in the social and behavioral sciences grew increasingly complex, the first comprehensive books were published (Hu and Bentler 1995). The advances in statistical software and the explicit interest of major journals, including the launching of a specialized journal in 1994 ('Structural Equation Modeling: A Multidisciplinary Journal') have certainly provided a fertile ground for the increased proliferation of SEM techniques.

Before we go into more technical details, a brief comment on what we understand by causality in the context of SEM seems appropriate because we will recurrently use the term 'causal'. In social sciences it is very rare to have data that have been obtained as a result of randomized trials where participants were randomly assigned to 'treatment' groups or where it was possible to control the exposure to the independent variables (Acock 2013, p. 59). Consequently, even if a model fits the data very well, the term 'causality' has to be used prudently because in most cases, the statistical results would be equally good if the relationship between two variables was reversed. Moreover, as discussed elsewhere, our primary objective in this volume is not to explain causal relationships between variables but to shed light on the relation – in the sense of 'common configurations' – of three manifest outcomes of economic vulnerability and the risk constellations that are typical for these configurations.

Following the logic outlined by Rick H. Hoyle (1994) and Alan C. Acock (2013), we will consider separately the two components that make up SEM, the measurement component based on latent variables and the structural component stemming from causal path models.

As discussed in the theoretical part, latent variables are variables that are unobservable and are not directly measurable. They are used for analyzing concepts that are somewhat vague and can therefore only be circumscribed in an approximate manner. In SEM it is common to distinguish between two types of latent variables

[4]According to Hoyle (1994), factor analysis can be considered a special case of the general structural equation model.

that are different with regard to their hypothesized relationship with the observed variables, which are called 'indicators'. Most latent variables used in social and psychological research are presumed to cause the observable indicators, that is, a latent construct makes participants respond in a certain way to the survey questions (Edwards and Bagozzi 2000). Because these observable variables reflect the latent variable, they are called *reflective indicators* (see Fig. 8.1). To take the example of measuring stress, a latent construct could be built upon a series of survey questions measuring the typical symptoms of stress, all of which tend to be highly correlated.

Sometimes it is more sensible to assume that the causal flow goes from the observed variables to the latent variable. Here, the former are called *formative indicators* (Fig. 8.2) because they cause the latent variable, which is referred to as a formative construct or a composite latent variable (Acock 2013). Typically, this is the case when each item provides a relatively independent piece of information about a broad construct, for example a deprivation index. Contrary to reflective indicators, which should all load very strongly on a single dimension, formative indicators vary in how strongly they are correlated with each other and, if a simple explanatory factor analysis was used, it could easily yield several factors (Acock 2013, p. 142).

Latent constructs such as the one depicted in Fig. 8.1 are usually based on questionnaires items that tap into the same construct, providing the basis for applying the method known as confirmatory factor analysis (CFA, for a detailed discussion refer to Brown 2006) on which the following discussion is primarily based). Factor analysis aims at discovering the latent variable that influences all (several) of these items and accounts for the correlation among them. The thus obtained scale score (factor) represents the shared meaning of the entire set of items, yielding a more parsimonious understanding of the underlying concept. Foundational for understanding factor analysis is the common factor model (Thurstone 1947), which posits

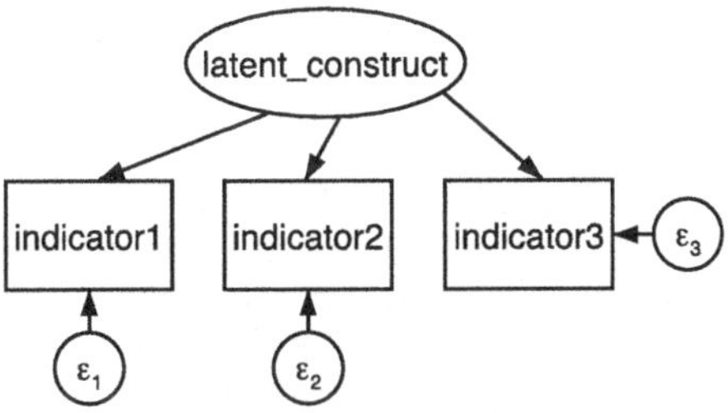

Fig. 8.1 Latent construct causing reflective indicators

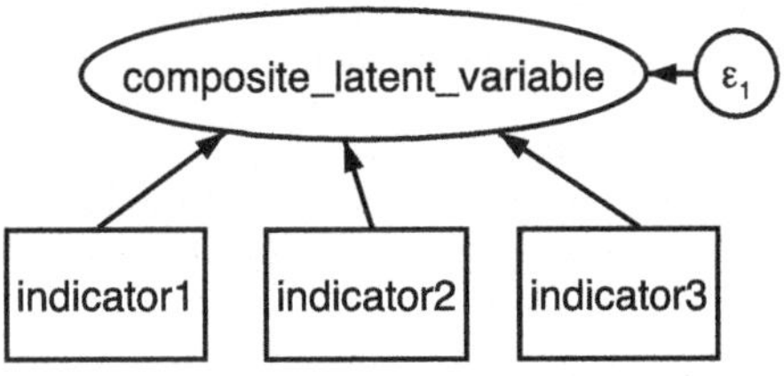

Fig. 8.2 Formative indicators causing composite latent variable

that each variable in a set of observed variables is a linear function of one or more common factors and one unique factor, thus, the variance deduced from the correlation matrix between these variables can be divided into the common variance shared by all variables (accounted for by the latent factor) and the variance that is unique to each variable, which includes the variance that is specific to this variable and the random error variance (Brown 2006). From this model, two general approaches are derived, exploratory factor analysis (EFA) and confirmatory factor analysis (CFA) (Jöreskog 1969). The difference between the two was succinctly stated by Bollen (2002, p. 615) in an article reviewing research on latent variables: 'in exploratory factor analysis, the factors are extracted from the data without specifying the number and pattern of loadings between the observed variables and the latent factor variables. In contrast, confirmatory factor analysis specifies the number, meaning, associations, and pattern of free parameters in the factor loading matrix before a researcher analyses the data' (cited in Marsh et al. 2014, p. 87). The CFA model is thus evaluated according to how well it reproduces the correlation matrix of the observed variables: the higher the shared variance (communality), the better the model fits the data.

Causal path model – the structural component of SEM – denotes a set of variables (latent or observed) that are in a particular relationship to each other. Two types of variables can be distinguished: *exogenous* variables are determined by factors outside the model and *endogenous* variables are explained within the model as function of other exogenous and endogenous variables (Pearl, 2009). Translated into the language of a diagram, an exogenous variable is any variable that has *no* arrow pointing to it (they never regress on other variables) whereas an endogenous variables is any variable that receives an arrow (other variables regress on it), independent from the fact whether arrows point away from it or not (Acock 2013). Consequently, in SEM it is possible that variables act as both, independent and dependent variables. This case of endogenous variables is usually referred to as mediator variables because they intervene between an exogenous and an endogenous variable and allow to measure the *indirect effect* of an independent variable on a dependent variable (Hoyle 2011).

The construction of *confirmatory* structural equation models, which is the type we are going to use, starts out by outlining the theoretical model in the sense of stating the relationship among variables. This allows visualizing the causal relationship: The arrows between the components (variables or set of variables) show the direction of the causal influence where a change in component A leads (all other components remaining equal) to a change in B. The specification of the causal model implies formulating statements about a set of parameters, which are constants that indicate the relationship between variables. The values of the free parameters[5], which may be factor loadings or regression coefficients, are then estimated:

[5]The pathways between two variables can either be left *free* to vary because the objective is to measure the relationship between the two, or it can be fixed, usually based on the estimations found in previous studies.

the matrix of covariance between the variables in the empirical data set is compared with the matrix of covariance calculated in the model and, using a 'fit criterion' (e.g. maximum likelihood estimation), the best fit is determined as the model that best represents the data (Hu and Bentler 1995). Besides the consideration for theoretical plausibility of the model, the researcher needs to ensure that the model contains at least as many pieces of information (observed variables and their covariances) as the parameters that need to be estimated[6].

Among the main advantages of SEM Chin (1998, p. vii) mentions the possibility to isolate errors in measurement for observed variables. Being able to measure and remove them, rather than assuming that they are random, as is the case in traditional regression models, increases the predictive power of the model (Acock 2013). Another advantage is the greater flexibility researchers have for the interplay between theory and data as compared to standing-alone techniques such as to factor analysis or multiple regression models: moving from variables to constructs, SEM allows to more closely align concepts with the corresponding statistical expression of the hypothesis (Hoyle 2011). In addition to afore mentioned capacity to model latent variables, this flexibility lies in the ability to model causal relationships between multiple independent and dependent variables. Given that in this volume we are dealing with three outcome variables and their interaction, this is a feature that makes SEM stand out as uniquely suitable to help in answering some of our research questions.

8.5 Weights

The survey design and data collection procedure of the VLV survey aimed at increasing the chance of accurately representing the surveyed regions and age groups of both sexes. For that reason enough observations for each subsets of the population in every region had to be collected in order to be able to make statistically significant inferences (Kish and Frankel 1974). Collecting strata of similar size (60, in our case) naturally means to over- or under-represent certain population groups, which, if analyzed without correction, could lead to biased or inaccurate point estimates and standard errors[7]. Furthermore, since the size of different age-sex population groups varies considerably, there exist differences in the probability of an individual

[6] If the model contains k observed variables, we have $\{k * (k + 1)\}/2$ pieces of information. A just-identified model (with 0 degrees of freedom) can be fitted but it cannot be tested by any measure of goodness of fit, which is why it should be avoided (Acock 2013, p. 43).

[7] In the case of the VLV survey, when we refer to the 'total population' we do not refer to all of Switzerland since the five sampling regions (cantons and parts of cantons) have not themselves been chosen in the context of a sampling process, rather, we refer to the total population aged 65 and older in all five sampling regions taken together. Because of the selection criteria described in 'Population surveyed' we believe that the survey is somewhat representative of the Swiss situation though not in a strictly speaking statistical sense.

to be sampled. We therefore had to calculate the sample weights[8] for each age-sex stratum in each sampling region, representing the inverse probability of an observation to be in the sample. This variable *weight* was calculated for each of the 60 strata, based on the stratifying variables *age* (65–69, 70–74, 75–79, 80–84, 85–89, 90 plus), *sex* and *canton* (Geneva, Valais, Berne, Basel, Ticino). The following is an example for the coefficient of stratum 1, where *n* denotes a subsample with the characteristics in parenthesis (for example: $\mathbf{n}_{(m,\,65-69,\,GE)}$=all male, aged 65–69, from Geneva) and where N is the total sample size and p the actual population size.

$$\frac{\mathrm{n}_{(\mathrm{m},65-69,\ \mathrm{GE})} \times \mathrm{N}}{\mathrm{p}_{(\mathrm{m},65-69,\mathrm{GE})} \times \sum n_{(\mathrm{m},65-69)}} = \quad \text{coefficient of stratum 1}\left(\text{male}, 65-69, \text{Geneva}\right)$$

For *descriptive statistics* with data obtained from surveys, it is necessary to take into account the survey design, because some observations represent a greater proportion of the underlying population than others and point estimates (e.g. comparing percentages or means) will be different for the sample left as such than if the actual composition of the sampling population is reconstructed with weights. We therefore used the standard procedures integrated in the Stata module for survey analysis to estimate uni- and bivariate statistics (Part III).

The decision whether or not to use sample weights for *analytical inference* about model parameters is more controversial. The spectrum of practices ranges from analysts who incorporate weights into every analysis to researchers who completely disregard survey design in their models (Pfeffermann 1993). Two sorts of questions can orient the researcher (Fuller 2009): the first one is substantial and pertains to the informativeness of the survey design for the model, that is, whether the design variables are statistically relevant to the subject matter parameters that are being estimated, but also when the distribution of the residuals is affected by the sampling design, resulting in biased parameters. In the second case, the weights have no content that pertains to subject matter but may be considered an artifact created by the sampler as a result of his/her knowledge of the sample population and, as a result, the estimated probabilities will be related to the errors in the model. If either of these scenarios has been confirmed, a methodological decision needs to be taken about how to consider the survey design in the analysis because instead of adding weights, the effect of design variables can be accounted for by specifying an extension of the model that incorporates their hypothetical influence (Holt et al. 1980). When it comes to analyzing survey data, the problems researchers are faced with when using regression models or structural equation models are very similar (Bollen et al. 2013) but we will address them separately in order to be able to exploit the more specialized literature.

[8]The method for calculating the sample weights was decided collectively by the researchers involved in the VLV study.

For *regression analysis*, homoscedasticity[9] is an important assumption on which statistical tests of significance are based. The assumption posits that the modeling errors are uncorrelated among themselves and normally distributed, and that their variances are not correlated with the effects being modeled, allowing to simplify the computations necessary to estimate model fit. Specialized statistical packages for survey data allow handling the heteroscedastic error terms, which could result in biased (overestimated) goodness of fit statistics as measured by the Pearson coefficient. The Stata module 'Svy' takes into account the heteroscedasticity of error terms by using adjusted Wald test statistic, which tends to increase statistical power of variables. Another drawback of using sample weights is that some post-estimation procedures, those that are based on maximum likelihood value, are inappropriate with survey estimation results (Thompson 1997). In this controversy, some authors propose to try both approaches, estimating the models with and without weights and preferring the non-weighted model if the difference between the estimated coefficients is considerably small (Dumouchel and Duncan 1983; Winship and Radbill 1994). We follow this recommendation, which is further justified by the fact that our stratifying variables (age, sex, canton) will be included in all regression models.

When it comes to *structural equation modeling* (SEM), despite the popularity of the approach, there is not yet much literature available on the use of weights (Bollen et al. 2013). In their recent review of the literature on SEM with complex survey designs, Bollen and colleagues explained the lack of attention to this methodological issue by the emergence of SEM within the psychometric and econometric tradition where simple random sampling is wide spread. They found a broad consensus for the use of pseudo-maximum likelihood[10] when estimating SEM with complex survey data. Their research showed that studies that did not apply this method tended to report biases in the parameters and wrong standard errors. They conclude that the decision on whether to include weights depends on the informativeness of the sampling design for the model parameters because if the sample design is not informative it can in fact be disadvantageous to use weights: 'when unequal selection probabilities do not produce bias, analyzing with weights can inflate standard errors' (Bollen et al. 2013, p. 1237). In our structural equation models, we will heed their advice and refrain from using weights because preliminary analysis indicates that in the case of our structural equation models, the sample variables are not informative.

---

[9]The assumption that model errors are independent with mean 0 and homogeneous variance (Kleinbaum et al. 2013).

[10]Pseudo Maximum Likelihood (PML) for linearization estimation of asymptotic covariance consists of two components: (1) replacing sample covariances by weighted sample covariances, and (2) replacing inverse Fisher information with a sandwich estimator of variance (Bollen et al. 2013, p. 1236). As alternative approaches a variety of resampling methods (jackknife repeated replication, balanced repeated replication, and bootstrapping) have been introduced in the estimation of SEM with complex survey data (Stapleton 2008).

References

Acock, A. C. (2008). *A gentle introduction to Stata* (2nd ed.). College Station: Stata Press.

Acock, A. C. (2013). *Discovering structural equation modeling using Stata*. College Station: Stata Press.

Agresti, A. (2012). *Analysis of ordinal categorical data*. Hoboken: Wiley.

Agresti, A. (2013). *Categorical data analysis*. Hoboken: Wiley.

Benninghaus, H. (2007). *Deskriptive Statistik*. Wiesbaden: Springer.

Blaikie, N. (2003). *Analyzing quantitative data: From description to explanation*. SAGE.

Bollen, K. A. (2002). Latent variables in psychology and the social sciences. *Annual Review of Psychology, 53*(1), 605–634.

Bollen, K. A., Tueller, S. J., & Oberski, D. (2013). Issues in the structural equation modeling of complex survey data. In *Proceedings 59th ISI World Statistics Congress*, 25–30 August, Hong Kong (S. 1235–1240).

Brown, T. A. (2006). *Confirmatory factor analysis for applied research*. New York: Guilford Press.

Chabert, J.-L. (1989). Gauss et la méthode des moindres carrés. *Revue d'histoire des sciences, 42*(1), 5–26.

Chin, W. W. (1998). Commentary: Issues and opinion on structural equation modeling. *MIS Quarterly, 22*(1), vii–xvi.

Clason, D. L., & Dormody, T. J. (1995). Analyzing data measured by individual Likert-type items. *Journal of Agricultural Education, 35*, 31–35.

Cliff, N. (2014). *Ordinal methods for behavioral data analysis*. New York: Psychology Press.

Cohen, J. (1988). *Statistical power analysis for the behavioral sciences* (2nd Ed., Auflage: Revised.). Hillsdale: Lawrence Erlbaum Assoc Inc.

Cramer, D. (1998). *Fundamental statistics for social research: Step-by-step calculations and computer techniques using SPSS for windows*. New York: Psychology Press.

Dumouchel, W. H., & Duncan, G. J. (1983). Using sample survey weights in multiple regression analyses of stratified samples. *Journal of the American Statistical Association, 78*(383), 535–543.

Edwards, J. R., & Bagozzi, R. P. (2000). On the nature and direction of relationships between constructs and measures. *Psychological Methods, 5*(2), 155–174.

Fuller, W. A. (2009). *Sampling statistics* (1st ed.). Hoboken: Wiley.

Goldberger, A. S. (1972). Structural equation methods in the social sciences. *Econometrica, 40*(6), 979–1001.

Healey, J. (2009). *The essentials of statistics: A tool for social research*. Boston: Cengage Learning.

Holt, D., Smith, T. M. F., & Winter, P. D. (1980). Regression analysis of data from complex surveys. *Journal of the Royal Statistical Society. Series A (General), 143*(4), 474–487.

Hoyle, R. H. (2011). *Structural equation modeling for social and personality psychology*. London: Sage.

Hu, L.-T., & Bentler, P. M. (1995). Evaluating model fit. In R. H. Hoyle (Hrsg.), *Structural equation modeling: Concepts, issues, and applications* (S. 76–99). Thousand Oaks: Sage.

Jamieson, S. (2004). Likert scales: How to (ab)use them. *Medical Education, 38*(12), 1217–1218.

Jann, B. (2005). *Einführung in die Statistik*. Wiesbaden: Oldenbourg Verlag.

Jöreskog, K. G. (1969). A general approach to confirmatory maximum likelihood factor analysis. *Psychometrika, 34*(2), 183–202.

Kish, L., & Frankel, M. (1974). Inference from complex samples. *Journal of the Royal Statistical Society, Series B, 36*, 1–37.

Kleinbaum, D., Kupper, L., Nizam, A., & Rosenberg, E. (2013). *Applied regression analysis and other multivariable methods*. Boston: Cengage Learning.

Kohler, U., & Kreuter, F. (2008). *Data analysis using Stata* (2nd Ed., 2. Aufl.). College Station: Stata Press.

Likert, R. (1932). A technique for the measurement of attitudes. *Archives of Psychology, 22*(140), 1–55.

Long, J. S. (1997). *Regression models for categorical and limited dependent variables*. Thousand Oaks: Sage.
Long, J. S., & Freese, J. (2006). *Regression models for categorical dependent variables using Stata* (Second ed.). Stata Press.
Marsh, H. W., Morin, A. J. S., Parker, P. D., & Kaur, G. (2014). Exploratory structural equation modeling: An integration of the best features of exploratory and confirmatory factor analysis. *Annual Review of Clinical Psychology, 10*(1), 85–110.
McCullagh, P. (1980). Regression models for ordinal data. *Journal of the Royal Statistical Society. Series B (Methodological), 42*(2), 109–142.
Michell, J. (1986). Measurement scales and statistics: A clash of paradigms. *Psychological Bulletin, 100*(3), 398–407.
Norman, G. (2010). Likert scales, levels of measurement and the "laws" of statistics. *Advances in Health Sciences Education, 15*(5), 625–632.
Pearl, J. (2009). Causality: Models. In *Reasoning and inference* (2nd ed.). Cambridge/New York: Cambridge University Press.
Pearson, E. S. (1967). Studies in the history of probability and statistics. XVII: Some reflexions on continuity in the development of mathematical statistics, 1885–1920. *Biometrika, 54*(3/4), 341–355.
Pfeffermann, D. (1993). The role of sampling weights when modeling survey data. *International Statistical Review/Revue Internationale de Statistique, 61*(2), 317–337.
Rhemtulla, M., Brosseau-Liard, P. É., & Savalei, V. (2012). When can categorical variables be treated as continuous? A comparison of robust continuous and categorical SEM estimation methods under suboptimal conditions. *Psychological Methods, 17*(3), 354–373.
Stapleton, L. M. (2008). Variance estimation using replication methods in structural equation modeling with complex sample data. *Structural Equation Modeling: A Multidisciplinary Journal, 15*(2), 183–210.
Stevens, S. S. (1946). On the theory of scales of measurement. *Science, 103*(2684), 677–680.
Thompson, M. (1997). *Theory of sample surveys* (Softcover reprint of the original 1st ed. 1997 edition). London: Chapman and Hall/CRC.
Thurstone, L. L. (1947). *Multiple-factor analysis; A development and expansion of The Vectors of Mind* (Bd. xix). Chicago: University of Chicago Press.
Toutenburg, H., & Heumann, C. (2008). *Deskriptive Statistik: eine Einführung in Methoden und Anwendungen mit R und SPSS*. Berlin/Heidelberg: Springer.
Weinberg, S. L., & Abramowitz, S. K. (2002). *Data analysis for the behavioral sciences using SPSS*. Cambridge: Cambridge University Press.
Winship, C., & Radbill, L. (1994, November 1). Sampling weights and regression analysis. *Sociological Methods & Research, 23*(2), 230–257. https://doi.org/10.1177/0049124194023002004.

Chapter 9
Choice of Variables

All of our research questions taken together, they span a wide array of life domains, requiring that a large number of covariates be considered in the empirical analyses. For each independent variable, a brief review of the literature will be given along with an introduction to the hypothesized association with the three dependent variables; the specific hypotheses will be explicated in more detail when introducing the respective analysis. Several variables needed to be transformed in order to be able to perform certain statistical methods. In these cases, details will be provided on the distributional characteristics (for a theoretical discussion of our choices concerning measurement levels see Chap. 2). In general, the descriptive analysis of the variables introduced here will be presented in Part III.

9.1 Dependent Variables: Measures of Economic Vulnerability

The concepts underlying the three indicators of economic vulnerability has already been discussed at considerable length in the theoretical part. In the following sections, the focus will therefore be on the specificity of operationalizing these measures with VLV data. A general word to the implications of the methodological choice to use binary indicators for measuring economic vulnerability is in order. As argued previously, we believe that, at the individual level, there is an absolute core to experimental economic vulnerability. The challenge now is to capture this threshold for the two subjective measures and this endeavor promises to be more successful, if the survey question provides an inbuilt threshold. In the case the self-assessed difficulties in making ends meet, we can suppose such an underlying dichotomy, indicated by the wording of the response items; the perceived worries about not having enough money, on the other hand, does not allow such an assumption. The questionable relevance of forcing all individuals into two income groups from the

J. Henke, *Revisiting Economic Vulnerability in Old Age*, Life Course Research and Social Policies 11, https://doi.org/10.1007/978-3-030-36323-9_9

perspective of the classified individual was one of our main reservations regarding the use of the monetary indicator, though we admit that it is relevant from a societal point of view. For the purpose of operationalizing the vulnerability typology, we are required to use binary measures; otherwise, the number of categories would explode as a result of all the combinations that are possible between three measures, each with several items. Combining the three binary measures into one is an attempt to preserve and interpret information that is lost if one looks at economic vulnerability through the lens of just one binary indicator. In order to prevent the loss of information, binary analysis among the three measures of economic vulnerability and between them and each of the independent variables will be performed using the original survey items, before dichotomization. This allows having a more precise measure of the strength of the association between each pair of variables; moreover, it prevents us from underestimating the variability in outcomes between groups and from overlooking non-linearity in the relationship between the variables (Altman and Royston 2006).

The *Objective Measure* of economic vulnerability (*obj_ev*) consists of an absolute poverty line based on the recommendations by the Swiss Conference for Public Welfare (CSIAS/SKOS). For the year 2010, the threshold was at CHF 2450.- per adult equivalent[1] (SKOS 2013). Household income being a categorical variable in the VLV questionnaire,[2] we recoded all responses in the lowest two categories (less than CHF1200/between CHF 1200 and 2400.-) as *income poor* after accounting for household size. Household equivalence was calculated based on the OECD scale.[3] The question was asked in a face-to-face interview setting, several interview techniques were applied in order to encourage a response to this delicate question: respondents were shown the nine income categories on a show card and asked to indicate which one most closely corresponded to their average monthly household income.[4] If respondents were very hesitant they were allowed to tick the box on the computer screen themselves so that the interviewer was kept from seeing their response. In the end, the percentage of respondents who chose not to answer this question was 14.5%, corresponding very closely to the proportion of nonresponse

---

[1] In January 2014, the most up-to-date recommendations available from CSIAS/SKOS are based on data from 2010.

[2] The scale consists of the following income categories: 'Less than 1200 CHF'/'Between 1200 and 2400 CHF'/'Between 2400 and 3600 CHF'/'Between 3600 and 4800 CHF'/'Between 4800 and 6000 CHF'/'Between 6000 and 7200 CHF'/'Between 7200 and 10000 CHF'/'Between 10000 and 10000 CHF'/'More than 15000 CHF'

[3] The OECD scale attributes a factor of 1 to the first adult in a given household and a factor of 0,5 for each additional adult. The OECD scale is among the most widely used equivalence scales. Given the categorical nature of our data on monthly income, this scale was more adapted than more fine-tuned versions of equivalence scales.

[4] The wording in the original questionnaire was as follows, in German and in French:

German:' Aufgrund des Einkommensmassstabes, den ich Ihnen zeigen werde, können Sie mir sagen, was das monatliche Brutoeinkommen Ihres Haushaltes ist?'

French: 'Sur la base de l'échelle de revenus que je vais vous énoncer, pouvez-vous me dire à combien se monte le revenu mensuel total brut de votre ménage?'

recorded in the 2nd wave (1994) of the survey (Lalive d'Epinay and Hofstetter-Bétemps 1999, p. 93).

The *Self-Assessed Measure* of economic vulnerability (*sa_ev*) is based on the following question: 'Thinking of your household's total monthly income, would you say that you are able to make ends meet?'[5] The response scale consists of four categories: with great difficulties, with some difficulties, quite easily, easily. This question was placed immediately after the question on income and wealth in the face-to-face interview. In order to obtain a binary variable, the response items were recoded into 1 representing vulnerable respondents ('with great difficulties' and 'with some difficulties') and 0 for non-vulnerable respondents ('easily' and 'very easily). Thus, the Self-Asssessed Measure meets the desired requirement of providing an inbuilt qualitative cut-point (see p. 34).

The *Perceived Measure* of economic vulnerability (*perc_ev*) was constructed using one item out of a set of questions on different sources of worry: 'To what degree do the following situations worry you today? – Not having enough money to pay for current expenses, for example for bills, rent or food.'[6] The response scale proposed five degrees of preoccupation: this doesn't worry me at all, this worries me a bit, this worries me somewhat, this worries me strongly, this does not concern me. The two subjective measures found in the VLV-survey are ideal for the purpose of assessing whether it makes sense to distinguish the self-assessed and the perceived dimension with regard to economic vulnerability because, while the question on financial worries taps into the emotional dimension of economic vulnerability, both survey items explicitly ask respondent to focus on the current financial situation (as opposed to a long-term perspective on provisions), and on the ability to cover basic needs, suggesting (or positing, in the case of sa_ev) that the frame of reference is to be their monthly household budget. Traditionally, the study of worries has been associated with anxiety, depression and general psychosomatic syndroms (Spielberger u. a. 1980). More recently, a non-clinical literature has emerged, conveying an understanding of worries that is no longer dominated by a pathological view. Some types of worries may even be associated with positive mental health,

[5] The wording in the original French questionnaire was as follows, in German and in French:

German:,Würden Sie sagen, dass Sie mit Ihrem monatlichen Haushaltseinkommen über die Runden kommen? Sehr schwer, Ziemlich schwer, Ziemlich einfach, Einfach'

French:,Compte tenu du revenu mensuel de votre ménage, diriez-vous que vous arrivez à joindre les deux bouts: Très difficilement, Assez difficilement, Assez facilement, Facilement'

[6] The wording in the original questionnaire was as follows, in German and in French:

German:,In welchem Ausmass belasten Sie die folgenden Situationen heute? Nicht über genügend Geld zu verfügen, um die laufenden Kosten zu begleichen, z.B. Rechnungen, die Miete und Nahrungsmittel zu bezahlen: Überhaupt nicht, Ein wenig, Ziemlich, Stark, Trifft nicht zu'

French:,Dans quelle mesure les situations suivantes constituent-elles une préoccupation pour vous aujourd'hui? Ne pas avoir suffisamment d'argent pour couvrir les dépenses courantes, par exemple pour payer les factures, le loyer ou la nourriture: Cela ne me préoccupe pas du tout, Cela me préoccupe un peu, Cela me préoccupe assez, Cela me préoccupe fortement, Cela ne me concerne pas'

specifically, if the focus of worries concerned wider sociopolitical problems (Boehnke et al. 1994). Boehnke and his colleagues found that while the *content* of worries is often related to the level of expectation and aspiration, the *intensity* of worries and their *duration* is often assciated with being personally confronted with the problem (for example, loss of income) (Boehnke et al. 1998). Our operationalization of the Perceived Measure of economic vulnerability taps into the content aspect, but also captures levels of intensity.

This survey question was found in the paper-pencil questionnaire. The response item 'this does not concern me' was designed to provide an opportunity for indicating that a given question among the list of worries is irrelevant to the personal situation of the respondent. For example, the question on worry about 'having to face serious conflicts with my partner or spouse' obviously only concerned individuals, who were in a relationship at the time of the interview. This response item seems to have been misunderstood by some respondents, who confused it with the response item 'this does not worry me at all'. However, for the construction of a binary indicator, these irregularities in response patterns did not cause any biases because both of the concerned response items were grouped into the same category, so that those who responded 'this worries me somewhat' and 'this worries me strongly' were categorized as perceived vulnerable (1) and those who said 'this does not worry me at all', 'this doesn't worry me at all' or 'this worries me a bit' were considered non-vulnerable. Contrary to the self-assessed difficulties in making ends meet, this variable does not provide clear qualitative criterion for subjective evaluation; consequently, the cut-point is less neat. Whether the proposed cut-point between the categories 'worries me a bit' (0) and 'worries me somewhat' (1) is relevant for distinguishing qualitatively distinct degrees of intensity of financial worries, remains to be shown empirically. Compared to other possible alternative, this cut-point seemed to represent a functional solution as it identified a group of a similar size as the two groups identified by the Objective and the Self-Assessed Measures.

9.2 Explanatory Variables

Following the plan of analysis outlined by our hypotheses, we classified our covariates according to four categories: *Background Characteristics, Economic Resources, Financial Needs and Expectations* and *Psychosocial Consequences and Symptoms*. The first block of variables simply refers to relatively permanent characteristics that count among the risks of exposure in the vulnerability framework, that is, demographic characteristics, marital status and educational attainment. We found it helpful to categorize our covariates in these four groups, corresponding to the components of our analytical model. However, it is not always possible to isolate the hypothesized influence of a predictor on the dependent variables in a neat way. For example, it would have been ideal to divide the covariates into a category 'increases needs' and 'diminishes resources', thus connecting the analysis more directly with the reviewed theory. The variable *marital status* provides a good illustration, why such a rigorous distinction can be complicated: Marital status is conventionally classified as Background Characteristic, but it can also influence Economic Resources through life course savings and economies of scale for cohabiting couples. It could also be considered within the frame of Psychosocial Consequences and Symptoms because being married or in a long-term relationship may represent a coping resource mediating the influence of self-assessed economic strain on perceived economic stress. The four blocks in which covariates have been categorized must therefore be seen as a framework for facilitating the analytical process by directing the focus on specific hypotheses. In what follows, a summary of the literature about the relationship between each covariate and economic quality of life or low levels thereof will be provided. Because of the novelty of the distinction between the Self-Assessed and the Perceived Measure, it will in most cases not be possible to cite literature addressing the two subjective dependent variables in a differentiated manner. The operationalization of each independent variable will be explained and, for variables requiring substantial recoding, the corresponding procedure will be outlined.

9.2.1 Background Characteristics

By 'background' we refer to a diverse set of variables, some of which are control variables and others that are thought to have an influence on the outcome variables without being part of any specific analytical component of our model. 'Sex' and 'age' were included as they are stratifying variables, though in some models, they are of analytical interest, too. As mentioned before, 'marital status' is a variable that may reflect psychosocial as well as economic resources. The variable 'canton' accounts for the varying socioeconomic contexts; it is the third stratifying variable of our sample. The variables 'origin' and 'educational attainment' were included to assess the influence of social background on our three measures of economic vulnerability. Lastly, we controlled for the influence of 'personality'. The decision to consider personality as a part of Background Characteristics has to do with the relatively permanent nature of these traits and the assumption that they could affect various model components such as the Self-Assessed and the Perceived-Measure as well as variables summarized in Psychosocial Symptoms and Consequences.

9.2.1.1 Sex

As explained before, women and men were sampled separately in our study because of the differences in response rates and the unequal sex-ratio among the oldest old. Due to the traditionally different life-courses of men and women in Switzerland, a gendered-pattern of economic vulnerability is likely (Budowski et al. 2002), though inequalities between the sexes have been reduced considerably by the 10th reform of the old age pensions in 1997 which introduced mandatory contributions independent of employment status: the previously cited study mandated by the Swiss

Federal Office for Social Security revealed that among the retired population, being a woman constitutes no risk in and of itself when controlling for age, education and level of old age provision (Wanner et al. 2008). A recent study with VLV data on the impact of life trajectories on poverty in old age confirms that gender differences are largely accounted for by differentials in economic resources and education (Gabriel et al. 2015). For the purpose of our study, the variable is particularly interesting because a series of studies suggest important gender differences with regard to emotional well-being (Weidekamp-Maicher 2008), which is relevant in the context of appraisal. While some authors attribute these differences to a generally more intense affect in women (Diener et al. 1985), others found that women experience only those negative emotions more strongly than men that have an internal focus, such as, for example, fear and nervousness (Noelen-Hoeksema and Rusting 1999).

9.2.1.2 Age

The relationship between the variable 'age' and measures of economic vulnerability is shaped by the triad 'cohort, period and age' – three effects that are not always easily disentangled. Studying older people from a life course perspective means to systematically place them into three temporal locations, a chronological age, a membership in family and generation and a historical period (Hagestad 2003). With regard to the objective financial situation of an individual, 'History' leaves its mark primarily through the age-specific potential of accumulation of economic resources at a given moment in the life cycle and the socioeconomic contexts an individual is placed at successive stages. For example, the observed decline in income with advancing age has commonly been explained by a gradual depletion of resources. However, in a study on tax data from the canton of Zurich, this trend is shown to have its roots in the historic development of the economy and the pensions system (Moser 2006). Age is also among the most discussed predictors of subjective economic vulnerability because this relationship has been shown to remain significant even when controlling for income and psychosocial variables (Litwin and Sapir 2009). Previous research has consistently found that older adults tend to be more satisfied with their income and that in fact, the predictive power of income decreases with advancing age (Weidekamp-Maicher and Naegele 2010). The age-gradient in subjective appraisals of adequacy at similar income levels has been suspected to reflect the improvements in living conditions of successive birth cohorts (George 1992). However, as mentioned above, this has not been confirmed for the Swiss context, suggesting, that psychological mechanisms related to individual age may have an effect in their own right.

9.2.1.3 Marital Status

As mentioned before, marital status may have a variety of direct and indirect effects on both Objective and Self-Assessed Measures of economic vulnerability. In the Swiss context today, the state of widowhood is no longer considered to have a generally negative impact on the financial situation (Bíró 2013; Wanner and Fall 2012), and these findings were confirmed by a study using VLV data (Gabriel et al. 2015). The state of being divorced, on the other hand, has been shown to remain a risk factor in Switzerland, especially for women (Falter 2009). The variable marital status was coded in the following way: married,[7] never married, separated/divorced, widowed.

9.2.1.4 Canton

The variable canton, another stratifying variable, reflects regional differences in standard of living and living costs as well as the variety of institutional tissue that is typical for the Swiss federal system. The designation 'canton' nonetheless contains a few variations in terms of geographical boundaries: only in the case of Geneva and Ticino, the population register of the entire canton was used for sampling, in Berne, it was limited to three out of six regions ('Mittelland' with the city of Berne, the mountainous 'Oberland' and 'Seeland', characterized by lakes); for Basel, both semi-cantons ('City' and the surrounding 'Countryside') were included and for Valais, only the region 'Central Valais' was covered. In the interest of clarity, we will use the names of the cantons as labels for these sampling regions: Basel, Berne, Geneva, Ticino and Valais.

9.2.1.5 Education

Educational attainment is a widely used measure to capture human capital, usually achieved in the first half of life. According to theory and the reviewed literature, low educational levels have to be considered a risk of exposure to future economic vulnerability, mediated via employment opportunities. If education is taken as a proxy for social background, it may furthermore be informative about the living standard respondents were exposed to when growing up, which in turn may reflect in the way they subjectively evaluate their current economic situation (Grusky 2001). The educational attainment scale features the following categories, representative of the Swiss educational system: Primary School (grades 1–6), Lower Secondary School (grades 7–9), Professional Training (apprenticeship, lasting 3 years on average), High School (grades 10–12), University of Applied Sciences, University. The

[7] Includes remarried couples and registered partnerships of homosexual couples.

distribution being approximately symmetric, this variable was treated as interval-level data (summary statistics in Appendix Table 9.6).

9.2.1.6 Personality

In the context of our research objectives, controlling for personality is important because it may directly influence the subjective evaluation of vulnerability. There is a wide agreement among scholars that the fundamental structure of the human personality is composed of five major dimensions (John and Srivastava 1999). This parsimonious framework classifies individuals in terms of their degree of Extraversion (energetic and sociable), Agreeableness (considerate and kind), Conscientiousness (hard-working and orderly), Neuroticism (nervous and tense), and Openness to experience (artistic and creative).

Specifically respondents who have a 'neurotic' personality may be more prone to feeling economically vulnerable (perc_ev) because of the propensity of this personality type to worry and be anxious. In previous research, personality profiles with high Neuroticism and low Conscientiousness have been positively associated with self-assessed economic vulnerability (Alley and Kahn 2012). Furthermore, some personality traits that could affect our subjective dependent variables are known to change with age, a phenomenon that needs to be controlled for in our models: previous research suggests that levels of Agreeableness and Conscientiousness tend to be higher at a more advanced age, whereas Extraversion and Openness are negatively associated with age (Donnellan and Lucas 2008).

We used a standard psychometric inventory for assessing personality, the so called 'Big Five' (Rammstedt and John 2007). The questions asked to respondents were introduced by 'I see myself as someone who', followed by a list of 10 attributes.[8] Participants were then scored (0–10) according to their responses (strongly agree, agree a little, neither agree nor disagree, disagree a little, disagree strongly) along five dimensions of personality. Since these scores are combined into (five)

---

[8]The wording of the list in English is:

1. is reserved
2. is generally trusting
3. tends to be lazy
4. is relaxed, handles stress well
5. has few artistic interests
6. is outgoing, sociable
7. tends to find fault with others
8. does a thorough job
9. gets nervous easily
10. has an active imagination

The coding was done as follows: Extraversion: 1R, 6; Agreeableness: 2, 7R; Conscientiousness: 3R, 8; Neuroticism: 4R, 9; Openness: 5R; 10. (R = Items 1, 3, 4, 5 and 7 were coded in reverse order.)

likert scales, they are conventionally treated as interval data (Boone and Boone 2012). In our sample population, the distributions of these scores follow an approximately normal curve; only in the case of Conscientiousness, the distribution is very skewed to the right and truncated (see Appendix Tables 9.1 and 9.2).

9.2.2 *Economic Resources*

The second set of explanatory variables represent various types of Economic Resources, in particular wealth and home ownership. For the descriptive analysis in Part III, we were moreover interested in the sources of income: occupational pension and financial support by family, State and private organizations as well as income stemming from occupational activities. These variables, describing the composition of household income, were not included in the inferential statistics because there, the aggregation of the sources – income, wealth and home ownership – was the focus of interest.

9.2.2.1 Wealth

The important role of wealth has been discussed in much detail in the theoretical part of this volume. The question was elicited by showing respondents a range of categories of wealth in CHF (nothing or hardly any, less than 60,000, 60,000–150,000, 150,000–500,000, 500,000–onc million, more than one million) asking them to point to the category that approximately represented their wealth.[9] The highest two categories were collapsed into a single category (500,000 and higher). The distribution being very skewed to the right and truncated, this variable was treated as ordinal-level data.

9.2.2.2 Owner

Home ownership plays a key role in the household portfolio of Europeans above age 50, 70% of whom are owners (Angelini et al. 2013). Living in one's own house or apartment was included as covariate because it may have an effect on reducing living costs, especially if the mortgage is completely or nearly paid off (Heady et al. 2009). The effect of home ownership may vary by region, depending on the average value of the real estate and on how common ownership is in a region: for example, it is known to be much more common in the canton of Valais than in Geneva. In Switzerland, the rate of home ownership among the population aged 50 and older is below the European average, at 56% (Kolb et al. 2013). The overwhelming effect of

[9]The wording of the question was « Parmi les catégories que je vais vous énoncer, où situeriez-vous approximativement votre wealth?'

home ownership for economic quality of life in Switzerland has been confirmed by Wanner and colleagues: they found that living in one's own home as compared to renting had the effect of dividing the risk of having low levels of economic resources by four, even after controlling for a number of demographic and contextual variables (Wanner et al. 2008). A comparative study by Cavasso and colleagues using SHARE data[10] confirmed this relationship: they found that the risk of financial distress[11] was lower among home owners compared to tenants (12% and 20% respectively) (Cavasso and Weber 2013). In the VLV survey, home ownership is operationalized by a binary variable.

9.2.2.3 Types of Sources of Income

The variables introduced in this section all recorded, whether a respondent benefits from a specific type of income and were therefore coded as dummy variables, 1 representing the existence of a given source of income.

Among these sources of income, two variables concern granted financial assistance: the legally granted *supplementary benefits* to the mandatory old age pensions, and the variable *support*, which combines assistance by family members, cantonal assistance or by a non-governmental institution. These three variables were combined into one because of the small number of respondents concerned. We hypothesize that if any one of the three types of financial support has been granted, this means, first of all, that the person has recognized her or his own financial need. Referring back to our theoretical discussion: the 'adversity' has been recognized and coping strategies have been embraced. This, in turn means that we expect the variable support to be positively associated with self-assessed economic vulnerability. Secondly, the presence of financial support or supplementary benefits means that someone with an outsider-view has come to the same conclusion as the respondent, which confirms the subjective appraisal that resources are inadequate.

Using Amartya Sen's terminology, both of these indicators contribute to a 'trans-positional' assessment (see Chap. 3) of economic vulnerability. While there is a strong probability that for both variables, the subjective appraisal of the economic situation will result in self-declared vulnerability, the association with the Objective Measure is more difficult to predict. As far as supplementary benefits are concerned, the threshold for eligibility was at CHF 2688.-[12] in 2011, meaning that beneficiaries are in fact likely to be above our threshold of CHF 2400.-.

---

[10]The Survey of Health, Ageing and Retirement in Europe (SHARE) is a multidisciplinary and cross-national panel database of micro data on individuals aged 50 and older.

[11]Financial distress was observed if financial wealth net of non-mortgage debt was less than the equivalent of 3 months of income (provided that income was not in the top 1/3 of the income distribution).

[12]In 2011, the eligibility threshold for a pensioner living alone was composed of CHF 19,050.- for general living costs plus up to CHF 13,200 to cover rent and utilities (Informationsstelle AHV/IV 2010).

Further types of economic resources include income from *occupational pensions* and *continued professional activity*. While the basic old age pension (AHV/AVS) can be taken for granted, occupational pensions are not yet universal. Information about continued professional activity was included, distinguishing between two binary variables of regular or irregular work. We expect *regular work* to be a protective factor against both, subjective and objective economic vulnerability. Besides the direct effect on monthly income, those pensioners who continue working on a regular basis tend to be highly qualified and well-earning. *Irregular work*, on the other hand, could just as well indicate the need for additional finances, in which case a positive association with self-assessed economic vulnerability would be expected.

9.2.3 Financial Needs and Expectations

The model component Financial Needs and Expectations was operationalized by a series of proxy variables. We included the 'socio-professional status' respondents attained prior to retirement in order to capture the social class they associated with, and which we expect to affect their expectations regarding a minimal level of economic quality of life. 'Health' is among the most important issues that preoccupy older adults (Idler 1993) and often comes with increased expenditures. At last, a lack of 'social participation' is what income-poor individuals in Switzerland seem to deplore mostly according to Pro Senectute's expert report.

9.2.3.1 Former Socio-Professional Status

Classifications by social class intend to capture the interwoven pattern of security, stability and prospects of an individual (Goldthorpe 2002). Literature reports several mechanisms through which social position can influence economic vulnerability in old age. Most obviously on a purely material level, a higher social position is associated with higher levels of income throughout the occupational life. This causal pathway has been confirmed for the Swiss general population where low social position as measured by a scale of socio-professional positions was the strongest predictor of income poverty (Budowski et al. 2002). Furthermore, social class may affect economic vulnerability indirectly through behavioral or lifestyle factors that influence physical and psychological health (Halleröd 2009), thus affecting the potential to accumulate resources throughout the life course. Another possible pathway that concerns subjective outcomes of economic vulnerability is the influence on expectations, for example, the minimum levels of self-assessed economic quality of life via the mechanisms of intra-biographical referencing and social comparison (Hazelrigg and Hardy 1997). Controlling for these effects is crucial for a responsible use of subjective economic measures because they are numbered among the core weaknesses of the subjective approach.

In the VLV survey, social position was operationalized based on the last pre-retirement socio-professional category (SPC). The SPC is a classification scheme that integrates both educational attainment and leadership position or employment status in a way that is adapted to the Swiss context (Joye and Schuler 1996). The result is a classification with some order between the sequential levels though statistically, it has to be treated as nominal-level data, since the ranking is not linear. The educational dimension specifically records training that is relevant for the exercised profession, allowing to distinguish between seemingly equivalent professions by differentials in the level of formal qualification (Joye and Bergmann 2005). In the VLV survey the original categorization scheme described by Joye and Bergmann (2005, p. 28) was extended by three more categories, singling out the farmers among the self-employed and distinguishing among the unskilled between the manual and non-manual workers. Furthermore, a category 'not active' was offered as option for those who had never worked (Appendix Table 9.3).

Socio-professional Categories:

1. Top executives and academic professions
2. Self-employed and farmers
3. Intermediary professions
4. Skilled, non-manual
5. Skilled, manual
6. Unskilled, manual and non-manual

Because the actual living standard as well as expectations thereof are likely to be forged by the life style shared within a household a 'household dominance model' (Grundy and Holt 2001) was applied, attributing the relatively higher SPC between the spouses to both of them.

9.2.3.2 Health

The variable 'health' was included because of its demonstrated relationship with economic vulnerability. Within our theoretical framework, we are interested in the variable health for two main reasons: as a marker for socio-economic standing throughout the life course and as a predictor for increased financial needs. Most research has focused on the effects of low socio-economic status on health. Many epidemiological studies use income as a marker of social standing and access to health-relevant resources across the life course (Newman and Cauley 2012). In the European context, poor health and health decline among older adults is associated more strongly with poor material conditions and subjective poverty than with income (Adena and Myck 2014). In late life, the mediation effect of income via the individual experience of economic strain or well-being on health seems to be particularly pronounced (Arber et al. 2014). There is furthermore evidence that the causal links between the material life situation and health outcomes are best accounted for by the psychological burden inflicted by persistent material hardship over the life course (Kahn and Pearlin 2006). Poor health may also be the reason why more financial resources are needed to maintain a given level of quality of life (Stoller and Stoller 2003). In this case, the pathway through which poor health may negatively affect the financial situation is via exploding medical expenditures (Chan et al. 2002) or costly adaptations of the living environment and investments in improved mobility. In the Swiss context, health problems are indeed counted among the main reasons why pensioners seek individual financial assistance (Seifert and Pilgram 2009).

We considered two different ways of operationalizing the variable health: the activities of daily living (ADL) scale (Katz S 1963) and the self-assessed general health status. The ADL scale was originally developed to evaluate treatments and make prognosis among the elderly and chronically ill. Based on an index that summarizes over-all performance in bathing, dressing, going to the toilet, transferring, continence and feeding, it is focused on the physiological aspects of health and the consequences of ill-health for individual autonomy. We only found a relatively small correlation between ADL and our dependent variables, suggesting that this conceptualization of health may be too narrow. Self-assessed health status is a construct that has been widely used in socio-economic research because it encompasses the bio-physical as well as the contextual factors that are relevant for the health experience of older adults (Jylhä 2009). The question 'How would you evaluate your current health status' was followed by a likert-type scale with five response items. Items were recoded into a variable with three categories (very bad and bad, satisfactory, good and very good).

In a cross-sectional design, it is difficult to establish the direction of the causal relationship between health and economic vulnerability. Previous research based on longitudinal data from the United States was not able to establish that episodes of illness predicted periods of economic hardship, defined by income at or below the poverty line (Lynch et al. 1997). However, the risk of reversed causation cannot be excluded with certainty.

9.2.3.3 Social Participation

The model component of social participation has been included because economic deprivation among retired people in Switzerland is essentially experienced as exclusion from social participation and as an inability to fulfill social roles (Seifert 2009). The importance that is ascribed to social participation for the general quality of life of older adults is reflected in a policy framework developed by the World Health Organization (WHO 2002): the response of the international community to the challenges and opportunities of population ageing is to propose a paradigm referred to as 'active aging' and it places a continued participation in social and cultural affairs at the center of its agenda.

Despite the considerable attention in policy literature and in empirical social gerontology, there is no agreement on the definition nor on the most important dimensions of social participation (Levasseur et al. 2010; Baeriswil 2016). Starting point for the concept of social participation used in this volume is the definition of

Table 9.1 Variables operationalizing the concept of social participation

Variable name	Description	Min.	Max.
Restaurant	Frequency of going to a restaurant, coffee shop or tea room	(1) never	(5) every day
Cinema	Frequency of going to see a movie, theater play or concert		(4) at least once a week
Trip	Frequency of taking a trip of at least 1 day		

a socially acceptable minimum level of participation in the Swiss society. Analogous to the discussion on the definition of poverty, the challenge consists in defining the components and standards that should be considered 'normal' in a multipolar society. A widely accepted approach to measuring socioeconomic participation is the deprivation approach. For the purpose of explaining the operationalization of our variable block 'social participation' it is helpful to briefly introduce this approach because it explicitly counts on social comparison processes. The concept of relative deprivation emerged in the late 1970s when a British sociologist demonstrated that poverty cannot be equated to a lack of finances but that it has to be seen in terms of the general standard of living of the society in question (Townsend, 1979). The approach has gained considerable support and a refined version has been adopted in the European survey on income and living conditions (SILC) and is thus also represented in the Swiss poverty report (OFS 2012). Deprivation is defined as a lack of consumer- or living-standard-items, which the majority of the population deems as essential. This list is established prior and respondents are then asked, which items they do not possess and the reason why they do not own it. Thus, for calculating the individual level of deprivation, the indicator only takes into account those items, which the respondent says to be lacking for financial reasons (that is, not by choice). The nine items featuring in the Swiss version of the indicator exemplify, what is considered minimal in this society today: to be able to effect a sudden expenditure of CHF 2000.-, to be able to take an annual week of vacation, having no arrears in payments (insurance primes, rent, etc.), to be able to enjoy a meal with meat, fish or a vegetarian equivalent every other day, to be able to afford adequate heating in the home, owning a washing machine, owning a color TV set, owning a phone and owning a car (Table 9.1).

The VLV questionnaire features an inventory of household and consumer items[13] that allows establishing a detailed description of the material living standard of Swiss pensioners. However, there is no information that would permit to differentiate between the reasons why items are lacking – because of a life-style choice or because of financial limitations. Moreover, the ownership of several consumer goods that are considered in the deprivation indicator described above, such as owning a TV or a phone, has become virtually universal. We argue that these items have

[13] The items include: hot water, toilet (WC), bath room, central heating, refrigerator, freezer, stove, micro wave, dish washer, tumbler, telephone (land line), cell phone, remote alarm, television, radio, CD player, DVD player, PC, internet.

therefore lost their relevance for a deprivation measure. Granted the specific requirements of a transnational database aiming at facilitating comparisons such as the SILC survey, there is reason to believe that socioeconomic participation in the Swiss context is more accurately captured with items that are higher up on the scale between 'basic' and 'luxurious'.

Given the data situation of the VLV survey and heeding the diagnostic of Pro Senectute concerning the unfulfilled social desires of many of their clients, we decided to focus our operationalization of social participation on a more restricted understanding, using the following three indicator variables:

Eating out in a restaurant is generally considered a luxury because, independent of the income level or the price segment, it is always relatively more expensive than a meal prepared at home. The frequency of going out to eat is thus considered to reflect a long-term strategy of managing according to a given household budget and can thus be used as a measure for assessing living standards (Angelini u. a. 2013). The variable 'restaurant' is an indicator based on the highest frequency of either one of two survey items: going to a restaurant or going to a coffee shop/tea room. The rationale for integrating these two survey items lies in the similarity of these activities in terms of social participation. There may in fact be a substitution effect between going out for a complete meal and consuming only a beverage and aggregating the two is a concession to a less luxury-bound measure, resulting in a more conservative indicator of social participation.

Among the three indicators, 'cinema' is the one that most clearly captures a life style that is interested in culture. It encompasses the options of seeing a movie or going to a concert; again, the item with the higher frequency was chosen. Interest in and enjoyment of cultural activities express a social habitus that is often rooted in socioeconomic position and educational attainment (Bennett u. a. 2009).

The variable 'trip' is again based on the highest frequency of two strongly correlated survey questions: taking a trip of 1–2 days by car, bus or train, and taking a trip of at least 3–4 days (to parents, friends or on holidays). Among the three variables, this one can be expected to be more strongly affected by ill-health and advancing age. Moreover, this activity being more costly than the others, it might be most strongly affected by scarcity of financial resources.

Following Scherger and colleagues (Scherger et al. 2011), the hypothesized association with our measures of economic vulnerability is thought to be influenced by several factors that are assumed to shape patterns of social participation in older age: firstly, there is the idea of continuity to pre-retirement as far as social capital and 'taste' is concerned because these things are the result of a life time of experiences and are not likely to change in later life. Secondly, leaning on Laslett's concept of a 'third age' marked by opportunities for new leisure and social activities (Laslett 1991), there is the expectations that the frequency of these activities is more likely to be intensified rather than decreased, all other things being constant.[14] The

[14] By this we understand that leisure activities in retirement are expected to be constant or more frequent, provided that important factors such as the financial situation and the health status allow for it.

view that older adults generally tend to 'disengage' from society is no longer a common view among gerontologists and specific reasons for withdrawal such as a decrease in general well-being or depressive symptoms as a consequence of retirement seem to affect only a small group for a limited duration; at least for the age group represented in our sample[15] (Scherger et al. 2011; (Fonseca et al. 2013). In the light of these considerations, the most obvious reason for a lower frequency of social participation is either economic strain or the reflection of a pre-retirement life style marked by low levels of participation. Thus, keeping economic resources constant, the frequency of participating in these socio-cultural activities is expected to be negatively associated with the Self-Assessed Measure of economic vulnerability.

Since the item 'restaurant' has five response categories (never, at least once a year, at least once a month, at least once a week, every day) and its distribution is close to symmetric, we treated it as an interval variable. The variables 'cinema' and 'trip' were treated at the ordinal level (Appendix Table 9.6) unless indicated otherwise.

9.2.4 Psychosocial Consequences and Symptoms of Economic Vulnerability

The third set of variables focuses on possible consequences and symptoms of economic vulnerability. Based on Pro Senectute's qualitative survey and other findings from poverty literature, three areas were identified in which the subjective experience of economic stress is most notable: a lack of social participation resulting in *social isolation* and *loneliness*, the inability to fulfill social roles and expectations, resulting in a *sense of diminishment* and a low *sense of mastery*.

9.2.4.1 Social Isolation

Social isolation is a condition that is frequently associated with old age. Two streams dominate the sociological debate about social isolation pertaining to older adults (Burns 2010). The first one suggests that older people are more vulnerable to social isolation because modern society tends to exclude them, particularly because of their withdrawal from the productive life. Furthermore, the risk of a reduced social network is suspected to increases with old age because loved ones of similar age are gradually passing away (Oris and Lerch 2009). Based on a life course perspective, the second debate views social isolation as risk that is independent from age but that increases with previous spells of isolation (Victor et al. 2008).

[15] Our sample population does not include individuals aged 85 and older.

A person's ability to participate in society and in networks of families and friends is affected by his or her financial situation. On a structural level, inadequate economic resources can constitute an obstacle to social participation because many social activities require spending money, such as entry fees, transportation or paying for a meal in a restaurant (Stewart et al. 2008). Economically vulnerable pensioners are also more likely not to be able to visit or receive visits than better-off pensioners because of reduced mobility due to poor health (Patsios 1999). Thus, economic vulnerability can go hand in hand with social isolation.

Already in the 1950s, Townsend distinguishes in ‚The Family Life of Old People' (1957, p. 166) between ‚social isolation', which is an (objectively) measurable lack of contact with other people and ‚loneliness' which is a (subjective) feeling of lack or loss of companionship. The distinctiveness but relatedness of the two concepts have been confirmed as being relevant for older adults by subsequent research (Havens et al. 2004; Cornwell and Waite 2009). Typical measures of social isolation include counting social contacts, social activities and describing social networks (Victor et al. 2008). The measure of social isolation used in this research is based on measures of frequency of contact with family and friends, either through visits or by phone calls (Table 9.2).

The two variables based on the frequency of visits are based on a combination of two survey questions, 'receiving visits (by family, or by friends respectively)' and 'visiting (family, or friends respectively)': among the two variables, the higher score was chosen in order to avoid underestimating the frequency of social contacts, for example of respondents with limited mobility who may receive more visits than they visit people themselves. The response options are 'no or hardly ever', 'at least once a year', 'at least once a month', 'at least once a week', 'almost every day or every day'. The distribution of the cases by category is perfectly symmetric for 'visiting friends' and approximately symmetric for the other three variables; thus all four variables pertaining to social isolation were consequently treated as interval-level data (see Appendix Table 9.6).

Table 9.2 Indicator variables of social isolation

Variable name	Description	Min.	Max.
Call friends	Frequency of phone calls with friends	(1) no or hardly ever	(5) almost every day / every day
Call family	Frequency of phone calls with family		
Visit friends	Frequency of visits to or by friends		
Visit family	Frequency of visits to or by family		

9.2.4.2 Loneliness

Previous research has shown that the lack of financial resources is among the most important predictors of loneliness (Fokkema et al. 2012) even when taking into account environmental resources, health, and psychological factors (Cohen-Mansfield and Parpura-Gill 2007). The feeling of loneliness was included in order to tap into the subjective dimension of the social situation. The wording of the survey question was ‘I am feeling a bit isolated and alone, even among friends’ and corresponds to an item of a well-established depression scale (Wang et al. 1975). Four response categories were offered: always, often, rarely, never. Because few respondents felt lonely ‘always’ or ‘often’, the two categories were united, resulting in an ordinal variable with three categories.

9.2.4.3 Sense of Diminishment

The concept of a sense of diminishment was operationalized using three variables: a self-worth scale, a question on self-confidence and on the worry about not being sufficiently recognized.

(a) *Self-worth*

The self-worth scale was constructed from a heuristic instrument called ‘assumptive world views’ (Janoff-Bulman 1989). It was designed to assess the basic schema people have about themselves and the world, and the impact of stressful life events on these basic assumptions. One of the dimensions focuses on respondents ‘worth of self’ and is a summary composition of four items shown in Table 9.3. The response items ranged from ‘strongly disagree’ to ‘strongly agree’ and were reverse coded for items 1, 3 and 4 (‘having a low opinion of self’, ‘satisfaction with self’, ‘feeling ashamed of self’ and ‘feeling no good’) before adding up the points to compute the scale.

Because the scale turned out to be highly skewed to the right (Table 9.4) a monotonic[16] transformation proved necessary: the ten lowest categories (6–15), containing very few observations, were recoded into one category; the subsequent categories were left unchanged. The resulting scale with ten scores now contains at least 5% of the observations in each category. The scale not being distributed normally, it was treated as an ordinal variable.

(b) *Self-confidence and recognition*

In addition to the self-worth scale, the concept of ‘a sense of diminishment of self’ was operationalized by the variables self-confidence and the worry about not being sufficiently recognized (Table 9.5). We expect these two questions to tap into

[16] Monotonic transformations only maintain the ranks but change the distances between points (Weiner et al. 2003).

Table 9.3 Items of the self-worth scale

Variable name	Survey question	Min.	Max.
1. Low opinion	I have a low opinion of myself.	(1) strongly disagree	(5) strongly agree
2. Satisfaction	I am very satisfied with the kind of person I am.		
3. Ashamed	I have reason to be ashamed of my personal character.		
4. No good	I often think I am no good at all.		

Table 9.4 Summary statistics of self-worth scale before and after recoding

Scale version (min.-max.)	25th perc.	Mean	75th perc.	Skewness	Kurtosis
Self-worth (6–24)	18	19.8	22	−0.82	3.6
Self-worth (1–10)	4	5.9	8	2.1	−0.3

Table 9.5 The variables self-confidence and worries about lack of recognition

Variable	Description	Min.	Max.
Self-confidence	To what degree are the following affirmations true for you? – I am self-confident.	(1) never or rarely	(3) always
Recognition	To what degree do the following situations worry you today? – Not getting sufficient recognition for what I do.	(1) strongly	(4) not at all worried/ doesn't concern me

the concept from different but complementary angles: a low level of self-confidence indicates a diminished sense of self, just as the worry about not being sufficiently recognized speaks about a deep insecurity that goes to the core of an individual's identity.

Both variables have likert-like response items that were coded in such a way that the vulnerability measures are expected to be associated with the lower scores. For 'self-confidence', the lowest two response options 'never' and 'rarely' were collapsed into one because there were very few respondents who answered 'never'. For 'recognition' the response options 'does not worry me at all' and 'does not concern me' were collapsed because of the previously noted ambiguity of meaning between these two response items; moreover, the distinction between the two items did not seem relevant for our research objectives. The question on worries about not getting sufficient recognition was part of the same survey question as the item used for constructing the Perceived Measure (worries about not having enough money for current expenses).

(c) *Mastery*

Mastery has been defined as 'the extent to which one regards one's life chances as being under one's own control in contrast to being fatalistically ruled' (Pearlin & Schooler 1978). Mastery or sense of control has been associated with self-assessed

Table 9.6 Summary statistics of the mastery scale, before and after coding

Variable (min.-max.)	25th perc.	Mean	75th perc.	Skewness	Kurtosis
Mastery (1–16)	11.0	12.5	14.0	−0.9	4.0
Mastery (1–10)	5.0	6.6	8.0	−0.5	2.5

financial difficulties in various studies (Alley and Kahn 2012; Caplan 1981; Krause and Baker 1992; Pearlin 1981). Previous studies report moreover that mastery can play a mediating effect between economic hardship and low levels of health (Mirowsky 2013) or poor mental health outcomes (Pearlin and Skaff 1996). The mediation effect comes about if ongoing financial strain leads to a diminished sense of mastery, which in turn, predicts deterioration of psychological functioning (Thoits 2013).

In the VLV questionnaire, the 4-item version of Cohen's perceived stress scale (PSS) comes closest to capturing this concept (Cohen et al. 1983; Cohen 1986). The measure is designed to 'tap into the degree to which respondents find their lives unpredictable, uncontrollable, and overloading' (Cohen et al. 1983, p. 387). The name of the scale seems to indicate an undesirable closeness to one of our three outcomes, economic stress. A close look at the items shows, however, that they circumscribe the very issues we hypothesize to be relevant for our research question and the wording of most questions is very similar to the items featuring in Pearlin's scale of mastery.[17] The items in Cohen's scale are as follows: In the last month, how often have you felt

..that you were unable to control the important things in your life?
..confident about your ability to handle your personal problems?
..that things were going your way?
..difficulties were piling up so high that you could not overcome them?

The response options encompassed five categories, ranging from never, almost never, sometimes, fairly often, to very often. A sum of the scores was calculated and because of the small number of very low scores, the lowest scores (0–7) were collapsed, resulting in a scale of 10 scores. Missing values were replaced using a prorating approach (see Chap. 10). The final scale consists of eight categories, that are normally distributed (Table 9.6).

An overview of all explanatory variables with summary information on number of categories and measurement levels is found in Appendix Table 9.5.

[17] Pearlin's scale of mastery (1981, p. 353) asks the following questions: How strongly do you agree or disagree with these statements about yourself,

1. There is really no way I can solve some of the problems I have. (1.00)
2. Sometimes I feel that I am being pushed around in life. (.75)
3. I have little control over the things that happen to me. (.995)
4. I can do just about anything I really set my mind to. (.55)
5. I often feel helpless in dealing with the problems of life. (.92)
6. What happens to me in the future mostly depends on me. (.47)
7. There is little I can do to change many of the important things in my life. (.90)

Appendix

Appendix Table 9.1 Distribution of scores of personality dimensions "Big Five"

Score	Percentage distribution of scores by personality dimension				
	Neurot	Agreeable	Extraver	Conscient	Open
2	10	0	1	0	2
3	10	1	5	0	3
4	14	4	12	1	6
5	14	6	12	2	10
6	25	19	25	10	21
7	11	19	16	14	14
8	10	22	12	16	16
9	4	17	9	19	15
10	2	11	8	37	13

Appendix Table 9.2 Mean and Standard Deviation of personality dimensions "Big Five"

Personality dimension	Neuroticism	Agreeable	Extraversion	Conscientious	Openness
Mean	5.4	7.4	6.5	8.5	7.1
St. deviation	2.0	1.7	1.9	1.6	2.0

Appendix Table 9.3 The Swiss socio-professional categorization operationalized in the VLV survey

<table>
<tr><th>Education →
Position ↓</th><th>University</th><th>Technical and professional</th><th>Apprenticeship</th><th>Compulsory education at most</th></tr>
<tr><td>Top executives</td><td colspan="4">1. Top executives (directors of medium or large company of at least 20 people)</td></tr>
<tr><td rowspan="2">Self-employed</td><td rowspan="2">2. Liberal professions (medical doctor, lawyer)</td><td colspan="3">3. Self-employed (CEO of a small company with less than 20 people)</td></tr>
<tr><td colspan="3">4. Farmers</td></tr>
<tr><td rowspan="2">Wage-earners</td><td rowspan="2">5. Academic professions (professor, engineer)</td><td rowspan="2">6. Intermediary professions (technician, social worker)</td><td>7. Skilled, non-manual (sales clerk, secretary)</td><td>9. Unskilled, non-manual (cleaner, security guard)</td></tr>
<tr><td>8. Skilled, manual (mechanic, electrician)</td><td>10. Unskilled, manual (agricultural laborer, construction worker)</td></tr>
</table>

Based on Joye and Bergmann (2005, p. 28)

Explanations to Appendix Table 9.3: The main focus of this research being on the lower end of the social stratification, the ten categories shown in Appendix Table 9.3 were regrouped into 7 categories, giving preference to maintaining a detailed idea of the less qualified categories while making sure that the frequency of observations in each category does not fall below a certain level (approx. 5%) of the total sample population. Appendix Table 9.4 shows the final operationalization of the variable 'former socio-professional status' in this volume. Top executives, liberal professions and academic professions were collapsed into category 1, self-employed and farmers, of whom there were less than 3%, into category 2; at last, unskilled workers – manual and non-manual – were combined in category 6 because each category separate would have represented less than 3% of the sample population

Appendix Table 9.4 Operationalization of former socio-professional status

New categories	Original categories Joye & Bergmann
1. Top executives and academic professions	1, 2, 5
2. Self-employed and farmers	3, 4
3. Intermediary professions	6
4. Skilled, non-manual	7
5. Skilled, manual	8
6. Unskilled, manual and non-manual	9, 10

Appendix Table 9.5 Overview of explanatory variables

Highest Measurement Level	Interval		Ordinal		Nominal
Model component	normally distributed	not normally distributed	> = 5 categories, normally distributed	< 5 categories	
Background characteristics	Neuroticism (2–10)	Conscientious (2–10)	Educational attainment (1–6)	Age group (1–4)	Sex (bi)
	Openness (2–10)				Canton (6)
	Extraversion (2–10)				Marital status (4)
	Agreeable (2–10)				
	Mastery (1–10)				
Economic resources			Wealth (1–5)		Assets (bi)
					Regular work (bi)
					Irregular work (bi)
					Occupational pension (bi)
					Support (bi)
					Supplement. Benefits (bi)
					Owner (bi)
Needs and expectations			Restaurant (1–5)	Cinema (1–4)	Last socio-professional category (7)
				Trip (1–4)	Health status (3)
Psychosocial symptoms and consequences		Self-worth (1–10)	Visit family (1–5)	Recognition (1–4)	
			Visit friends (1–5)	Lonely (1–3)	
			Call friends (1–5)		
			Call family (1–5)		

Appendix Table 9.6 Summary Statistics of interval-level and ordinal-level (5+ categories) variables

	Min.	Max.	25th perc.	Mean	75th perc.	St.dev.	Skewn.	Kurtosis
Openness	2	10	6	7.1	9	1.9	2.4	−0.3
Extraversion	2	10	5	6.5	8	2.0	2.4	0.0
Agreeable	2	10	6	7.3	8	1.7	2.8	−0.4
Neuroticism	2	10	4	5.4	7	2.0	2.4	0.1
Conscientiousness	2	10	7	8.4	10	1.6	2.9	−0.8
Mastery	1	10	5	6.6	8	2.3	−0.5	2.5
Self-worth	1	10	4	5.9	8	2.6	2.1	−0.3
Wealth	1	5	2	3.5	5	1.4	2.0	−0.5
Education	1	6	3	3.6	5	1.5	2.3	0.0
Restaurant	1	5	3	3.4	4	1.1	2.8	−0.6
Call friends	1	5	3	3.5	4	1.0	3.2	−0.6
Call family	1	5	3	3.8	4	1.0	3.8	−0.9
Visit friends	1	5	2	2.9	4	1.0	2.7	−0.1
Visit family	1	5	3	3.3	4	1.1	2.4	−0.2

References

Adena, M., & Myck, M. (2014). Poverty and transitions in health in later life. *Social Science & Medicine, 116*, 202–210.

Alley, D., & Kahn, J. R. (2012). Demographic and psychosocial predictors of financial strain in older adults. *Paper submitted at the 2012 annual meeting of the population Association of America.*

Altman, D. G., & Royston, P. (2006). The cost of dichotomising continuous variables. *BMJ: British Medical Journal, 332*(7549), 1080.

Angelini, V., Weber, G., & Brugiavini, A. (2013). Household consumption and the crisis. In A. Börsch-Supan, M. Brandt, H. Litwin, & G. Weber (Eds.), *Active ageing and solidarity between generations in Europe: First results from SHARE after the economic crisis* (pp. 19–26). Berlin/Boston: Walter de Gruyter.

Arber, S., Fenn, K., & Meadows, R. (2014). Subjective financial well-being, income and health inequalities in mid and later life in Britain. *Social Science & Medicine, 100*, 12–20.

Bennett, T., Savage, M., Silva, E., Warde, A., Gayo-Cal, M., & Wright, D. (2009). *Culture, class, distinction*. London: Routledge.

Bíró, A. (2013). Adverse effects of widowhood in Europe. *Advances in Life Course Research, 18*(1), 68–82.

Boehnke, K., Richmond, B. O., Chandra, S., & Stromberg, C. (1994). *Worries, values and worries, values and well-being: A comparison of East and West German, Nepalese and Fijian undergraduate well-being: A comparison of East and West German, Nepalese and Fijian undergraduate – Google scholar*. Pamplona: Gehalten auf der 12th Congress of the International Association for Cross-Cultural Psychology.

Boehnke, K., Schwartz, S., Stromberg, C., & Sagiv, L. (1998). The structure and dynamics of worry: Theory, measurement, and Cross-National Replications. *Journal of Personality, 66*(5), 745–782.

Boone, H. N., & Boone, D. A. (2012). Analyzing Likert Data. *The Journal of Extension (JOE), 50*(2). Retrieved:http://www.joe.org/joe/2012april/tt2.php

Budowski, M., Tillmann, R., & Bergman, M. M. (2002). Poverty, stratification, and gender in Switzerland. *Schweizerische Zeitschrift Für Soziologie, 28*(2), 297–318.
Burns, V. F. (2010). *Unpacking the Phenomenon of Social Isolation through the Unique Experiences of Autonomous Older Adults Living in Social Housing* (M.A.). McGill University (Canada), Canada.
Baeriswil, M. (2016). Participation et rôles sociaux à l'âge de la retraite: une analyse des évolutions et enjeux autour de la participation sociale et des rapports sociaux de sexe. Doctoral Thesis, University of Geneva.
Caplan, G. (1981, April). Mastery of stress: Psychosocial aspects. *The American Journal of Psychiatry, 138*(4), 413–420.
Cavasso, B., & Weber, G. (2013). The effect of the great recession on the wealth and financial distress of 65+ Europeans. In A. Börsch-Supan, M. Brandt, H. Litwin, & G. Weber (Hrsg.), *Active ageing and solidarity between generations in Europe: First results from SHARE after the economic crisis* (S. 27–36). Berlin/Boston: Walter de Gruyter.
Chan, A., Ofstedal, M. B., & Hermalin, A. I. (2002). Changes in subjective and objective measures of economic Well-being and their interrelationship among the elderly in Singapore and Taiwan. *Social Indicators Research, 57*(3), 263–300.
Cohen, S., Kamarck, T., & Mermelstein, R. (1983). A global measure of perceived stress. *Journal of Health and Social Behavior, 24*(4), 385–396. https://doi.org/10.2307/2136404.
Cohen, S. (1986). Contrasting the hassles scale and the perceived stress scale: Who's really measuring appraised stress? *American Psychologist, 41*(6), 716–718.
Cohen-Mansfield, J., & Parpura-Gill, A. (2007). Loneliness in older persons: A theoretical model and empirical findings. *International Psychogeriatrics, 19*(02), 279–294.
Cornwell, E. Y., & Waite, L. J. (2009, March 1). Social disconnectedness, perceived isolation, and health among older adults. *Journal of Health and Social Behavior, 50*(1), 31–48. https://doi.org/10.1177/002214650905000103.
Diener, E., Sandvik, E., & Larsen, R. J. (1985). Age and sex effects for emotional intensity. *Developmental Psychology, 21*(3), 542–546.
Donnellan, M. B., & Lucas, R. E. (2008). Age differences in the big five across the life span: Evidence from two National Samples. *Psychology and Aging, 23*(3), 558–566.
Falter, J.-M. (2009). Le divorce, facteur de pauvreté féminine. In *Transitions dans les parcours de vie et construction des inégalités* (S. 297–316). PPUR presses polytechniques.
Fokkema, T., De Jong Gierveld, J., & Dykstra, P. A. (2012). Cross-National Differences in older adult loneliness. *The Journal of Psychology, 146*(1–2), 201–228.
Fonseca, R., Kapteyn, A., Lee, J., Zamarro, G., & Feeney, K. (2013). Financial and subjective Well-being of older Europans. In A. Börsch-Supan, M. Brandt, H. Litwin, & G. Weber (Eds.), *Active ageing and solidarity between generations in Europe: First results from SHARE after the economic crisis* (pp. 137–148). Berlin/Boston: Walter de Gruyter.
Gabriel, R., Oris, M., Studer, M., & Baeriswil, M. (2015). The persistence of social stratification? A life course perspective on poverty in old-age in Switzerland. *Swiss Journal of Sociology, 41*(3), 465–487.
George, L. K. (1992). Economic status and subjective Well-being: A review of the literature and an agenda for future research. In N. E. Cutler, D. W. Gregg, & M. P. Lawton (Eds.), *Aging, money, and life satisfaction: Aspects of financial gerontology* (pp. 69–99). New York: Springer Publishing.
Goldthorpe, J. H. (2002). Occupational sociology, yes: Class analysis, no: Comment on Grusky and Weeden's "research agenda". *Acta Sociologica, 45*(3), 211–217.
Grundy, E., & Holt, G. (2001). The socioeconomic status of older adults: How should we measure it in studies of health inequalities? *Journal of Epidemiology and Community Health, 55*(12), 895–904.
Grusky, D. B. (2001). Social Stratification. In N. J. Smelser & P. B. Baltes (Hrsg.), *International Encyclopedia of Social & Behavioral Sciences* (1st ed, S. 14443–14452). Amsterdam/New York: Pergamon.

Hagestad, G. (2003). Interdependent lives and relationships. A life-course view of families and aging. In R. A. Settersten (Ed.), *Invitation to the life course: Toward new understandings of later life* (pp. 135–160). Amityville: Baywood Publication.

Halleröd, B. (2009). Ill, worried or worried sick? Inter-relationships among indicators of wellbeing among older people in Sweden. *Ageing & Society, 29*(04), 563–584.

Havens, B., Hall, M., Sylvestre, G., & Jivan, T. (2004). Social isolation and loneliness: Differences between older rural and urban Manitobans. *Canadian Journal on Aging/La Revue canadienne du vieillissement, 23*(Special Issue/Numéro special 02), 129–140. https://doi.org/10.1353/cja.2004.0022.

Hazelrigg, L. E., & Hardy, M. A. (1997). Perceived income adequacy among older adults issues of conceptualization and measurement, with an analysis of data. *Research on Aging, 19*(1), 69–107.

Heady, B., Krause, P., & Wagner, G. (2009). Poverty redifined as low consumption and low wealth, not just low income: Psychological consequences in Australia and Germany. In *Gehalten auf der joint OECD-University of Maryland conference, "measuring poverty, inequality and social exclusion: Lessons from Europe"*. Paris: OECD.

Idler, E. L. (1993). Age differences in self–assessments of health: Age changes, cohort differences, or survivorship? *Journal of Gerontology, 48*(6), S289–S300.

Informationsstelle AHV/IV. (2010). *Änderungen auf 1. Januar 2011 bei Beiträgen und Leistungen AHV – IV.*

Janoff-Bulman, R. (1989). Assumptive worlds and the stress of traumatic events: Applications of the Schema construct. *Social Cognition, 7*(2), 113–136.

John, O. P., & Srivastava, S. (1999). The big five trait taxonomy: History, measurement, and theoretical perspectives. In L. A. Pervin & O. P. John (Eds.), *Handbook of personality: Theory and research* (pp. 102–138). New York: Elsevier.

Joye, D., & Bergmann, M. M. (2005). Comparing social stratification schemas: CAMSIS, CSP-CH, Goldthorpe, ISCO-88, Treiman, and Wright. *Cambridge Studies in Social Research, 9*, 1–37.

Joye, D., & Schuler, M. (1996). *Sozialstruktur der Schweiz: Sozio-Professionelle Kategorien.* Bern: Bundesamt für Statistik.

Jylhä, M. (2009). What is self-rated health and why does it predict mortality? Towards a unified conceptual model. *Social Science & Medicine, 69*(3), 307–316.

Kahn, J. R., & Pearlin, L. I. (2006). Financial strain over the life course and health among older adults. *Journal of Health and Social Behavior, 47*(1), 17–31.

Katz, S. (1963). Studies of illness in the aged: The index of adl: A standardized measure of biological and psychosocial function. *JAMA, 185*(12), 914–919.

Kolb, K., Blossfeld, H.-P., & Skopek, N. (2013). The two dimensions of housing inequality in Europe. Are high home ownership rates an Indicator of low housing values? *Comparative Population Studies – Zeitschrift für Bevölkerungswissenschaft, 38*(4), 1009–1040.

Krause, N., & Baker, E. (1992). Financial strain, economic values, and somatic symptoms in later life. *Psychology and Aging, 7*(1), 4–14.

Lalive d'Epinay, C., & Hofstetter-Bétemps, C. (1999). *Vieillesses Au Fil Du Temps, 1979–1994: Une Révolution Tranquille: Santé, Situations De Vie, Formes De Participation Et Visions Du Monde Des Personnes âGées En Suisse*. Lausanne: Réalités sociales.

Laslett, P. (1991). *A fresh map of life: The emergence of the third age.* Cambridge: Harvard University Press.

Levasseur, M., Richard, L., Gauvin, L., & Raymond, É. (2010). Inventory and analysis of definitions of social participation found in the aging literature: Proposed taxonomy of social activities. *Social Science & Medicine (1982), 71*(12), 2141–2149.

Litwin, H., & Sapir, E. V. (2009). Perceived income adequacy among older adults in 12 countries: Findings from the survey of health, ageing, and retirement in Europe. *The Gerontologist, 49*(3), 397–406.

Lynch, J. W., Kaplan, G. A., & Shema, S. J. (1997). Cumulative impact of sustained economic hardship on physical, cognitive, psychological, and social functioning. *The New England Journal of Medicine, 337*(26), 1889–1895.

Mirowsky, J. (2013). Analyzing associations between mental health and social circumstances. In C. S. Aneshensel, J. C. Phelan, & A. Bierman (Hrsg.), *Handbook of the Sociology of Mental Health* (S. 143–165). Dordrecht: Springer Netherlands.

Moser, P. (2006). Einkommen und Vermögen der Generationen im Lebenszyklus. In *Eine Querschnitts-Kohortenanalyse der Zürcher Stuerdaten 1992–2003*. Zürich: Statistisches Amt des Kantons.

Newman, A., & Cauley, J. A. (2012). *The epidemiology of aging*. Dordrecht: Springer Science & Business Media.

Noelen-Hoeksema, S., & Rusting, C. L. (1999). Gender differences in well-being. In D. Kahneman, E. Diener, & N. Schwarz (Hrsg.), *Well-being: The foundations of hedonic psychology* (S. 330–350). New York: Russell Sage Foundation.

OFS. (2012). *Pauvreté en Suisse: concepts, résultats et méthodes. Résultats calculés sur la base de l'enqête SILC 2008 à 2010*. Neuchâtel: Office fédéral de la statistique.

Oris, M., & Lerch, M. (2009). La transition ultime. Longévité et mortalité aux grands âges dans le bassin lémanique. In M. Oris, E. D. Widmer, A. de Ribaupierre, D. Joye, D. Spini, G. Labouvie-Vief, & J.-M. Falter (Hrsg.), *Transitions dans les parcours de vie et construction des inégalités* (S. 407–433). Lausanne: PPUR presses polytechniques.

Patsios, D. (1999). Poverty and social exclusion amongst the elderly. In *Poverty and social exclusion survey of Britain working paper 20*. Bristol: Avon.

Pearlin, L. I., & Schooler, C. (1978). The structure of coping. *Journal of Health and Social Behavior, 19*(1), 2–21.

Pearlin, L. I., & Skaff, M. M. (1996). Stress and the life course: A paradigmatic Alliance. *The Gerontologist, 36*(2), 239–247.

Rammstedt, B., & John, O. P. (2007). Measuring personality in one minute or less: A 10-item short version of the big five inventory in English and German. *Journal of Research in Personality, 41*(1), 203–212.

Scherger, S., Nazroo, J., & Higgs, P. (2011). Leisure activities and retirement: Do structures of inequality change in old age? *Ageing & Society, 31*(01), 146–172.

Schweizerische Konferenz für Sozialhilfe (SKOS). (2013). *Armut und Armutsgrenzen*.

Seifert, K. (2009). Altersarmut: Wird das Rad der Geschichte zurückgedreht? In Caritas Schweiz (Hrsg.), *Sozialalmanach 2010* (S. 89–98). Luzern: Caritas.

Seifert, K., & Pilgram, A. (2009). *Vivre avec peu de moyens. La pauvreté des personnes âgées en Suisse*. Zürich: Edition Pro Senectute.

Spielberger, C. D., Gonzalez, H. P., Taylor, C. J., Anton, W. D., Algaze, B., Ross, G. R., & Westberry, L. G. (1980). *Preliminary professional manual for the test anxiety inventory: TAI*. Palo Alto: Consulting Psychologists.

Stewart, M., Reutter, L., Makwarimba, E., Veenstra, G., Love, R., & Raphael, D. (2008). Left out: Perspectives on social exclusion and inclusion across income groups. *Health Sociology Review, 17*(1), 78–94.

Stoller, M. A., & Stoller, E. P. (2003). Perceived income adequacy among elderly retirees. *Journal of Applied Gerontology, 22*(2), 230–251.

Thoits, P. A. (2013). Self, identity, stress, and mental health. In C. S. Aneshensel, J. C. Phelan, & A. Bierman (Hrsg.), *Handbook of the Sociology of Mental Health* (S. 357–377). Dordrecht: Springer.

Townsend, P. (1957). *The family life of old people: An inquiry in East London*. London: Routledge & K. Paul.

Townsend, P. (1979). *Poverty in the United Kingdom: A survey of household resources and standards of living*. Berkeley: University of California Press.

Victor, C., Scambler, S., & Bond, J. (2008). *The social world of older people: Understanding loneliness and social isolation in later life: Understanding loneliness and social isolation in later life*. Maidenhead: McGraw-Hill International.

Wang, R. I. H., Treul, S., & Alverno, L. (1975). A brief self-assessing depression scale. *The Journal of Clinical Pharmacology, 15*(2–3), 163–167.

Wanner, P., Gabadinho, A., & Pecoraro, M. (2008). *La situation économique des actifs et des rentiers. Rapport technique et tableaux commentés*. Berne: OFAS.

Wanner, P., & Fall, S. (2012). *La situation économique des veuves et des veufs* (Vol. Bd. 5). Berne: Office fédéral des assurances sociales.

Weidekamp-Maicher, M., & Naegele, G. (2010). Economic resources and subjective well-being in old age. In H. Mollenkopf & A. Walker (Hrsg.), *Quality of Life in Old Age: International and Multi-Disciplinary Perspectives* (Softcover reprint of hardcover 1st ed. 2007, S. 65–84). Springer.

Weidekamp-Maicher, Manuela. (2008). *Materielles Wohlbefinden im späten Erwachsenenalter und Alter* (Text).

Weiner, I. B., Schinka, J. A., & Velicer, W. F. (2003). *Handbook of psychology, research methods in psychology*. Hoboken: Wiley.

World Health Organization. (2002). *Active aging: A policy framework*. Geneva: WHO.

Chapter 10
Missing Values

The validity of the results found in the presented analysis critically depends on the representativeness of the research design and this question needs to be assessed in light of the difficulties of surveying the specific group of respondents we were interested in: economically vulnerable households could potentially be underrepresented in the sample because they are less likely to participate in studies (Budowski et al. 2002). Previous research on the VLV survey suggests that the economically vulnerable population has been successfully captured within the sample population as the recorded poverty rates are equivalent to those reported in other studies as well as in official statistics (Gabriel and Oris 2013; Oris and Nicolet 2016). Still, it was important to investigate the pattern of non-responses, in particular for the variables income and wealth, in order to assess any potential biases.

Missing values are an inevitable problem with survey data. Ignoring them can yield distorted results if values are not missing at random or if comparisons are based on samples of varying size. Depending on the kind of missing data one is dealing with, different options for resolving the problem can be considered. Rubin famously distinguished between data that is missing completely at random (MCAR), at random (MAR) or missing not at random (MNAR) (Rubin 1976). Whether missing values are independent of both observable variables and of unobservable parameters of interest (MCAR) can – by definition – not be examined; this type of missing values is thought to be extremely rare. The MAR is of an unproblematic type because the fact that a certain response is missing is not associated with a specific value of the variable (Polit and Beck 2011).

In our treatment of missing values, we proceeded as follows: First, we attempted to fill-in missing data in cases where we had sufficient knowledge about other items that were related to the mission questionnaire item. Basically only psychometric scale qualified for this method (pro-rating). Second, we assessed the pattern of missing values of those variables that had a considerable amount of missing values in order to detect whether they were missing in a non-random manner.

J. Henke, *Revisiting Economic Vulnerability in Old Age*, Life Course Research and Social Policies 11, https://doi.org/10.1007/978-3-030-36323-9_10

10.1 Pro-rating for Validated Psychometric Scales

In the case of a validated scale or sub-dimension of an internally consistent measurement instruments we applied a pro-rating approach to recover as many values as possible: we substituted the missing variable of a respondent by his or her mean score on the other items. This procedure is sometimes done by using regression estimates (Heckman 1977) but in the case of a scale with known psychometric characteristics, it has been argued that an arithmetic mean is a valid approximation to a linear regression.[1]Specifically, we used this method to treat missing values in the 'big five' inventory of personality for the variables *openness, extraversion, neuroticism, agreeableness* and *conscientiousness*), Cohen's Perceived Stress Scale (variable *mastery*) and the self-worth scale taken from Assumptive World-View model (variables *shame, opinion, satisfy* and *no good*). For these variables we were able to impute 1.9% of missing values on average for variables on personality, 2.5% in self-worth and 6% in mastery, all of which is considerably below the recommended maximum threshold of 20%[1].

10.2 Patterns of Missing Values

Using a program written by Long and colleagues specifically for the analysis of missing data (Long 2006) we first assessed the frequency of missing values and only conducted an analysis of the distributional pattern if more than 5% were missing, which concerned the following seven variables[2] (Table 10.1):

The variables with the highest proportion of missing values all concerned questions on personal finances: wealth and income (obj_ev), followed by worry on

Table 10.1 Variables with more than 5% missing values (N = 3080)

Variable	Description	N Missing	% Missing
Wealth	Wealth	640	20.8
obj_ev	Income below poverty line	449	14.6
perc_ev	Worries about not having enough money	401	13.0
Self-worth	Scale based on 4 items	234	7.6
Extravertion	Personality dimension, based on 2 items	221	7.2
Openness		215	7.0
Conscientiousness		199	6.5
Neuroticism		198	6.4
Agreeableness		195	6.3

[1] http://www.leehw.com/pro-rating-for-missing-data (2.9.14).

[2] All other variables had less than 3.3% missing values.

finances (perc_ev). The Self-Assessed Measure 'making ends meet' (sa_ev) is not represented in the table as only 3.3% of its observations were missing. All three of these variables are very important to our analysis, two of them representing outcome measures of economic vulnerability. The other variables with a high rate of missing values are items featuring in psychometric scales. While the resistance against disclosing information on finances is well known, the clustering of missing values for the personality items may have to do with their format: responding to a long list of similar questions can sometimes be experienced as burdensome, causing some respondents to skip questions.

We proceeded to analyze the clustering of missing values by order of frequency (> = 1%) in order to identify combination patterns across those variables with more than 5% missing values (Appendix Table 10.1). The most important cluster is found between wealth and income (9.1%) followed by perc_ev and the psychometric items (2.2). In 1.4% of the cases, at least one of the five dimensions of personality is missing.

The emerging pattern clearly indicates that respondents who did not answer the question on income also tended not to provide information about wealth, and vice versa. Similarly but less important in number, there are some respondents who are generally reluctant to respond to psychological questions. As many as one third of the missing values concerns both sets of variables, indicators of economic vulnerability and psychometric items. However, apart from the combinations mentioned above, no pattern suggesting a concentration of missing values emerges.

The high number of missing values among these variables necessitates that we analyze the probability distribution of missingness. In the quest to recover as many of thc 583 missing values as possible, we will heed the following caution that highlights the balance we are trying to strike: 'With or without missing data, the goal of a statistical procedure should be to make valid and efficient inferences about a population of interest, not to estimate, predict, or recover missing observations nor to obtain the same results that we would have seen with complete data. Attempts to recover missing values may impair inference.' (Schafer and Graham 2002, p.149).

10.3 Determinants of Missing Values in Financial Worry, Income and Wealth

Questions on income and wealth are known to be delicate and prone to non-response in surveys. Not surprisingly, the rate of missing values for the variable income (on which the Objective Measure is based) is at a high rate of 14.6%, and for the variable wealth it is at 20.8%. These rates are almost identical with the share non-responses recorded in the survey of 1979 (14% and 22%, respectively). Also, non-response to the question about financial worry (on which the Perceived Measure is based) is surprisingly high at 10.4%. Maybe this is again due to the fact that this question was part of a scale of as many as 17 questions, causing some people to skip this question, all the more as it was administered by a paper-pencil questionnaire. Indeed, among the 17 items addressing various burdensome situations across life-domains, an average rate of 14% of missing values was recorded.

In order to examine the potential bias that might be caused by deleting all cases with missing values in perc_ev, obj_ev and wealth from our dataset, we proceeded to verify if we had to deal with a bias of over- or under-representation of certain population groups (MAR) or whether it is likely that the missingness is associated with a value of the variable itself, for example, wealthy respondents are less likely to report income (MNAR). For this purpose, we created binary indicators (MV perc_ev, MVobj_ev, MV wealth), which take 1 if the value is missing for the observation and 0 if the value is observed. Firstly, we wanted to be able to exclude the possibility of MNAR by testing if the *type of economic resources* respondents have access to influences the propensity to respond to questions on financial worry, income and wealth. In the context of economic vulnerability, we furthermore examined, whether the subjective evaluation (sa_ev and perc_ev) has an influence on the missingness of obj_ev because this kind of a relationship among the three measures of economic vulnerability would represent a serious risk of distorting the planned analyses. Secondly, we tested if certain population characteristics such as *educational attainment* and last *socio-professional category,* may have influenced the response behavior for income and wealth. The results of the regression analyses "Type of economic resources as determinants of missing values in financial worry, income and wealth" are found in Appendix Table 10.2.

In all models we controlled for age, sex, marital status and linguistic region. Though not all of these covariates are expected to have the same influence on missingness, for the sake of comparability the same independent variables were included in all models and only complete sets of observations were used.[3]

In the first model, old age turned out to be the most important factor for increasing the odds of a non-response to the question on *MV perc_ev*. Among the types of economic resources, only occupational pension had a significant negative effect (OR = 0.7, $p < 0.05$). This indicates that based on the present data set, the analysis of our Perceived Measure will yield slightly biased results because the probability of non-response to this question is higher among those who do not have an occupational pension. Receiving supplementary benefits to old age pensions had no statistically significant influence on missingness even when the variable pension was excluded from the model (results not shown).

For *MV obj_ev* and *MV wealth*, several explanatory variables were significant and the odd ratios were mostly congruent in explaining the missingness in both as was to be expected given the similar pattern of missing values. The single most influential variable was supplementary benefits, which reduced the probability of non-response by a factor of 2.7 and 2.8 respectively (reversed odds). The effect of occupational pensions was also negative but less strong. The demographic factors

[3] This means that the data sets contained no missing values among all covariates used for regression analysis on outcome variables MV perc_ev, MV obj_ev and MV wealth.

were all statistically significant: (for obj_ev, Italian-speakers were double as likely not to respond compared to German-speakers), age (respondents older than 85 were double as likely not to respond compared to those aged 65–74) and sex (women were twice as likely not to respond). The variable assets only had an effect on missing values in wealth: owning a third pillar of a life insurance caused the probability of non-response to decrease.

We performed the same analysis on a variable that recorded whether there was a non-response to both income *and* wealth. This concerned 12% of the sample (372 individuals). The objective of this analysis was to examine whether the trends identified for missing values in income or wealth are exacerbated when neither of the two items were answered. The coefficients were between those found for MV obj_ev and MV wealth (results not shown), which was to be expected given the large overlap. It also means that there is no exacerbation of the observed trends when both variables are missing at the same time.

Focusing this time only on MV obj_ev, we did not find that the *subjective evaluation of economic vulnerability* (sa_ev and perc_ev) had any statistically significant influence on missingness in the income-based measure (Results not shown). This finding is important as it gives the all-clear signal that there is no hidden associations between the objective and the subjective measures of economic vulnerability in terms of their patterns of missing values, which could have led to wrong interpretations in subsequent analysis.

Concerning the models testing for the influence of *educational attainment* and last *socio-professional category* (results not shown) no significant effects were found, with one exception: respondents who only attended primary school were significantly more likely not to respond to the question about financial worries ($OR = 1.65$; $p < 0.01$).

In summary, we observe that the odds of encountering missing values in variables pertaining to economic vulnerability are significantly and strongly related to being very old (85+). As we will see in the next section, this is a more general phenomenon that has to do with older adults getting tired more easily. For the variables obj_ev and wealth there is furthermore a risk of producing biased results due to the excessive exclusion of female and Italian-speaking respondents. As we will see in Part III, these population groups are characterized by a higher than average rate of economic vulnerability measured as by objective indicators. However, the association remains even when controlling for sources of income, though occupational pensions and supplementary benefits significantly reduce the risk of non-response. The latter can be read as an encouraging sign for the representativeness of our data since respondents who benefit from supplementary benefits can doubtlessly be considered economically vulnerable. Another factor to be considered here is the effect of going through the administrative process of placing the demand for supplementary benefits: this process involves a willingness to disclose financial information, a predisposition that is also required for participating in a survey.

The ambiguity of our findings thus far requires that we attempt to circumscribe more closely the profile of those respondents who would be excluded if we dropped the missing values, especially those in income and wealth. We created a cross tabulation with a binary variable that takes 1 if both, income and wealth feature a missing value, and the variables educational attainment and last socio-professional category (Appendix Table 10.3).

The distribution of missing values in income and wealth by education does reveal minor differences between attainment levels but no congruent pattern is discernable: the highest rates are recorded for absolvents of 'lower secondary school', followed by 'High school' graduates. Respondents with primary level education have close to the same rate of missing values in income and wealth as university graduates (11.3%, 10.2%). Reflecting the response behavior of survey participants, the category 'missing' in the variable education corresponds to the highest rate of missing values in the financial items; the same is true for the last socio-professional status of the household. Missing values by last socio-professional status of the household are highest in the category 'unskilled' (15.3) and 'not active' (20%), followed by 'top executive (12.9). Again, the results are such that no unidirectional factor (low or high social status) can be identified for explaining the propensity not to respond to questions about income and wealth.

We conclude that the missing values in the three variables examined, income, wealth and the perceived measure of economic vulnerability, are not missing at random, though their missingness cannot be easily attributed to a specific social background. It is not possible to recover missing values in these variables by using multiple imputation methods (Rising 2010). The fact that we are interested in the difference between several measures on economic vulnerability means that we cannot draw on information from similar variables by imputation without taking an important risk of reducing the variation. This means that we have to drop cases that have missing values in these variables, which causes our sample to shrink by some 30% to a sample size of 2137 (Appendix Table 10.4).

In the subsequent analysis we therefore must do with a loss in power and the fact that the exclusion of sets with missing values causes some bias in our results: there is a more than proportionate loss of (a) respondents with certain demographic characteristics (very old, women, Italian-speakers) and, (b) in a more ambiguous manner, respondents who are at the lower (though not the lowest end) of the economic spectrum. We now proceed to conducting the same analysis for the psychometric scale self-worth.

10.4 Determinants of Missing Values in Self-worth

The self-worth scale featured 7.6% missing values. Like before, a regression model was estimated on a dummy variable (1 = missing in self-worth, 0 = not missing in self-worth). As dependent variables we included basic demographic characteristics, personality traits that are suspected to influence response behavior and our binary measures obj_ev, sa_ev and perc_ev. Three models were estimated, each one featuring one of the measures of economic vulnerability. Using the original sample, we omitted all sets of observations that had at least one missing value among the explanatory variables in order to ensure comparability of the models. Results of these regression models are found in Appendix Table 10.5.

The analysis shows that 'age above 85' is the only explanatory factor that clearly and significantly increases the odds of non-response for self-worth. Neither the personality traits nor any of the three measures of economic vulnerability was significantly associated to non-response in the four items constituting 'self-worth'.

10.5 Conclusion: Handling Missing Values and Final Sample

Based on the analysis of those variables with more than 5% missing values we decided to eliminate those sets of observation that have missing values among obj_ev, perc_ev and wealth because these missing values are not randomly distributed (MNAR) and the variables are directly related to our research objective, which means they cannot be imputed without considerable risk.

The sample we are left with after dropping those variables with the highest number of missing values continues to have good-sized strata for individuals up to the age group 80–84. However, the strata of the older cohorts have shrunk to such a small sub-sample that it is no longer possible to make statistically significant inferences. In fact, the older age groups have shrunk considerably, given that many a set of observations were based on proxy interviews and were thus excluded at an earlier stage. For the sake of statistical quality we therefore decided to limit our multivariate analysis to four age groups 65–69, 70–74, 75–79, 80–85, dropping age groups 85 and older from our sample. Below is the final sample that will be used for multivariate analysis (Table 10.2).

Concerning the handling of missing values in multivariate analysis, we will have to use listwise deletion of observations, meaning that if there is a single variable with a missing value among all the variables being used in the model, the observation is ignored (StataCorp 2013). The drawback of an analysis using only complete cases is that it reduces statistical power because of a smaller sample size though with the thorough preparations undertaken here, this danger has largely been avoided: with the exception of the psychometric items discussed above, the risk of biases in estimates is limited in our case because we are dealing with very small numbers of missing values per variable after having deleted perc_ev, obj_ev and wealth because all other variables had less than 3.3% missing values.

Table 10.2 Final sample with respondents aged 65–84

Age groups	Geneva		Valais		Berne		Basel		Ticino		Total
	Women	Men	Women	Men	Women	Men	Women	Men	Women	Men	
65–69	45	49	42	49	42	59	54	52	33	39	464
70–74	39	44	34	40	50	54	40	45	34	47	427
75–79	39	42	32	50	49	54	35	42	29	38	410
80–84	28	34	20	43	39	50	30	33	20	33	330
Total	151	169	128	182	180	217	159	172	116	157	**1631**

Appendix

Appendix Table 10.1 Clustering of missing values in variables with more than 5% missing values, by frequency of combination

Freq.	%	wealth	obj_ev	perc_ev	self-worth	neurot	open	agreeable	extraver	conscient
2035	66.1									
279	9.1	x	x							
216	7.0	x								
110	3.6			x						
67	2.2			x	x	x	x	x	x	x
61	2.0		x							
42	1.4					X Missing in any of the 5 personality dimensions				
41	1.3				x					

Appendix Table 10.2 Type of economic resources as determinants of missing values in financial worry, income and wealth

Dependent variable:		MV perc_ev	MV obj_ev	MV wealth
Variable (Ref.)		*Odds ratios*		
Age (65–74)	75–84	1.3*	ns	ns
	85+	1.6**	2.0***	1.7***
Sex	Male	ns	0.6***	0.6***
Linguistic Region (German-speaking)	French-speaking	ns	ns	ns
	Italian-speaking	ns	2.0***	1.7***
Economic resources	Owner	ns	ns	ns
	Occup. pension	0.7*	0.6***	0.6***
	Assets	ns	ns	0.7***
	Supplement. benefits	ns	0.4***	0.4***
Constant		0.2***	0.2***	0.4***
Key: *** <0.001; ** <0.01; * <0.05; *ns* not statistically significant				
N =		3013	3013	3013

Appendix Table 10.3 Rate of missing values in wealth and income, combined, for educational attainment and previous socio-professional status of the household

Educational attainment	% of MVs in wealth and income	previous socio-professional status of household	% of MVs in wealth and income
Primary	11.3	Top executive	12.9
Lower secondary	17.0	Liberal	11.1
Professional training	10.3	Intermed.prof./farmers	12.0
High school	14.7	Non-manual	11.0
University of applied sciences	11.5	Skilled manual	9.3
University	10.2	Unskilled	15.2
Missing	20.2	Not active	20.0
		Missing	38.5
Total	12.1	Total	12.1
N =	372.0	N =	372.0

Appendix Table 10.4 Missing values in wealth, the objective, and the perceived measure of economic vulnerability

Missing for how many economic variables (wealth, obj_ev, perc_ev)?	Frequency	%	Cumulative %
0 (= new sample)	2′137	69.4	69.4
1	528	17.1	86.5
2	357	11.6	98.1
3	58	1.1	100
Total	3′080	100	

Appendix Table 10.5 Determinants of missing values in self-worth

Dep. Var. = MV self-worth		Model 1	Model 2	Model 3
Variable (Ref.)		Odds ratios		
Age (65–74)	75–84	ns	ns	ns
	85+	1.969∗	2.029∗	2.050∗
Sex	Male	ns	ns	ns
Linguistic region (German-speaking)	French-speaking	ns	ns	ns
	Italian-speaking	ns	ns	ns
Personality	Extraversion	ns	ns	ns
	Agreeableness	ns	ns	ns
	Conscientiousness	ns	ns	ns
	Neuroticism	ns	ns	ns
	Openness	ns	ns	ns
Economic vulnerability	**obj_ev**	**ns**	–	–
	sa_ev	**–**	**ns**	**–**
	perc_ev	**–**	**–**	**ns**
	Constant	0.031∗∗		
Key: ∗∗∗ < 0.001; ∗∗ < 0.01; ∗ < 0.05; *ns* not statist. significant				
N =		2300	2300	2300

References

Budowski, M., Tillmann, R., & Bergman, M. M. (2002). Poverty, stratification, and gender in Switzerland. *Schweizerische Zeitschrift Für Soziologie, 28*(2), 297–318.

Gabriel, R., & Oris, M. (2013). *Poverty and inequality amongst the elderly population in Switzerland, 1979–2011*. Bern, Switzerland.

Heckman, J. J. (1977). *Sample selection bias as a specification error (with an application to the estimation of labor supply functions)* (Working Paper No. 172). National Bureau of Economic Research. Abgerufen von.

Long, J. S. (2006). *Regression models for categorical dependent variables using stata* (2nd ed.). College Station: Stata Press.

Oris, M., & Nicolet, M. (2016). Mesures et capture de la vulnérabilité dans une enquête sur les conditions de vie et de santé des personnes âgées. L'expérience de VLV (Vivre-Leben-Vivere) en Suisse. In Oris, Michel & Cordazzo, Philippe & Bellis, Gil & Brown, Elizabeth & Parant, Alain. Les populations vulnérables. Actes du XVIe colloque national de démographie (207–224). Aix-en-Provence - 28-31 mai 2013 - Bordeaux: CUDEP (Conférence Universitaire de Démographie et d'Étude des Populations).

Polit, D. F., & Beck, C. T. (2011). *Nursing research: Generating and assessing evidence for nursing practice* (Auflage: Ninthtion.). Philadelphia: Lippincott Raven.
Rising, B. (2010). *Multiple imputation*. Stata Corp LP. Abgerufen von.
Rubin, D. B. (1976). Inference and missing data. *Biometrika, 63*(3), 581–592.
Schafer, J. L., & Graham, J. W. (2002). Missing data: Our view of the state of the art. *Psychological Methods, 7*(2), 147–177.
StataCorp. (2013). *Stata 13 structural equation modeling reference manual*. College Station: Stata Press.

Part III
Prevalence of Economic Vulnerability Among Swiss Pensioners

This part presents a descriptive analysis of our sample population. It provides the background for more complex multivariate analyses by systematically exploring the relationship between the three measures of economic vulnerability and all potential explanatory variables. The covariates are presented according to the section order outlined previously, corresponding to the components of our theoretical model: Background Characteristics, Economic Resources, Financial Needs and Expectations, and Psychosocial Consequences and Symptoms.

Examining the association between each dependent and independent variable separately is a necessary step for gaining a thorough understanding of the nature of the statistical relationship. The primary objective being the verification that there is indeed a statistically significant relationship, these analyses pave the way for defining the operationalization of our more complex models by uncovering unexpected patterns of association. For categorical variables, contingency tables with cell frequency and percentage distributions are shown; for interval variables, uni- and bivariate statistics are presented. The strength of the relationship between each of the three measures of economic vulnerability and the covariates is assessed using different measures of effect size and corresponding tests of significance (see also Chap. 8). For those covariates that have a statistically significant relationship of at least moderate strength, an analysis of the congruency/discrepancy between the three measures is presented. Interpreting the discrepancies at the binary level yields provisionary elements of response to some of our hypotheses, indicating potential directions that will be put to the test once we proceed to multivariate analyses. For this purpose, cross-tabulations based on the covariate and the Vulnerability Typology are shown. In some instances, type AAA (total absence of vulnerability, representing 69% of the sample population) will not be considered in the analysis in order to facilitate the interpretation of those results focusing on vulnerable groups. Some auxiliary variables that will not feature in the more advanced models are drawn on because they help in describing the composition of our sample population and allow contextualizing our findings. Most of the analysis will be performed on the sample shown in Table 10.8; only the first section on socio-demographic characteristics is based on the larger sample (age groups '65–69' to '90 plus', N = 3080). In this way, we are able to provide a description of the total population '65 plus' in Switzerland, allowing us to get a better idea of the degree to which the older cohorts diverge from the population aged 84 and younger, thus allowing us to know whether we can generalize our findings. Before delving into the descriptive exploration of our hypotheses, the next chapter provides an overview of some basic socio-demographic characteristics of the sample population.

Chapter 11
Socio-demographic Characteristics of the Sample Population

The socio-demographic characteristics analyzed in this chapter have been selected to provide a general idea of the social situation, including the social standing, of the surveyed population by taking into account age and sex. The variables include marital status,[1] educational attainment[1] and the last pre-retirement socio-professional category of the household.

Table 11.1 shows the distribution of marital status by age-sex groups, all cantons taken together.

The largest share of the population 65 and older has the status 'married or remarried' (57%),[2] followed by 'widowed' (27%). Naturally, the rate of married people declines with advancing age when widowhood becomes more prevalent, but the sex-gap remains strong. While for women, starting at age group 75–79 and all subsequent age groups, widowhood is more common than being married, this ratio is never reversed for men: at all ages, the predominant share of men is married. Analogous, for any age group except those younger than 70, singleness is twice as common among women than among men. The sample proportion of pensioners who are divorced is at 9%. Being divorced is a more frequent state among women than among men and it becomes rarer in older birth cohorts, with 12.5% for the age group 65–69, compared to only 5.5% for the age group 90 plus.

As Table 11.2 shows, educational attainment varies strongly by sex and – to a lesser degree – by cohort. There is a u-shaped gradient in the differences between men and women, with greater gaps at the lower and higher levels of education: compared to men, women are two and a half times as frequently found among those who only attended primary or lower secondary school (17.7% versus 7.8% for primary school and 11.6% versus 4.7% for lower secondary school). In contrast, the relative proportion of men who attended university is three times higher than the proportion of women

[1] Later in this chapter, the variables marital status and educational attainment will be discussed more thoroughly in view of economic vulnerability outcomes.

[2] Our data indicates that cohabitation is even more frequent: 63% of the population share their lives with someone (45% women, 79% men).

J. Henke, *Revisiting Economic Vulnerability in Old Age*, Life Course Research and Social Policies 11, https://doi.org/10.1007/978-3-030-36323-9_11

Table 11.1 Sample distribution of marital status by sex and age group

	65–69		70–74		75–79		80–84		85–89		90 +		
	Women	Men	Women	Men	Women	Men	Women	Men	Women	Men	Women	Men	Total
Single	8.3	6.1	5.8	3.3	14.5	5.9	10.0	4.3	8.0	3.8	9.8	2.5	6.8
Married/rem.	67.3	77.1	54.9	80.1	35.0	76.8	30.3	72.5	23.6	65.7	11.8	53.0	57.0
Divorced/sep.	13.9	10.6	15.0	11.0	12.4	8.0	7.1	8.9	4.7	2.9	7.2	3.5	9.3
Widowed	10.6	6.1	24.2	5.6	38.2	9.3	52.7	14.3	63.7	27.6	71.2	40.9	26.9
Total (N = 3080)	*100*	*100*	*100*	*100*	*100*	*100*	*100*	*100*	*100*	*100*	*100*	*100*	*100*

Table 11.2 Sample distribution of educational attainment by sex and age group

	65–69		70–74		75–79		80–84		85–89		90 +		
Educational attainment	Women	Men	Women	Men	Women	Men	Women	Men	Women	Men	Women	Men	Total
Primary	9.4	4.3	14.2	9.2	17.0	8.4	18.3	9.2	25.0	8.8	22.0	7.2	12.1
Lower secondary	7.4	1.6	10.4	3.4	10.7	4.0	15.3	5.6	10.4	7.9	15.3	5.6	7.7
Professional training	39.1	45.6	32.2	36.7	27.4	44.5	27.9	33.5	26.4	37.7	22.0	35.4	34.9
High school	20.4	8.2	22.1	14.6	25.6	12.8	20.1	17.5	17.9	9.6	20.7	17.4	17.1
University of applied sciences	15.4	16.1	12.8	18.0	11.9	13.1	14.0	14.7	14.2	15.8	14.7	12.3	14.5
University	8.4	24.3	8.3	18.0	7.4	17.2	4.4	19.5	6.1	20.2	5.3	22.1	13.8
Total ($N = 3080$)	*100*	*100*	*100*	*100*	*100*	*100*	*100*	*100*	*100*	*100*	*100*	*100*	*100*

(20.2% compared to 6.7%). The most important conclusion is that women of all age groups are overrepresented among the lower levels of educational attainment.

As expected, the gender gap is less pronounced among younger cohorts. The changing trend is most clearly visible in the evolution of the share of primary level education (9% for 65–69 year old women compared to 22% for women aged 90 and older). The comparison of proportions of attainment levels across age groups is to be done with caution as the younger cohorts are more likely to accurately reflect the 'original composition' of educational attainment, meaning that these proportions are closer to those that would have been observed at age 25, whereas the composition of older cohorts has gone through increasing transformations due to differentials in mortality rates by educational attainment (Oris and Lerch 2009; Spoerri u. a. 2006).

For university-level degrees, the rate of attainment among the youngest female pensioners is noticeably higher than among the very old women; however, the attainment rate for university of applied sciences reveals no change between younger and older female cohorts. Among younger male cohorts, there is a noticeable trend towards higher educational levels, except for the highest levels, which remain fairly constant. The weak or stagnant trends towards a university education and the downwards slope in male attainment of a university degree can probably at least partially be explained by the above mentioned selection bias due to an educational mortality gradient (Schumacher and Vilpert 2011): this phenomenon, where individuals with a higher education tend to live longer, causing their relative share to increase as a given cohort grows older, has previously been observed in the Swiss context (Oris and Lerch 2009).

Not shown in the table above but noteworthy are the cantonal differences in educational attainment. Ticino and Valais have the highest percentage of people who only went to primary school or lower secondary school (30%, 27%). The sample population from Basel and Geneva feature a very similar educational profile: in both cantons, 35% have attained some kind of a university degree (including applied sciences). This is roughly one third higher than in Valais and double the rate observed in Ticino while Berne tends to be in a middle range.

Social position is not only determined by educational attainment but also by post-educational experience and work career.

Table 11.3 shows the distribution of socio-professional categories among the population 65–84 by canton. For an overview of the sample characteristics in terms

Table 11.3 Last socio-professional category of household by canton, population 65–84

Socio-professional category	Geneva	Valais	Berne	Basel	Ticino	Total
Executives, liberal professions	45	35	32	36	33	36
Self-employed	8	15	15	8	8	11
Intermediary professions	15	14	15	17	14	15
Skilled non manual	13	13	15	21	15	16
Skilled manual	14	19	17	13	24	17
Unqualified	3	5	5	4	6	5
Not active	1	0	1	0	1	1
Total (N = 1622)	*100*	*100*	*100*	*100*	*100*	*100*

of their last (pre-retirement) socio-professional category, it makes less sense to look at the distribution by sex and age, since the household dominance model was applied in the coding of this variable. The most frequent category in all five regions is 'executives and liberal professions' (36% on average), Geneva reporting a higher than average prevalence of this category. The less prestigious categories 'skilled manual' and 'unqualified' are most present in Ticino and Basel (24%, 29%).

Before describing the economic situation in more specific terms, the next chapter presents the distribution of our sample population according to each one of the three measures of economic vulnerability.

References

Oris, M., & Lerch, M. (2009). La transition ultime. Longévité et mortalité aux grands âges dans lebassin lémanique. In M. Oris, E. D. Widmer, A. de Ribaupierre, D. Joye, D. Spini, G. Labouvie-Vief, & J.-M. Falter (Eds.), *Transitions dans les parcours de vie et construction des inégalités (407–433)*. Lausanne: PPUR presses polytechniques.

Schumacher, R., & Vilpert, S. (2011). Gender differences in social mortality differentials in Switzerland (1990-2005). *Demographic Research, 25*(8), 285–310.

Spoerri, A., Zwahlen, M., Egger, M., Gutzwiler, F., Minder, C., & Bopp, M. (2006). Educational inequalities in life expectancy in the German speaking part of Switzerland between 1990 and 1997: Swiss National Cohort. *Swiss Medical Weekly, 136*, 145–148.

Chapter 12
Prevalence and Overlap of Three Measures of Economic Vulnerability

12.1 Economic Vulnerability Rate by Measure

The rate of economic vulnerability as measured by the Objective, Self-Assessed and Perceived Measure among the population 65–84. The proportion of the economically vulnerable within the total population is close to the same according to the Objective Measure (14.8%) and the Self-Assessed Measure (14.2%); it is 20% higher for the Perceived Measure (17.8%).

We will now turn to the prevalence of economic vulnerability in terms of the intersection between the measures, yielding the vulnerability types defined in Part I. Table 12.1 shows that the overlap between the three population groups identified as vulnerable by one of the three measures is very small: only 3.4% are vulnerable on all three dimensions (Type BBB), 8.7% are vulnerable on exactly two dimensions (Types BBA, BAB, ABB) and 19.3% are vulnerable on only one dimension (Types AAB, ABA, BAA). So, while as many as 31.4% of the sample population are vulnerable according to *at least one* measure, by far the largest share of the population is not vulnerable by any of the proposed standards (Type AAA).

According to our theoretical considerations, some of the vulnerability types are more likely to occur than others. Earlier we stated that vulnerability types AAB and BAB are not likely to be found according to theory. Empirically we find, however, that type AAB occurs in 8.2% of our sample population, which corresponds to the second most frequent combination (after AAA, which is the absence of vulnerability and thus can only theoretically be considered a 'type' of vulnerability). Put differently, they represent a share of pensioners in Switzerland who are worried about not having enough money to pay their current expenses despite the fact that they are finding it easy (or very easy) to make ends meet with their monthly household income. In Chap. 13 we will examine whether this constellation is associated with specific personality traits, such as neuroticism, which could explain an above average propensity to worry. The second type that we were not able to derive theoretically from any risk constellation was type BAB because of the hypothesized

J. Henke, *Revisiting Economic Vulnerability in Old Age*, Life Course Research and Social Policies 11, https://doi.org/10.1007/978-3-030-36323-9_12

Table 12.1 Rates of economic vulnerability by vulnerability type

Type	obj_ev	sa_ev	perc_ev	Frequency	%
AAA	non-vuln.	non-vuln.	non-vuln.	1'114	68.3
AAB	non-vuln.	non-vuln.	vulnerable	133	8.2
ABA	non-vuln.	vulnerable	non-vuln.	62	3.8
ABB	non-vuln.	vulnerable	vulnerable	74	4.5
BAA	vulnerable	non-vuln.	non-vuln.	119	7.3
BAB	vulnerable	non-vuln.	vulnerable	28	1.7
BBA	vulnerable	vulnerable	non-vuln.	40	2.5
BBB	vulnerable	vulnerable	vulnerable	55	3.4
N				*1'625*	*100*

pathways to stress, leading from 'objectivity' towards increasing degrees of 'subjectivity'. Indeed, type BAB turns out to be the least frequent constellation (1.7%).

Let us now briefly comment on the remaining vulnerability types, in the order of frequency. Type BAA takes the 3rd rank (7.3%). As depicted previously (Table 5.3) we hypothesize that these respondents are in fact unaware of their own vulnerability, not recognizing the scarcity of their resources as 'adversity'. We expect this group to be better endowed with economic resources other than monthly income, particularly in comparison with group BBX (BBA or BBB),[1] being the counterfactual of BAX. Another likely scenario is that these individuals have adjusted to living with scarce resources and are probably long-standing members of the lower social classes. In this case, reflecting the mechanisms of 'adaptive preferences' and 'downward social comparison', we would furthermore expect to find evidence thereof in the variables educational attainment, socio-professional category and among the psychosocial variables.

Vulnerability type ABB represents 4.5% of the sample population. According to our theoretical schema we posited that, by an objective standard, individuals in this group have been able to ensure more than the minimum income such as defined by the poverty line. However, they are finding it insufficient for making ends meet and the strategies for ensuring their monthly income may not be very sustainable, causing role strains, which results in the experience of financial stress. Among this group we would expect to find more individuals belonging to the middle class but with low levels of savings, and a higher rate of individuals with poor health. Furthermore, these respondents are likely to have a relatively high frequency of social participation compared to other groups. The comparison with group AAX will be especially relevant though the most important difference between these two groups will most certainly be a difference in (mean) income. In order to obtain further nuances, it would therefore be interesting to single out respondents with the vulnerability profile of AAX who are in the income category just above the poverty line and compare them to ABX.

[1] In our typology, the letter X denotes a placeholder for either A or B wherever a specific measurement angle is irrelevant for the hypothesis under examination.

Type ABA is a very similar type to the one just discussed and only slightly less frequent (3.8%). In our theoretical considerations we described this group as successfully coping with their (subjectively) precarious income situation. The combination of insufficient self-assessed income (though above the poverty line) yet without financial worries lets us conclude that they have found coping strategies that seem to work out well for them. It will be particularly interesting to compare this group to ABB with regard to the variables block Psychosocial Consequences and Symptoms. We expect to find less evidence of role strain and financial stress among respondents with the ABA profile compared to those belonging to group ABB.

The most vulnerable type imaginable in our typology is evidently type BBB, ranking second to last in terms of frequency: only 3.4% of the sample population are affected by all three dimensions of economic vulnerability. In our analysis, we expect to find high risks of exposure combined with insufficient coping resources for dealing with the situation practically, but also signs of downward adaptations in terms of expectations. This population group is likely to include many respondents of low-income-earning socio-professional category; moreover, it is expected to be affected by a lower than average level of health. In comparison with group BBA, we expect these respondents to show lower levels of mastery and a diminished sense of self as well as symptoms of social isolation.

Type BBA (2.5%) represents those individuals who, so we suspect, are choosing not to mobilize their coping resources to change the situation. While they do recognize the adversity of their situation by admitting their financial difficulties, they either have intentionally embraced a modest lifestyle or they are focusing on other things they deem as more important. It seems plausible that this group includes a more than proportionate share of respondents who live in remote rural areas as well as very old individuals, for whom material things have ceased to be (or have never been) among the highest priorities.[2] It is very likely also that members of this group benefit from well-developed social networks (i.e. low levels of social isolation), providing stability and moral support.

12.2 Congruence Between Measures

As Table 12.1 has shown, there is relatively little overlap between the three measures. In terms of proximity between angles of measurement, Table 12.2 shows that among the three indicators, the Perceived (perc_ev) and the Self-Assessed Measure (sa_ev) achieve the highest total congruence in allocating a respondent as either 'vulnerable' or 'non-vulnerable (83.8% of sample population). Breaking these percentages down into those that are congruently 'vulnerable', by adding up the percentage points of BBB and ABB, we see that 7.9% of the sample population was

[2]As a result of methodological decisions described on p. 101, our final sample does not include very old individuals, which makes it difficult to verify this assumption.

Table 12.2 Congruence between measures in percent of sample population

Measures	Congruently vulnerable	Congruently non-vulnerable	Total congruence
perc_ev / sa_ev	7.9	75.9	83.8
sa_ev / obj_ev	5.8	76.7	82.6
perc_ev / obj_ev	5.1	72.4	77.5

Table 12.3 Odds of being economically vulnerable by another measure

	obj_ev	sa_ev	perc_ev
obj_ev		4.9***	1.6*
sa_ev	4.9***		8.5***
perc_ev	1.6*	8.5***	

Key: *** < 0.001, ** < 0.01, * < 0.05; *ns* not statistically significant

congruently classified as both, vulnerable in terms of 'difficulties in making ends meet' and 'worrying about not having enough money for current expenses'. Naturally, since there is a majority of 'non-vulnerable' respondents, the proportion of congruently classified 'non-vulnerable' respondents is also very high (75.9% of the sample population). As was to be expected, the congruent classification of the Perceived and the Objective Measures (obj_ev) is least frequent, but occurs nonetheless in 77.5% (total congruence).

In order to obtain the probability of correctly predicting one type of vulnerability, taking into account the other measures, a logistic regression[3] was estimated, using successively each of the measures as outcome variable. Table 12.3 shows that the risk (odds) of those who are economically vulnerable by any one of the measures to also be vulnerable by the other two measures is significantly higher.

As we have seen already, the strongest relationship exists between the two subjective measures, sa_ev and perc_ev: even after controlling for low income, the odds of being worried about not having enough money for current expenses are 8.5 times higher for people who responded that they are having difficulties in making ends meet. Also in agreement with theory, the Objective Measure is more closely associated with the Self-Assessed than with the Perceived Measure: after controlling for financial worries, respondents with less than CHF 2400 a month are 4.9 times more likely to say that they have difficulties in making ends meet, whereas they are only 1.6 times more likely to worry about not being able to cover their current expenses. In the context of our regression model, these probabilities do obviously not represent causal linkages but are to be interpreted as strength of association. This result can nonetheless be interpreted as supporting our hypotheses that there is a clear difference between the two subjective measures. It furthermore suggests that the Self-Assessed Measure, while being more closely associated with the Perceived Measure, may indeed have a mediating role between the Objective and the Perceived Measure.

[3] Based on a similar analysis in Bradshaw and Finch (2003).

12.3 Associations at the Ordinal Level

The dependent variables this volume focuses on are binary measures of economic vulnerability. In order to gain a thorough understanding of the relationship between the angles of measurement that underlie the three indicators, it is relevant to also study the association between the ordinal variables from which the indicators were computed.

Table 12.4 shows cell frequencies and percentage distribution of 'monthly household income', the basis of the Objective Measure, and the variable 'difficulties in making ends meet', which is the stem of the Self-Assessed Measure.

Looking first at the percentage distribution of income categories by level of difficulties in making ends meet, we see that, while most respondents who are finding it very difficult to make ends meet are found in the lowest income category (50%), this conclusion cannot be reversed: among the 242 respondents (14.9%) whose equivalent household income is less than CHF 2400 per month, 88 (36%) are in category 3 finding it 'quite easy' to make ends meet. 11.6% and 27% are finding it '1 very difficult' and '2 difficult' to make ends meet. The mode (underlined in the

Table 12.4 Cell frequencies and percentage distribution of monthly household income and difficulties in making ends meet

Monthly household income	Difficulties in making ends meet				
(equivalized, in CHF)	1 very difficult	2 quite difficult	3 quite easy	4 very easy	Total
Count					
<2400	28	67	88	59	*242*
2400–3600	21	68	129	104	*322*
3600–4800	5	27	139	346	*517*
4800–6000	2	10	56	218	*286*
> 6000	0	3	21	234	*258*
Total	*56*	*175*	*433*	*961*	*1625*
Percent within household income					
<2400	11.6	27.7	<u>36.4</u>	24.4	*100*
2400–3600	6.5	21.1	<u>40.1</u>	32.3	*100*
3600–4800	1.0	5.2	26.9	<u>66.9</u>	*100*
4800–6000	0.7	3.5	19.6	<u>76.2</u>	*100*
> 6000	0.0	1.2	8.1	<u>90.7</u>	*100*
Total	*3.5*	*10.8*	*26.7*	*59.1*	*100*
Percent within difficulties in making ends meet					
<2400	<u>50.0</u>	38.3	20.3	6.1	*14.9*
2400–3600	37.5	<u>38.9</u>	29.8	10.8	*19.8*
3600–4800	8.9	15.4	<u>32.1</u>	<u>36.0</u>	*31.8*
4800–6000	3.6	5.7	12.9	22.7	*17.6*
>6000	0.0	1.7	4.9	24.4	*15.9*
Total	*100*	*100*	*100*	*100*	*100*

χ^2 (12) = 447.20, $p < 0.001$; gamma = 0.62; z = 28.14, $p < 0.001$; Kendall's tau-b = 0.43, z = 25.14, $p < 0.001$

Table 12.5 Cell frequencies and percentage distribution of difficulties in making ends meet and worry about finances

	Worried about not having enough money				
Difficulties in making ends meet	1 strongly	2 somewhat	3 a little	4 not at all/does not concern me	Total
Count					
1 very difficult	31	13	4	8	*56*
2 quite difficult	42	43	52	37	*174*
3 quite easy	30	54	109	229	*422*
4 very easy	33	44	110	741	*928*
Total	*136*	*154*	*275*	*1015*	*1580*
Percent within difficulties in making ends meet					
1 very difficult	55.4	23.2	7.1	14.3	*100*
2 quite difficult	24.1	24.7	29.9	21.3	*100*
3 quite easy	7.1	12.8	25.8	54.3	*100*
4 very easy	3.6	4.7	11.9	79.9	*100*
Total	*8.6*	*9.8*	*17.4*	*64.2*	*100*
Percent within worry about finances					
1 very difficult	22.8	8.4	1.5	0.8	*3.5*
2 quite difficult	30.9	27.9	18.9	3.7	*11*
3 quite easy	22.1	35.1	39.6	22.6	*26.7*
4 very easy	24.3	28.6	40	73	*58.7*
Total	*100*	*100*	*100*	*100*	*100*

χ^2 (9) = 464.0, p < 0.001; gamma = 0.63, z = 25.23, p < 0.001

table) within the next higher income category CHF 2400–3600 remains stable: 129 out of 322 respondents (36.4%) at this income level are making ends meet '3 quite easily'. From the third income class onwards, the mode is in the highest category of self-assessment: respondents found in income categories 3600–4800, 4800–6000 and > 6000 are predominantly finding it '4 very easy' to make ends meet.

The strength of association is positive, strong and highly significant as measured by both, gamma and Kendall's tau-b: $\gamma = 0.62$, z = 25.23, p < 0.001 and tau-b = 0.43, z = 25.14, p < 0.001.

Table 12.5 shows the observed cell frequencies and percentage distribution of the ordinal-level variables 'difficulties in making ends meet' and 'being worried about not having enough money to cover current expenses', which is the variable underlying the Perceived Measure.

In absolute terms, among the 56 respondents who are finding it '1 very difficult' to make ends meet, 31 are in category '1 strongly worried' (mode, underlined in the table) and 13 are in category '2 somewhat worried'. For category '2 quite difficult' the mode clearly shifts higher up on the scale to category '3 a little worried', but here the distribution is more balanced: among the 174 respondents 42 are '1 strongly worried', 43 are '2 somewhat worried' and 52 are '3 a little worried'. The frequency distribution indicates a very strong congruence between the measurement angles for the most vulnerable respondents: among those respondents who are most vulnerable in terms of self-assessment, 55% are also in the most vulnerable category according to the perceived measurement angle. On the less vulnerable side of the contingency table, there is also a very strong concordance: among the 928 respondents who are making ends meet '4 very easily', 741 (79.9%) answered '4 not at all worried/does not concern me'. When looking at the relationship from the perspective of financial worry, a diagonal pattern emerges: the mode of '1 strongly worried' is found in '2 quite difficult' to make ends meet and moves up by one category as respondents worry less.

The effect size of this relationship is strong, positive and statistically highly significant with gamma $\gamma = 0.63$, $z = 25.23$, $p < 0.001$. The gamma coefficient seems to be more indicated for measuring the effect size of this particular relationship because of the more pronounced correlations in the extremes of the contingency table.

In Table 12.6 we see the observed cell frequencies and percentage distribution of 'monthly household income' (variable underlying the Objective Measure) and

Table 12.6 Cell frequencies and percentage distribution of monthly household income and worry about finances

Monthly household income	Worried about not having enough money				
(equivalized, in CHF)	1 strongly	2 somewhat	3 a little	4 not at all/does not concern me	Total
Count					
<2400	46	37	64	89	*236*
2400–3600	38	44	75	159	*316*
3600–4800	34	49	88	335	*506*
4800–6000	14	13	30	220	*277*
>6000	5	11	21	214	*251*
Total	*137*	*154*	*278*	*1017*	*1586*
Percent within income					
<2400	19.5	15.7	27.1	<u>37.7</u>	*100*
2400–3600	12.0	13.9	23.7	<u>50.3</u>	*100*
3600–4800	6.7	9.7	17.4	<u>66.2</u>	*100*
4800–6000	5.1	4.7	10.8	<u>79.4</u>	*100*
>6000	2.0	4.4	8.4	<u>85.3</u>	*100*
Total	*8.6*	*9.7*	*17.5*	*64.1*	*100*
Percent within worried about finances					
<2400	<u>33.6</u>	24.0	23.0	8.8	*14.9*
2400–3600	27.7	28.6	27.0	15.6	*19.9*
3600–4800	24.8	<u>31.8</u>	<u>31.7</u>	<u>32.9</u>	*31.9*
4800–6000	10.2	8.4	10.8	21.6	*17.5*
>6000	3.7	7.1	7.6	21.0	*15.8*
Total	*100*	*100*	*100*	*100*	*100*

$\chi^2 (12) = 185.89$, $p < 0.001$; gamma = 0.43; $z = 15.45$, $p < 0.001$; Kendall's tau-b = 0.28, $z = 15.04$, $p < 0.001$

'worry about not having enough money' (variable underlying the Perceived Measure).

The relationship between the objective and the perceived angle is far from linear: For each income class, the proportion of respondents is highest in the category '4 not at all worried/does not concern me'. Reversing the perspective, those who are '1 strongly worried' are most frequently found in the lowest income categories: 33% have to do with less than CHF 2400 a month, followed by the second largest group of 27% with an income between CHF 2400 and 3600. For all other levels of worry ('2 somewhat' to '4 not at all/does not concern me') the mode is in the income group CHF3600 to 4800, representing one third of the sample population.

The relationship between the objective and perceived base variable, too, is positive and highly significant though the effect size is moderate: $\gamma = 0.43$, $z = 15.45$, $p < 0.001$ and tau-b = 0.28, $z = 15.04$, $p < 0.001$. Among the three pairs of measures, the correlation between these two is clearly the least strong.

This in-depth analysis of the relationship between the three angles of measurement of economic vulnerability revealed that the association between each one of the base variables is statistically highly significant. The effect size is particularly strong between, firstly, the objective and the self-assessed angle and, secondly, between the self-assessed and the perceived angle. These results are encouraging in view of our hypothesis, that the self-assessed angle may play a mediating role between the objective and the perceived angle. Now, the foundation has been laid to move on to the descriptive analysis of these three distinct but overlapping groups of economically vulnerable respondents.

Reference

Bradshaw, J., & Finch, N. (2003). Overlaps in dimensions of poverty. *Journal of Social Policy, 32*(04), 513–525.

Chapter 13
Background Characteristics

13.1 Sex

The percentage distribution by sex (Appendix 13.1) confirms that the rate of economic vulnerability differs considerably between men and women: women, who make up 45% of the sample population, have a high and constant rate of vulnerability across the three measures (18.5%, 17% and 17.4%). Among men, the most important type of vulnerability is being worried about finances (18.2%), self-assessed and objective vulnerability being less frequent (10.8% and 10.6%).

A phi coefficient was calculated to assess the strength of the relationship between sex and measures of economic vulnerability, which is generally weak: the Objective Measure reports the highest coefficient ($\Phi = -0.10$), the other coefficients being even weaker. The association of sex with obj_ev is highly significant (χ^2 (1) = 16.6; $p < 0.001$) and somewhat significant for sa_ev (χ^2 (1) = 9.2; $p = 0.002$).

Given the gender-difference in vulnerability rates at the sample population level, it is of interest to examine how the three measures combine at the individual level by means of the Vulnerability Typology (Appendix 13.2).

It is noteworthy that the proportions of women are higher among all types characterized by an objective vulnerability (=types BXX). BAB is the exception, with male respondents representing the higher percentage share; however, women are still overrepresented compared to their sample proportion (46.4% vs. 44.9%). The most important sex difference is found for type BBB, where women represent close to double the proportion of men (65.5% vs. 34.6%). A very significant majority of women exists as well in type BAA, which we qualified as the group that does not recognize their own vulnerability, possibly because of adaptive preferences as a result of long-term spells of poverty in the past. For some of the types marked by subjective vulnerability, the trend is inversed and women are underrepresented: three fifth of ABB are men and close to two thirds of AAB are men.

J. Henke, *Revisiting Economic Vulnerability in Old Age*, Life Course Research and Social Policies 11, https://doi.org/10.1007/978-3-030-36323-9_13

The effect size of the relationship between sex and vulnerability type is moderate with Cramér's V = 0.14 but statistically highly significant with χ^2 (1) = 30.6; $p < 0.001$. We conclude that the nature of the relationship goes in the direction of women being more frequently economically vulnerable than men, with the exception of perceived vulnerability, which is the vulnerability type men are more frequently affected by than women.

13.2 Age

For the distribution by age we first looked at the larger sample that includes the older cohorts in order to be better prepared for examining the relationship of our measures with this important covariate.

With regards to age, objective and subjective measures seem to be following opposite trends (Appendix 13.3): the two subjective indicators describes an overall descending gradient from younger to older age, while according to the Objective Measure, we find that very old retirees are more vulnerable than younger ones (with the exception of age group 90 plus). The differences between the youngest and the oldest cohorts are striking for the Perceived Measure: as many as 18% of those aged 65–69 are worried about not having enough money for current expenses, compared to only 8.8% of those who are older than 90. The same trend is observed for the Self-Assessed Measure, though it is not linear, peaking at age group 70–74 (18.3%).

In contrast, the poverty rate (obj_ev) is clearly higher among the very old, at 18% for those aged 80–89, compared to 13% for the recently retired. The drop in rate for the Objective Measure among respondents of 90 and older must be thought of as an interaction of cohort and period effects: on the one hand, it is certain that the very old age groups, especially those older than 90, are made up of the 'survivors' of their cohort, and as such, their current composition is the result of processes of selection. One of these well-known selection processes is reflected in a socio-economic gradient of mortality, meaning that the financially well-off tend to live longer on average (Mackenbach et al. 1997; Oris and Lerch 2009). On the other hand, it is well established for the Swiss context, that older generations of pensioners tend to have a lower income than households composed of recently retired adults. It has been found that lower income in old age groups is not as much the result of a diminishment of income over time as it is the reflection of a less favorable institutional context – i.e. average level of income and the extension of old age pension schemes – reflected in less financial cushioning of the older generations (Moser 2006; Gabriel et al. 2015). Bearing in mind that these figures focus on the lower end of the income spectrum, we nonetheless must reckon with conflicting causal factors: in the present case, it looks as if cohort effects favoring the survival of the wealthy overrides the period effect, which would have made a lower economic status more likely among very old adults.

These preliminary findings seem to confirm previous research on the well-established income satisfaction paradox that the evaluation of economic resources tends to be more positive at older age for a given level of economic quality of life (Weidekamp-Maicher and Naegele 2010). Anticipating a more detailed investigation, possible explanations are likely to be found within a framework of reference utility, include adaptive preferences[1] and aspiration theory.[2]

The logical next step is to consider income and its appraisal at the individual level, facilitated by the Vulnerability Typology. For the analysis of the age distribution by vulnerability type we now return to the base sample, which only includes individuals who are younger than 85 (Appendix 13.4).

For the percentage distribution by age groups a clear gradient can readily be identified for the two vulnerability types that represent opposing sides of the typology: the frequency of type AAB decreases with advancing age, whereas BAA is more prevalent among the older cohorts. For the other types, no clear pattern can be discerned. A look at the most frequent vulnerability types within each age group (lower part of the table) shows that type AAB is – after the majority type AAA – most frequent among age groups 65–59 (10%) and 70–74 (9.1%). For age groups 75–79 and 80–84, type BAA is more prevalent (7.3% and 11.3%). The effect size of the relationship between age group and vulnerability type is small with Cramér's $V = 0.08$ but statistically significant at $p < 0.05$.

Given the clear sex differences that appeared for certain types, a distinction by age *and* sex will shed more light on those trends that may run in different directions for men and women (Appendix 13.5).

Combining both demographic variables shows that age is associated with different vulnerability profiles for women than for men: In the case of BBB, with increasing age the proportion stagnates for women (5.1%) but is cut in half for men (from 2.8% to 1.4%). For BAA, the proportion increases with age but it increases by a factor of 2.3 for men (to 8.3% at age 75–84) and, being high already for women, it increases by a factor of 1.3 (to 10.7% at age 75–84). For type ABA, the trend is opposite: for men, members of this type tend to be older, while for women the proportion sinks as age increases. Type AAB is less frequent among the older, for both sexes, but the gap is very substantial for women, where the proportion is cut in half (to 4.5% at age 75–84). For BBA, BAB and ABB the proportion is lower at higher age for both sexes.

We conclude that the interaction effect between age and sex is relevant for the analysis of economic vulnerability with regard to selected vulnerability types. The effect is most clearly seen in the mirroring of the two types BAA and AAB.

---

[1] Hazelrigg and Hardy (1997) found that the likelihood of assessing income as being adequate increased, the longer the interval since retirement

[2] Stoller and Stoller (2003) mention how an individual's aspirations may be influenced by the experience of one's own parents' living standard.

13.3 Canton

There exist considerable differences in the prevalence of economic vulnerability between the cantons (Appendix 13.6). In terms of income poverty (obj_ev) close to one in four residents in Valais and Ticino are vulnerable, while in Basel and Geneva the share of those who have to do with less than CHF 2400 per month is closer to one tenth of the population.

As for the subjective measures, there is what could be called a 'latin gap', with close to 10% of French- and Italian-speaking respondents (Geneva, Valais and Ticino) who evaluate themselves clearly as more vulnerable than the predominantly German-speaking cantons (Basel and Berne). The strength of the relationship as measured by Cramér's V is weak to moderate for all three measures: the effect size for obj_ev is $V = 0.12$, which is a statistically significant association with $\chi^2 (4) = 25.4$, $p < 0.001$, for sa_ev we find the exact same effect with $V = 0.12$ and $\chi^2 (4) = 22.6$, $p < 0.001$. The association of canton is strongest with perc_ev: $V = 0.15$ and $\chi^2 (4) = 38.4$, $p < 0.001$.

Geneva has among the lowest rate of economic vulnerability as measured by obj_ev (12%), standing in contrast to the high rate of individuals who self-assess themselves as vulnerable (17%) and who are worried about their financial situation (19%). Keeping in mind that we are looking at population-level statistics, part of the explanation for the considerable disparities may lie in the exceptionally high living costs, which elevates Geneva even above the other Swiss cities of comparable size, such as Basel and Zurich.[3]

Ticino systematically records higher than average vulnerability rates on all measures (obj_ev = 22%, sa_ev = 20%, perc_ev = 30%), converging with findings by previous research based on the Swiss Household panel (Falter 2009). The discrepancy between obj_ev and perc_ev amounts to 7.3% but the balance is a completely different one for obj_ev and sa_ev: in Ticino, the share of the population who are vulnerable in monetary terms is higher than the proportion of those who self-assess themselves as vulnerable. It seems all the more paradoxical that the rate of respondents who are worried about not having enough money supersedes those who are having difficulties in making ends meet by 10% points. This result is unexpected and points to the possibility of cultural differences in perceiving and/or voicing dissatisfaction, a question that will be addressed later in this volume.

In both Geneva and Ticino, type AAB is most frequent (29.5%, 28.6%). In all other cantons, type BAA dominates. As was to be expected by the previously shown frequencies by measure, the most intense form of vulnerability – type BBB – is most prevalent in Ticino and Valais (15.1% and 14.8%). As far as the strength of this statistically significant relationship is concerned, Cramér's V indicates an effect size that is barely moderate $V = 0.11$ ($p < 0.001$).

[3] http://www.statistik-bs.ch/thema/wirtschaft/preisindex (14.01.2015)

13.4 Marital Status

Among our sample population, the majority (63%) are married or remarried, followed by those who are widowed (19%), divorced or separated (11%) and never married (7%). As far as the Objective Measure is concerned, our results reflect the same ranking of risk by marital status (Appendix 13.7): married, single, widowed, divorced/separated) that was confirmed for the population in Valais and Geneva in the previous wave of this study, the 1994-survey (Lalive d'Epinay and Hofstetter-Bétemps 1999). The group of respondents whose status was divorced or separated features the highest proportion of economic vulnerability by any measure (obj_ev = 23%, sa_ev = 28%, perc_ev = 27%). From among the 102 women and 83 men who were divorced or separated at the time of their interview, 27% of the women and 17% of the men lived below the poverty line. Pro Senectute's experience confirms this risk profile, highlighting how the legal situation continues to affect especially very old adults: many a female client ends up needing financial support to cope with the tight budget, especially if she got divorced before the coming into effect of the divorce law (in 2000) which regulated splitting the assets stemming from occupational pensions between the spouses.

It is noteworthy that divorced and separated individuals tend to be burdened much more by their circumstances compared to widowed individuals, whose proportions of objective vulnerability is close to the same but among whom subjective vulnerability is less frequent. The relationship between marital status and obj_ev has a small to moderate sized effect according to Cramér's $V = 0.11$ and is highly significant ($p < 0.001$). The effect on self-assessed economic vulnerability is of a medium size with $V = 0.16$ ($p < 0.001$) and of a small size for perceived vulnerability $V = 0.10$ ($p < 0.001$).

Considering how the three angles of economic vulnerability combine at the individual level adds to our understanding of typical economic situations that tend to accompany each marital status (Appendix 13.8). Looking through the lenses of the typology, several things are worth noting: among those who are single, as many as 38% are found among vulnerability type BAA. The most frequent vulnerability type among married people is type AAB (32%). In agreement with the figures shown above, those respondents whose current status is 'divorced or separated' record the highest share of the most pronounced vulnerability type BBB (21%). Among widows and widowers, type BAA is again most prevalent (26%), while type BBB is much lower (10%).

The overall strength of the relationship between marital status and the Vulnerability Typology can be qualified as low to moderate with Cramér's $V = 0.12$ and highly significant ($p < 0.001$).

13.5 Educational Attainment

Most pensioners aged 65–84 went through a professional training (38%), followed by three educational categories of about equivalent size: high school (16%), university of applied sciences (16%) and university (14%).

As far as the vulnerability rates within each educational level are concerned, we find that objective economic vulnerability yields a classical gradient (Appendix 13.9): the poverty rate among respondents who only went to primary school is seven times as high (35%) as for university graduates (5%). A gradient is also found for the two subjective measures, though the extremes are less pronounced and the slope is not monotonous: the highest rate of economic vulnerability by the Perceived Measure is at 38% within 'lower secondary school'.

The relationship between the variable educational attainment and all economic measures is highly significant ($p < 0.001$) for both measures of associations, gamma and Kendall's tau-b. However, the effect size varies considerably between the two coefficients, especially for obj_ev, where gamma shows a strong effect (−0.43) while tau-b only attests to a weak relationship (−0.19). The analysis of the percentage distribution having revealed a difference of a factor 7 between the vulnerability rate in the lowest versus the rate in the highest educational category and a consistent pattern of increasing risk from lowest to highest educational attainment level, it is justified to speak of a strong effect for obj_ev. The Self-Assessed and the Perceived Measure on the other hand can be said to have an association of a moderate strength with education (gamma is −0.32 for sa_ev and − 0.24 for perc_ev).

The distribution of types within each educational level is instructive (Appendix 13.10): among the higher educational levels (university, university of applied sciences, high school and professional training) the Perceived Measure standing alone is the most frequent type (AAB = 7%, 10%, 5%, 8%) after the majority type AAA, which is the most frequent type at any educational level. Among those whose highest attainment was professional training, an equally high proportion of respondents are objectively vulnerable, without self-assessing or perceiving themselves as such (BAA = 8%). This finding is an indication that in the context of the Swiss dual educational system, among the cohorts studied, only some absolvents of an apprenticeship have been able to secure a level of economic quality of life similar to those who are in possession of a university degree. A considerable share display the same features as the least educated groups, marked by downward adaptations of expectations and preferences. Indeed, type BAA represents the highest proportion among absolvents of primary school (17%) and there is also a relatively high rate among those who completed lower secondary (11.1%).

13.6 Personality

The bivariate analysis of the 'Big Five' dimensions of personality are included in the chapter on Background Characteristics because, while not at the center of interest for our research questions, they may have an association with the subjective measures of economic vulnerability and therefore need to be controlled for. A point biserial correlation coefficients was used for those four dimensions of personality that follow a normal distribution and were treated as interval data: neuroticism, openness, extraversion and agreeableness (Appendix 13.11). The distribution of conscientiousness, which does not follow a normal distribution, was treated at the ordinal level (Appendix 13.12).

None of the five personality dimensions has more than just a weak effect on any of the three measures of economic vulnerability. The strongest and highly significant effect is between *openness* and obj_ev ($r_{pb} = -0.10$, $p < 0.001$) meaning that low income is associated with a lower score on the openness scale. A weak positive and significant association exists between *neuroticism* and self-assessed economic vulnerability ($r_{pb} = 0.08$, $p = 0.002$). The coefficient of determination is, however, very small ($r^2_{pb} = 0.064$), variation in the score of neuroticism explaining less than 1% of the variation in sa_ev. It would therefore be a mistake to conclude that the subjective measurement angle is *per se* influenced by personality. In fact, it is noteworthy that despite the wording of the survey item, based on the question 'to what extent are you *worried*', there is no statistically significant correlation between perc_ev and neuroticism ($r_{pb} = 0.06$, $p = 0.011$). Furthermore, no association was detected between any of the measures and agreeableness, extraversion and conscientiousness.

These results signify that there exists a great diversity of psychological profiles for each vulnerable group as identified by the three measures. However, we cannot exclude that certain vulnerability types are associated with specific personality traits. It is for instance likely that the difference between types XXA and XXB can at least partly be explained by differentials in the levels of neuroticism and not only, as the vulnerability framework suggests, by differences in the successfulness of coping strategies and role strains (risk constellations 2a and 2b in Table 5.2). In order to be able to assess whether the neurotic personality type has an influence on the Perceived Measure in the case of specific risk constellations, we will test a series of specific hypotheses by comparing pairs of vulnerability types ending in either B or A.

We begin by conducting a Kruskal-Wallis test, a non-parametric omnibus test, in order to determine whether there is a statistically significant difference in the position (median) of neuroticism between at least two vulnerable population groups within the typology. After verifying the assumption that neuroticism follows the same distribution within each group of the Vulnerability Typology, we calculate the Kruskal-Wallis H, $\chi^2(7) = 23.8$, $p = 0.001$, based on which we reject the null hypothesis that all population groups have the same median score of neuroticism.

We proceed to examine the difference in medians of the specific population groups we are interested in. In each case, our null hypothesis is that there is no statistically significant difference in the median scores of neuroticism between the two compared groups ($H_0 = n_{XXA} = n_{XXB}$). For each test, we formulate the alternative hypothesis, that group XXA has a lower median score of neuroticism than group XXB. Table 13.1 (right column) shows the results of the Wilkoxon rank-sum test, which calculates the Mann-Whitney Test two sample statistic, an effect size mea-

Table 13.1 Wilkoxon rank-sum test for neuroticism and types of vulnerability

Alternative hypothesis (median group 1 < median group2)	z	Prob > |z|	P{median(group1) < median(group2)}
H_{a1} = nAAA < nAAB	1.51	0.132	0.54
H_{a2} = nABA < nABB	−2.11	0.035	0.40
H_{a3} = nBAA < nBAB	1.99	0.047	0.62
H_{a4} = nBBA < nBBB	1.43	0.154	0.59

sure that calculates the probability of an observation being in one group versus the other group (Goldstein 1997).

Two comparisons of population groups, H_{a2} and H_{a3}, yielded significant results ($p < 0.05$): for H_{a2}, we see that for 40% of the pairs, the score of neuroticism is lower for ABA than for ABB or, put differently: the prevalence of neuroticism is higher in group ABA than in group ABB. This finding deviates from the expected effect we predicted by our theoretical model, but is compatible with previous findings on the multifaceted construct of neuroticism: it not only captures a propensity to emotional lability and worry, but also reflects an absence of optimism (Scheier et al. 1994). Our results therefore suggest that in the case of economic vulnerability among Swiss pensioners, the 'pessimism component' of neuroticism is more clearly at work than the emotion-based components.

For H_{a1} the result[4] is non-significant but nonetheless interesting if we recall that type AAB is a type that we were not able to deduce logically from the risk constellations provided by the vulnerability framework (see Table 13.1). In fact, it seems illogical that someone who has no difficulties in making ends meet would still worry about not having enough money to cover current expenses. Differences in personality seemed a plausible explanation to account for the existence of this – theoretically – unlikely type but as we see, the personality trait neuroticism is not to be blamed for why respondents of type AAB are worried.

For H_{a3}, the effect goes in the expected direction: for 62% of the pairs, the score of neuroticism is lower for BAA than for BAB. This is the only type, where the Perceived Measure is significantly associated with higher scores of neuroticism, namely in the combination with objective vulnerability and a positive self-assessment.

In summary, we have found that the personality trait of neuroticism is the only personality dimension of theoretical relevance that is revealing a statistically significant relationship with economic vulnerability. Since it explains only a very small share of variance in the Vulnerability Typology, we conclude that this variable is of minor explanatory power and does therefore not need to be retained in the multivariate analysis.

[4]The score (e.g. 0.54 for Ha_1 = nAAA < nAAB) shows the proportion of 'pairs' for which the score of group 1 is smaller than the score of group 2. The number of pairs is defined by the number of people in the first group (e.g. AAA = 1099) times the number of people in the second group (AAB = 129). For Ha_1, among the 141,771 pairs, 54% were lower for AAA than for AAB, which is a non-significant result.

Appendices

Appendix 13.1: Sample Distribution by Sex and Measures of Economic Vulnerability

Sex	Sample distribution		obj_ev	sa_ev	perc_ev
	n	%	% vulnerable within sex		
Women	734	45.0	18.8	17.0	17.44
Men	897	55.0	11.6	11.8	18.17
Total	*1631*	*100.0*	*14.8*	*14.2*	*17.8*
Phi Coefficient			−0.10	−0.08	0.01
χ^2 (1)			16.59	9.19	0.15
p			<0.001	0.002	0.700

Appendix 13.2: Distribution of Vulnerability Types by Sex

	Vulnerability type								
Sex	BBB	BBA	BAB	ABB	BAA	ABA	AAB	AAA	*Total*
Percentage within typology									
Women	<u>65.5</u>	<u>52.5</u>	46.4	40.5	<u>57.1</u>	<u>61.3</u>	36.8	42.6	*44.9*
Men	34.6	47.5	<u>53.6</u>	<u>59.5</u>	42.9	38.7	<u>63.2</u>	<u>57.4</u>	*55.1*
Total %	*100*	*100*	*100*	*100*	*100*	*100*	*100*	*100*	*100*
Percentage within sex									
Women	4.9	2.9	1.8	*4.1*	<u>9.3</u>	5.2	6.7	<u>65.1</u>	*100*
Men	2.1	2.1	1.7	4.9	5.7	2.7	<u>9.4</u>	<u>71.4</u>	*100*
Total	3.4	2.5	1.7	4.6	7.3	3.8	8.2	68.6	
Count									
Women	36	21	13	30	68	38	49	475	*730*
Men	19	19	15	44	51	24	84	639	*895*
Total	*55*	*40*	*28*	*74*	*119*	*62*	*133*	*1114*	*1625*

Cramér's V = 0.14; χ^2 (7) = 30.6; p = < 0.001
The types are ordered so that types representing multiple vulnerabilities (i.e. featuring more 'Bs' than 'As') are on the left side and those that can be considered as less vulnerable are on the right side. For easier interpretation, the relatively higher frequency between women and men is underlined in the upper part of the table, and the highest and second to highest frequencies within sex are underlined in the middle part of the table (the most frequent category is underlined with a double line, the second most frequent category is underlined with a simple line)

Appendix 13.3: Sample Distribution by Age Group and Measures of Economic Vulnerability

Age	Sample distribution		obj_ev	sa_ev	perc_ev
	n	%	% vulnerable within age group		
65–69	613	19.9	12.5	12.3	18.1
70–74	594	19.3	15.7	18.3	21.1
75–79	572	18.6	13.9	14.2	18.1
80–84	499	16.2	18.2	11.5	13.0
85–89	451	14.6	18.4	13.1	12.4
90+	351	11.4	13.7	9.4	8.8
Total	*2467*	*100.0*	*14.8*	*14.2*	*17.8*
Cramér's V			0.09	0.08	0.09
χ^2 (5)			19.4	16.8	24.6
p			0.002	0.005	<0.001

Appendix 13.4: Distribution of Vulnerability Types by Age Groups

Age	Vulnerability type								
	BBB	BBA	BAB	ABB	BAA	ABA	AAB	AAA	*Total*
%									
65–69	25.5	30.0	28.6	21.6	20.2	24.2	34.6	29.4	*28.4*
70–74	36.4	32.5	21.4	33.8	23.5	32.3	29.3	24.8	*26.3*
75–79	18.2	20.0	32.1	35.1	25.2	22.6	21.8	25.4	*25.2*
80–84	20.0	17.5	17.9	9.5	31.1	21.0	14.3	20.5	*20.1*
Total %	*100*	*100*	*100*	*100*	*100*	*100*	*100*	*100*	*100*
%									
65–69	3.0	2.6	1.7	3.5	5.2	3.3	10.0	70.8	*100*
70–74	4.7	3.0	1.4	5.9	6.6	4.7	9.1	64.6	*100*
75–79	2.4	2.0	2.2	6.4	7.3	3.4	7.1	69.2	*100*
80–84	3.4	2.1	1.5	2.1	11.3	4.0	5.8	69.7	*100*
Total	3.4	2.5	1.7	4.6	7.3	3.8	8.2	68.6	
Count									
65–69	14	12	8	16	24	15	46	327	*462*
70–74	20	13	6	25	28	20	39	276	*427*
75–79	10	8	9	26	30	14	29	283	*409*
80–84	11	7	5	7	37	13	19	228	*327*
Total	*55*	*40*	*28*	*74*	*119*	*62*	*133*	*1114*	*1625*

Cramér's V = 0.08; χ^2 (21) = 33.5; p = 0.041

Appendix 13.5: Percentage Distribution of Vulnerability Types by Age-Sex Groups

Vulnerability	Age-sex group			
Type	Male 65–74	Female 65–74	Male 75–84	Female 75–84
BBB	2.8	5.1	1.4	5.1
BBA	2.4	3.4	1.9	2.2
BAB	1.9	1.2	1.4	2.5
ABB	5.1	4.2	4.7	3.6
BAA	3.6	8.6	8.3	10.7
ABA	2.3	5.9	3.2	4.2
AAB	10.6	8.3	7.9	4.5
AAA	71.4	63.4	71.2	67.2
Total	*100*	*100*	*100*	*100*

Appendix 13.6: Sample Distribution by Canton and Measures of Economic Vulnerability

	Sample distribution		obj_ev	sa_ev	perc_ev
Canton	n	%	% vulnerable within canton		
Geneva	320	19.6	11.6	16.9	19.1
Valais	310	19.0	19.4	16.8	18.7
Berne	397	24.3	12.9	9.3	12.9
Basel	331	20.3	10.3	10.3	12.4
Ticino	273	16.7	22.0	19.8	29.3
Total	*1631*	*100.0*	*14.8*	*14.2*	*17.8*
Cramér's V			0.12	0.12	0.15
χ^2 (4)			25.4	22.6	38.4
p			<0.001	<0.001	<0.001

Appendix 13.7: Sample Distribution by Marital Status and Measures of Economic Vulnerability

	Sample distribution		obj_ev	sa_ev	perc_ev
Marital status	n	%	% vulnerable within marital status		
Single	119	7.3	15.1	7.6	11.8
Married/remarried	1024	62.8	12.1	11.3	16.3
Divorced/separated	186	11.4	22.6	28.0	26.9
Widowed	302	18.5	19.2	17.9	19.9
Total	*1631*	*100.0*	*14.8*	*14.2*	*17.8*
Cramér's V			0.11	0.16	0.10
χ^2 (3)			19.4	43.5	15.8
p			<0.001	<0.001	0.001

Appendix 13.8: Percentage Distribution of Vulnerability Types by Marital Status, Without Type AAA

	Vulnerability type							
Marital status	BBB	BBA	BAB	ABB	BAA	ABA	AAB	*Total*
Single	6.9	6.9	10.3	10.3	37.9	6.9	20.7	*100*
Married	8.5	8.1	3.9	14.1	23.2	10.2	32.0	*100*
Divorced/separated	21.3	10.0	7.5	20.0	13.8	13.8	13.8	*100*
Widowed	10.2	5.9	6.8	12.7	26.3	17.0	21.2	*100*
Total percent	*10.8*	*7.8*	*5.5*	*14.5*	*23.3*	*12.1*	*26.0*	*100*

Cramér's V = 0.12; χ^2 (21) = 72.59; p < 0.001

Appendix 13.9: Sample Distribution by Education and Measures of Economic Vulnerability

Education	Sample distribution		obj_ev	sa_ev	perc_ev
	n	%	% vulnerable within education		
Primary	160	10.0	35.0	27.5	28.1
Lower secondary	99	6.2	26.3	28.3	38.4
Professional training	597	37.5	15.4	13.9	16.3
High school	261	16.4	10.7	11.5	13.0
University of applied sciences	248	15.6	9.7	9.7	17.3
University	228	14.3	4.8	7.9	11.0
Total	*1593*	*100*	*14.8*	*14.2*	*17.8*
Gamma			−0.43	−0.32	−0.24
z			9.5	6.4	5.1
p			<0.001	<0.001	<0.001
Kendall's tau-b			−0.19	−0.14	−0.12
z			9.1	6.3	5.1
p			<0.001	<0.001	<0.001

Appendix 13.10: Percentage Distribution of Vulnerability Types by Educational Attainment

Education	Vulnerability type							
	BBB	BBA	BAB	ABB	BAA	ABA	AAB	AAA	*Total*
Primary	27.3	22.5	17.9	18.1	23.7	11.7	9.5	6.5	10.0
Lower secondary	12.7	10.0	14.3	19.4	9.7	5.0	10.3	3.9	6.2
Professional training	27.3	45.0	42.9	31.9	41.2	45.0	37.3	37.1	37.4
High school	16.4	7.5	10.7	11.1	11.4	16.7	11.1	18.4	16.5
University of Appl. Sciences	11.0	12.5	14.3	11.1	7.9	8.3	19.1	16.9	15.5
University	5.5	2.5	0.0	8.3	6.1	13.3	12.7	17.1	14.4
Total	*100*	*100*	*100*	*100*	*100*	*100*	*100*	*100*	*100*
Primary	9.4	5.7	3.1	8.2	17.0	4.4	7.6	44.7	*100*
Lower secondary	7.1	4.0	4.0	14.1	11.1	3.0	13.1	43.4	*100*
Professional training	2.5	3.0	2.0	3.9	7.9	4.6	7.9	68.2	*100*
High school	3.5	1.2	1.2	3.1	5.0	3.8	5.4	77.0	*100*
University of Appl. Sciences	2.4	2.0	1.6	3.3	3.7	2.0	9.8	75.2	*100*
University	1.3	0.4	0.0	2.6	3.1	3.5	7.0	82.0	*100*
Total	3.4	2.5	1.7	4.6	7.3	3.8	8.2	68.6	

Cramér's V = 0.14; χ^2 (35)[5] = 146; p < 0.001

Appendix 13.11: Correlation Between Personality Traits Neuroticism, Openness, Extraversion, Agreeableness and Measures of Economic Vulnerability

Personality	obj_ev		sa_ev		perc_ev	
(Scores 2–10)	Point Biserial correlation	P > |t|	Point Biserial correlation	P > |t|	Point Biserial correlation	P > |t|
Neuroticism	0.04	0.126	0.08	0.002	0.06	0.011
Openness	−0.10	<0.001	0.03	0.302	−0.01	0.784
Extraversion	−0.01	0.637	0.00	0.930	−0.04	0.125
Agreeableness	0.04	0.123	0.06	0.012	0.02	0.466

[5] Given the small cell frequency it would have been better to calculate a Fisher's exact instead of a Chi-squared; however, given the large sample size, Stata was not powerful enough.

Appendix 13.12: Association Between Personality Trait Conscientiousness and Measures of Economic Vulnerability

Conscientiousness	Sample distribution		
Score	n	%	Cum. %
2	1	0.1	0.1
3	6	0.4	0.4
4	21	1.3	1.7
5	40	2.5	4.2
6	157	9.8	14.0
7	239	14.8	28.8
8	254	15.8	44.6
9	308	19.1	63.7
10	585	36.3	100.0
Total	1611	100	
	obj_ev	**sa_ev**	**perc_ev**
Gamma	0.05	0.02	−0.03
z	1.0	0.3	0.6
p	0.161	0.385	0.235
Kendall's tau-b	0.02	0.01	−0.01
z	1.0	0.3	0.6
p	0.158	0.382	0.239

References

Falter, J.-M. (2009). Le divorce, facteur de pauvreté féminine. In *Transitions dans les parcours de vie et construction des inégalités (S. 297–316)*. PPUR presses polytechniques.

Gabriel, R., Oris, M., Studer, M., & Baeriswil, M. (2015). The persistence of social stratification? A life course perspective on poverty in old-age in Switzerland. *Swiss Journal of Sociology, 41*(3), 465–487.

Goldstein, R. (1997). Immediate Mann–Whitney and binomial effect-size display. *Stata Technical Bulletin, 36*, 29–31.

Hazelrigg, L. E., & Hardy, M. A. (1997). Perceived income adequacy among older adults issues of conceptualization and measurement, with an analysis of data. *Research on Aging, 19*(1), 69–107.

Lalive d'Epinay, C., & Hofstetter-Bétemps, C. (1999). *Vieillesses Au Fil Du Temps, 1979–1994: Une Révolution Tranquille: Santé, Situations De Vie, Formes De Participation Et Visions Du Monde Des Personnes âGées En Suisse*. Lausanne: Réalités sociales.

Mackenbach, J. P., Kunst, A. E., Cavelaars, A. E., Groenhof, F., & Geurts, J. J. (1997). Socioeconomic inequalities in morbidity and mortality in western Europe. The EU Working Group on Socioeconomic Inequalities in Health. *Lancet, 349*(9066), 1655–1659.

Moser, P. (2006). *Einkommen und Vermögen der Generationen im Lebenszyklus. Eine Querschnitts-Kohortenanalyse der Zürcher Stuerdaten 1992–2003*. Statistisches Amt des Kantons: Zürich.

Oris, M., & Lerch, M. (2009). La transition ultime. Longévité et mortalité aux grands âges dans le bassin lémanique. In M. Oris, E. D. Widmer, A. de Ribaupierre, D. Joye, D. Spini, G. Labouvie-Vief, & J.-M. Falter (Eds.), *Transitions dans les parcours de vie et construction des inégalités (407–433)*. Lausanne: PPUR presses polytechniques.

Scheier, M. F., Carver, C. S., & Bridges, M. W. (1994). Distinguishing optimism from neuroticism (and trait anxiety, self-mastery, and self-esteem): A reevaluation of the life orientation test. *Journal of Personality and Social Psychology, 67*(6), 1063–1078.

Stoller, M. A., & Stoller, E. P. (2003). Perceived income adequacy among elderly retirees. *Journal of Applied Gerontology, 22*(2), 230–251.

Weidekamp-Maicher, M., & Naegele, G. (2010). Economic resources and subjective well-being in old age. In H. Mollenkopf & A. Walker (Hrsg.), *Quality of life in old age: International and multi-disciplinary perspectives* (Softcover reprint of hardcover 1st ed. 2007, S. 65–84). Dordrecht: Springer.

Chapter 14
Economic Resources

In this chapter, we consider the prevalence of various types of financial and economic resources and their association with the three measures of economic vulnerability. According to our theoretical framework, the entire model component of Economic Resources (of which income is but one element) needs to be taken into account in the explanation of economic strain (sa_ev). The main focus of the preliminary analysis presented here is therefore to examine the correlation between types of resources (other than income) with obj_ev and with sa_ev. The types of resources considered are financial *support* from either family, canton and/or non-governmental organizations, the State-funded *supplementary benefits* to old age pensions, *assets* such as personal savings and life insurance (3rd pillar), occupational *pensions* (2nd pillar), irregular and regular *work*, home *ownership* and *wealth*.

Both, low income (obj_ev) and economic strain (sa_ev) are associated with low levels of Economic Resources such as types of pensions, home ownership and wealth. But the relationship with economic strain is furthermore mediated by (higher) expectations of a certain living standard and (increased) financial needs, operationalized by a set of variables that will be discussed later. In our framework, the association of Economic Resources with financial stress (perc_ev) occurs via the Self-Assessed Measure but for reasons of completeness and rigor, the association between the Perceived Measure and each component of Economic Resources will nonetheless be examined.

14.1 Financial Support

Only a very small share of the population aged 65–84 benefit from *supplementary benefits* to old age pensions (6%) or from financial *support* by family, non-governmental organizations or canton (6%). Since the proportion of the two is exactly identical, it is worth noting that the overlap between these two sources of support, i.e. respondents who benefit from both, is less than 20%. Respondents who

J. Henke, *Revisiting Economic Vulnerability in Old Age*, Life Course Research and Social Policies 11, https://doi.org/10.1007/978-3-030-36323-9_14

indicated receiving support are expected to be more vulnerable in monetary terms, as is reflected in the very high rate of objective vulnerability (31% for support, 26% for supplementary benefits).

The fact that one in four recipients of supplementary benefits remains below the poverty line, a threshold that experts have established as the minimum level of existence, is disturbing and suggests that there is something out of joint in the Swiss pension plan. The Pro Senectute report identifies several types of disturbances that explain this reality, referred to as 'post-transfer poverty'[1] (Seifert and Pilgram 2009, pp. 59–66). The first one refers to the 14% who are poor (obj_ev) yet who do not receive supplementary benefits. This figure partly reflects a group of pensioners who is legally eligible for support in the form of supplementary benefits. The motives for refraining to make a claim when faced with economic strain range from modesty, a sense of responsibility, fear of stigmatization to a lack of information. Another type of malfunctioning is the result of a series of mechanisms that manifest in a so-called threshold effect and that are rooted in the high level of standardization of this type of financial transfer: individuals, whose income exceeds the legal minimum level of monthly income by very little and who are therefore not eligible to receive supplementary benefits may find themselves excluded from a series of additional privileges and preferential treatments (e.g. tax exemptions). This may end up positioning them in an economically worse-off place than individuals with a virtually identical economic standing but who were assessed as eligible. Another source of inequality has to do with the way the amounts to be disbursed are calculated: eligibility is assessed based on the one hand on available revenues, which includes 1/10 of the wealth exceeding the amount of exemption[2] and, on the other hand, on needs composed essentially of a lump sum for living costs, and a regionally varying allowance for rent and health insurance. Thus, depending on the amount factored in for rent and health insurance, the monthly amount at the disposition of an individual might be below the threshold of the poverty line.

Receiving support or supplementary benefits is expected to heighten the individual self-awareness of vulnerability, resulting in higher rates of sa_ev and perc_ev, because these transfers are usually received after voicing a need. Indeed, the rate of vulnerability is higher if measured by sa_ev than if measured by obj_ev: sa_ev = 34% for support and 47% for supplementary benefits (Appendix 14.1). Perception of vulnerability is less strongly associated with support but still, 29% of respondents who receive money from an alternative source of support are worried about their finances; among beneficiaries of supplementary benefits, 40% express this concern.

The strength of the relationship measured by the Phi Coefficient is moderately strong for support by family, NGOs and cantons and the Objective Measure ($\Phi = 0.11$) and the Self-Assessed Measure ($\Phi = 0.14$). For both indicators it is

---

[1] German: Nachtransferarmut.

[2] The amount of exemption is currently CHF 37,500 for individuals living alone and CHF 60,000 for married couples.

statistically highly significant (obj_ev: χ^2 (1) = 19.36; $p < 0.001$; sa_ev: χ^2 (1) = 31.97; $p < 0.001$). The effect size of support and the Perceived Measure is weak but significant ($\Phi = 0.07$; χ^2 (1) = 7.57; $p = 0.006$).

For supplementary benefits, the pattern of association changes in favor of the subjective angle: the strength of relationship with self-assessed vulnerability is also moderate (but clearly stronger than for support) at $\Phi = 0.24$ and highly significant (χ^2 (1) = 90.34; $p < 0.001$), followed by the moderate and significant association with the perceived indicator (χ^2 (1) = 32.40; $p < 0.001$). Objective vulnerability is only weakly associated with benefitting from supplementary benefits, though the relationship remains significant ($\Phi = 0.08$; χ^2 (1) = 9.39; $p = 0.002$).

The discrepancy between the measurement angles is particularly interesting in the case of supplementary benefits. In 2011, the minimum amount for living costs below which a person was eligible to receive supplementary benefits was at a monthly CHF 2700 for an individual person living alone[3] (BSV 2012). Since this amount ranges above the poverty line of CHF 2400 per month, we expect to detect a threshold effect, expressed in the share of respondents who, while receiving supplementary benefits, actually find themselves just above the poverty line and thus no longer 'economically vulnerable' by our Objective Measure. There is no statistic on the average income of beneficiaries of supplementary benefits to old age pensions; in 2011, the average monthly amount disbursed to individuals living alone (not in an institution) was CHF 888 (BSV 2012).

For both types of financial support, supplementary benefits and other sources of support, the mode lies in the income category CHF 2400–3600. As was to be expected, in the case of supplementary benefits some 54% of the benefitting population group is found in the aforementioned income category just above the poverty line. With the data at hand, it cannot be verified whether the monthly household income of these individuals is located just above 2400 or at the higher end of the income category, closer to 3600.

For the variable financial support, the share of benefitting individuals is more evenly distributed across the three lowest income categories (31%, 33%, 25%). If we assume that financial assistance from family and friends, from cantonal institutions or NGOs like Pro Senectute are also based on some kind of criterion, be it standardized or intuitive, the observation that some 70% of beneficiaries are *above* the poverty line is remarkable and deserves further discussion.

The interpretation of the prevalence of these sources of income according to vulnerability type has to be done with some reservations considering the very small number of respondents who benefit from them. Since the expected cell frequency was less than five for at least one cell, Fisher's exact instead of chi-square was calculated. For both variables, the one-sided test yielded $p < 0.001$, indicating a statistically highly significant relationship with the Vulnerability Typology.

Type BAA records the highest proportion of individuals receiving either type of financial support (24% supplementary benefits, 19% other types of support). As

[3] Eighty-six percent of beneficiaries live alone (BSV 2012).

previously mentioned, these figures reflect the post-transfer situation; it is likely that some of these individuals are only above the poverty line because of these measures of financial assistance.

Considering the difference between BBB and BBA is enlightening. For supplementary benefits, where eligibility is objectively established, there is no reason to expect a difference in share between the two vulnerability types, since the only difference is their response in terms of economic stress measured by perc_ev. This is confirmed in the very similar proportions: 18.2% for BBB, 17.5% for BBA, however, it makes sense that there would be a difference between two groups who distinguish themselves in the intensity of worry about finances, with regard to the proportions receiving support from 'less standardized' sources: we indeed find that the group who is more worried receives double the rate of support from canton, family, friends or NGOs (BBB = 14.6% compared to BBA = 7.5%).

14.2 Occupational Pensions

Table 14.1 shows the rate of respondents benefitting from occupational pensions in the sample population and among the members of vulnerable groups as defined by the three measures. Sixty-eight percent of pensioners aged 65–84 benefit from this type economic security. The Objective Measure is logically the measure that is most strongly associated with occupational pensions: among those who have to live with less than CHF 2400 per month, there are more than seven times as many who do not benefit from occupational pensions, compared to those who are above the poverty line (34% vs. 6%). The effect size measured by phi is strong with $\Phi = -0.37$ and statistically significant (χ^2 (1) = 220.94; $p < 0.001$). For both subjective measures we observe that vulnerability rates of individuals who receive occupational pensions (sa_ev = 10.4% and perc_ev = 14.7%) correspond to approximately half the rates of individuals who have to go without this type of pension (sa_ev = 22% and perc_ev = 25%). Accordingly, the strength of association is of a low to moderate strength but highly significant for both subjective measures (sa_ev: $\Phi = -0.16$ (χ^2 (1) = 41.35; $p < 0.001$; perc_ev: $\Phi = -0.12$ (χ^2 (1) = 23.34; $p < 0.001$).

Table 14.1 Sample distribution by occupational pension and measures of economic vulnerability

	Sample distribution		obj_ev	sa_ev	perc_ev
Occupational pension	n	%	% vulnerable within occupational pension		
No	517	31.7	34.0	22.2	24.6
Yes	1'114	68.3	5.9	10.4	14.7
Total	*1631*	*100.0*	*14.8*	*14.2*	*17.8*
Phi coefficient			−0.37	−0.16	−0.12
χ^2 (1)			220.94	41.35	23.34
p			<0.001	<0.001	<0.001

The share of respondents who do not benefit from occupational pensions decreases sharply for income classes above CHF 2400: while 73% of those who are vulnerable by objective standards do not benefit from this pension, it is merely a rate of 38% in the next higher income class CHF 2400–3600, and 23% for income class CHF 3600–4800. One reason for the slight increase in the proportion of non-beneficiaries among the highest incomes may be found in the Swiss legal framework: self-employed persons (some of whom are very wealthy, as we will discuss later) are not legally obliged to contribute to an occupational pension plan. This might explain why the proportion of people without work pensions records a slight increase for incomes higher than CHF 6000.

We conclude that the effect of occupational pensions is primarily and quasi directly reflected in the level of income and thus in the Objective Measure, having only a moderate though statistically significant association with the subjective measures.

14.3 Home Ownership

Close to two thirds of the sampled retirees own their own apartment or house (58%) as shown in Table 14.2. The relationship with low income is not statistically significant, however, there is a significant effect on the evaluation of adequacy of income and the propensity to worry frequently: both subjective measures record double the rate of economic vulnerability among respondents who do not own their home. The effect size as measured by the phi coefficient is moderate and negative with $\Phi = -0.16$ for sa_ev (χ^2 (1) = 42.95; $p < 0.001$) and $\Phi = -0.15$ for perc_ev (χ^2 (1) = 34.08; $p < 0.001$).

The lack of relationship between low income and home ownership triggers questions. A look at the proportions of pensioners who do not own a home by income category reveals that there is no linear relationship as their share is relatively stable for income categories up to monthly CHF 6000, above which the rate of non-home ownership drops below 20%.

Regarding the correlation between ownership and sa_ev, one possible explanation pertains to a differential in actual living costs and disposable income of

Table 14.2 Sample distribution by home ownership and measures of economic vulnerability

	Sample distribution		obj_ev	sa_ev	perc_ev
Home owner	n	%	% vulnerable within home owner		
No	647	40.6	17.3	21.0	24.6
Yes	948	59.4	13.1	9.4	13.2
Total	*1595*	*100.0*	*14.8*	*14.2*	*17.8*
Phi coefficient			Ns	−0.16	−0.15
χ^2 (1)			5.46	42.95	34.08
p			0.063	<0.001	<0.001

homeowners compared to renters. Furthermore, and this is even more plausible for explaining the effect size on perc_ev, there may be a psychological effect, stemming from the sense of financial stability and prosperity that comes with home ownership (though we cannot, with our data, verify whether individuals have actually paid off the mortgage of their homes). The protective effect of home ownership against subjective economic vulnerability seems to receive some confirmation by the exceptionally high proportion of individuals who do not own a home in type ABB (72% compared to 41% on the total sample).

14.4 Professional Activity

Among the pensioners in our sample, 8% continue to work on an *irregular* basis and 6% pursue a *regular* professional activity (Table 14.3).

The only statistically significant relationships are found between the Objective and the Self-Assessed Measure and regular work but the effect size is very small in both cases with $\Phi = -0.07$ for obj_ev (χ^2 (1) = 7.99; p = 0.005 and $\Phi = -0.07$ for sa_ev (χ^2 (1) = 7.26; p = 0.022. However, the majority of pensioners who work beyond retirement age do not appear to be motivated by a lack of income; they are in fact most frequently found in the socio-professional category 'top executives, liberal professions' that can look back on careers in the high-wage segments of the labor market.

Table 14.3 Sample distribution of irregular and regular professional activity and measures of economic vulnerability

Irregular work	**Sample distribution**		**obj_ev**	**sa_ev**	**perc_ev**
	n	%	**% vulnerable within irregular work**		
No	1499	91.9	15.0	14.1	18.0
Yes	132	8.1	12.9	15.2	16.7
Total	*1631*	*100.0*	*14.8*	*14.2*	*17.8*
Phi coefficient			Ns	Ns	Ns
χ^2 (1)			0.44	0.10	0.13
p			0.509	0.728	0.713
Regular work	n	%	**% vulnerable within regular work**		
No	1532	93.9	15.5	14.8	17.8
Yes	99	6.1	5.1	5.1	19.2
Total	*1631*	*100.0*	*14.8*	*14.2*	*17.8*
Phi coefficient			−0.07	−0.07	Ns
χ^2 (1)			7.99	7.26	0.13
p			0.005	0.022	0.717

14.5 Wealth

Table 14.4 shows the distribution of wealth among our sample population of pensioners aged 65–84 residing in Switzerland. One in four respondents owns less than CHF 60′000, while one in three respondents has more than CHF 500,000 in savings or investments. Among those who responded 'not having any wealth', 30% are income-poor, compared to 6% among the wealthiest category. In terms of self-assessment, the difference between the proportions of vulnerability in the lowest (47%) and the highest wealth category (4%) corresponds to a factor of 10. The perceived angle of vulnerability displays a disparity of almost identical size between the highest and the lowest category of wealth: 44% of those who own 'nothing' are worried about their finances, compared to 9% of those owning more than half a million Swiss francs.

As we could expect, the effect size of wealth is strong, negative and highly significant ($p < 0.001$) for all three measures of economic vulnerability. But interestingly, it is greatest for the Self-Assessed Measure (gamma = −0.63), followed by the Perceived Measure (gamma = −0.48) and lastly the Objective Measure (gamma = −0.43).

Figure 14.1 shows the prevalence of wealth categories by vulnerability type, which have been ordered by the size of proportion of the lowest wealth category. Type ABB takes the lead in this ranking with 54% of this type not having any savings, followed by type BBB (46%). Bearing in mind the strong correlation between income and wealth, a close look at all types featuring objective vulnerability (types BXX) is imperative: on the upper side of the graph we see that among type BAA, 16% own no wealth but a relatively high share (42%) are in wealth categories above 150,000. The important difference between type BAA and BBB seems to support our general hypothesis stating that the income-based measure does not sufficiently take into account other relevant resources and that, in contrast, the self-assessed

Table 14.4 Sample distribution by wealth category and measures of economic vulnerability

	Sample distribution		obj_ev	sa_ev	perc_ev
Wealth	n	%	% vulnerable within wealth		
(almost) nothing	195	12.0	29.7	47.2	43.6
<60,000	223	13.7	23.3	23.8	31.8
60,000–150,000	284	17.4	19.4	12.3	14.8
150,000–500,000	422	25.9	11.1	6.9	10.9
>500,000	507	31.1	5.9	4.3	9.3
Total	1631	100	*14.8*	*14.2*	*17.8*
Gamma			−0.43	−0.63	−0.48
z			10.7	17.6	12.3
p			<0.001	<0.001	<0.001
Kendall's tau-b			−0.19	−0.30	−0.24
z			9.1	−14.4	11.1
p			<0.001	<0.001	<0.001

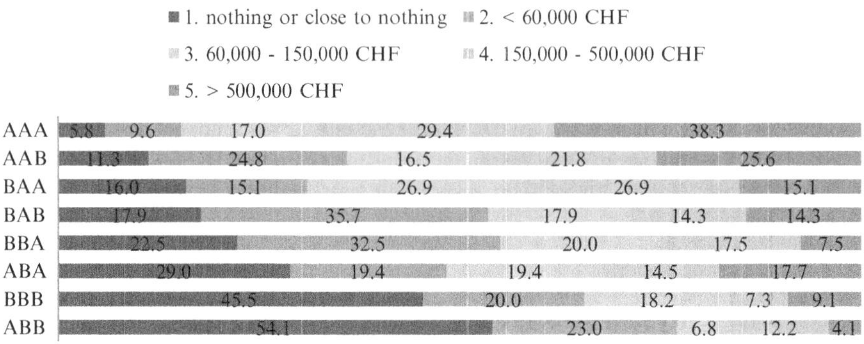

Fig. 14.1 Distribution of wealth categories, by vulnerability type

Table 14.5 Distribution of combinations between wealth and income categories

Wealth	Income				
Count	1. <2400	2. 2400–3600	3. 3600–4800	4. >4800	
1. <60,000	110	139	123	46	
2. 60,000–150,000	55	64	97	68	
3. 150,000–500,000	47	77	162	136	
4. >500,000	30	45	137	295	1631
%					
1. <60,000	6.7	8.5	7.5	2.8	
2. 60,000–150,000	3.4	3.9	5.9	4.2	
3. 150,000–500,000	2.9	4.7	9.9	8.3	
4. >500,000	1.8	2.8	8.4	18.1	
					100%

angle performs better at revealing the existence of coping resources (in this case, the availability of financial assets).

At this point, it is of general interest to look at the most frequent combinations of categories of wealth and income. Table 14.5 confirms a clustering of high-income categories with high-wealth categories at the lower right corner, peaking at the intersection of the highest income with the highest wealth category (18%).

At the opposite end of the spectrum, among the financially less privileged, the combinations of income and wealth are more diverse: while most individuals with modest monthly household incomes are found to have small levels of savings, low levels of wealth is similarly prevalent among middle income classes.

The next step naturally consists in taking a closer look at the income-wealth constellations by vulnerability type, in order to further examine how stock and flow of resources interact and reflect in subjective appraisals. Table 14.6 shows the frequency of all possible combinations between wealth and income categories in our sample population (N = 1631).

At first sight it stands out that several vulnerability types occur in a very concentrated manner, part of which is naturally the result of the coding of the lowest

Table 14.6 Absolute frequencies of wealth-income constellations, by vulnerability type

Fortune/income	BBB	BAA	ABB	ABA	AAB	BAB	BBA	AAA	Total	%
1. <60,000/<2400	36	37				15	22		110	6.8
2. 60,000–150,000/<2400	10	32				5	8		55	3.4
3. 150,000–500,000/<2400	4	32				4	7		47	2.9
4. >500,000/<2400	5	18				4	3		30	1.8
5. <60,000/2400–3600			39	23	15			60	137	8.4
6. 60,000–150,000/2400–3600			3	8	5			48	64	3.9
7. 150,000–500,000/2400–3600			6	5	8			57	76	4.7
8. >500,000/2400–3600			2	3	4			36	45	2.8
9. <60,000/3600–4800			15	5	24			79	123	7.6
10. 60,000–150,000/3600–4800			1	2	13			80	96	5.9
11. 150,000–500,000/3600–4800			2	2	14			144	162	10.0
12. >500,000/3600–4800			1	4	12			119	136	8.4
13. <60,000/>4800			3	2	9			32	46	2.8
14. 60,000–150,000/>4800			1	2	4			61	68	4.2
15. 150,000–500,000/>4800			1	2	7			126	136	8.4
16. >500,000/>4800				4	18			272	294	18.1
									1625	100

income category as 'BXX'. The fact that BAA is exclusively found among constellations 1 to 4 is an example of this; however, the concentration of BBB in the very lowest category (constellation 1. = 36 respondents) lends weight to the congruence of the three angles of measurement. Two other wealth-income constellations stand out: type ABB is highly concentrated in constellation 6 (39 respondents), and type AAB is clearly most often found for constellation 9 (24 respondents), both of which are combinations based on wealth category 'less than 60,000'. As for type ABB, the fact that as many as 50% of this group are in the low wealth-income constellation '<60,000/2400–3600' provides strong evidence for the potential *relevance* of the typology in terms of what it is able to reveal about coping resources: in the present constellation we can see the stylized example of someone with an income just above the poverty line, who experiences economic strain (sa_ev), resulting in stress (perc_ev) because low levels of savings do not permit to sustainably cope with the situation.

Regarding type AAB, we are now onto a possible explanation for this type that makes little sense theoretically (see Table 5.2) and for which the plausible explanation of higher levels of neuroticism was rejected. Indeed, it seems that for these 24 respondents, not having a lot of savings provokes financial worry, despite the fact that the wording of the survey question clearly states a very short time horizon.[4] Possibly, these respondents fear that they have to start consuming their wealth in order to be able to maintain their living standard, which seems at a comfortable level. Still within group AAB, we must not ignore another, clearly distinct subgroup: 18 respondents are worried or very worried about their current finances, while finding themselves in the highest possible wealth-income situation.

The analysis of the variable wealth, and particular in interaction with household income, confirms the overall importance of financial stocks that was discussed in the theoretical part. Firstly, the strong correlation between monthly household income and wealth was confirmed: the concordance of income and wealth is particularly marked in the wealthier segment while lower wealth levels are common throughout the lower and middle range of income categories. However, by far most respondents that are objectively vulnerable are also found in the lowest wealth category. Secondly, the presence of assets is exceedingly important for the subjective appraisal of economic vulnerability. This finding is in line with our theoretical model, which posits that the self-assessed angle takes into account economic resources other than monthly income, and that the perceived angle reflects whether coping resources (such as wealth consumption) are thought to be sustainable in the long run.

Appendix

Appendix 14.1: Sample Distribution by Financial Support and Supplementary Benefits and Measures of Economic Vulnerability

Support	Sample distribution		obj_ev	sa_ev	perc_ev
	n	%	% vulnerable within support		
No	1540	94.4	13.9	13.0	17.2
Yes	91	5.6	30.8	34.1	28.6
Total	*1631*	*100.0*	*14.8*	*14.2*	*17.8*
Phi coefficient			0.11	0.14	0.07
χ^2 (1)			19.36	31.97	7.57
p			<0.001	<0.001	0.006
Supplementary benefits			% vulnerable within supplementary benefits		
No	1540	94.4	14.2	12.2	16.5
Yes	91	5.6	25.8	47.3	39.8
Total	*1631*	*100.0*	*14.8*	*14.2*	*17.8*

[4] To what degree do the following situations worry you today? – Not having enough money to pay for current expenses, for example for bills, rent or food.

Support	Sample distribution		obj_ev	sa_ev	perc_ev
	n	%	% vulnerable within support		
Phi coefficient			0.08	0.24	0.14
χ^2 (1)			9.39	90.34	32.40
p			0.002	<0.001	<0.001

Reference

BSV. (2012). Statistik der Ergänzungsleistungen zur AHV und IV 2011.. Bundesamt für Sozialversicherungen BSV.

Chapter 15
Financial Needs and Expectations

According to our theoretical model the variable block referred to as Financial Needs and Expectations interacts with the totality of available economic resources to enhance or mitigate the experience of economic strain (sa_ev). By definition, the mediating effect of these variables can only be observed in multivariate analysis. The covariates selected for operationalizing diverse influences on what respondents feel they need and/or what they expect in terms of material living standard encompass the pre-retirement socio-professional category of the household, the self-assessed general health status and items of social participation, such as how often respondents go to a restaurant, to the movies or take a trip.

15.1 Socio-professional Status

The pre-retirement socio-professional category (SPC) contains layers of information that need to be disentangled carefully: reflecting a stratification of some sort, though without an underlying linear hierarchy, its being built on educational attainment, leadership position and employment status creates the basis for a strong correlation with financial assets. In the context of this research, we are interested in how former socio-professional status might shape people's perspective on their economic quality of life. However, at the bivariate stage of the analysis it is not possible to distinguish between the effects of financial means (for which SPC is a proxy-indicator) and how a certain social standing shows up in the perception of needs and expectations with regard to living standard. What can be ascertained is that this variable focuses more on capturing the 'expectations' than the 'needs' aspect of our model component.

The most frequent SPC in our sample population is 'top executives and academic professions' (36%) followed by skilled manual workers (17%), non-manual workers (16%) and intermediary professions (15%), then the self-employed and farmers (11%) and lastly the unskilled workers (5%). When looking at the distribution of

J. Henke, *Revisiting Economic Vulnerability in Old Age*, Life Course Research and Social Policies 11, https://doi.org/10.1007/978-3-030-36323-9_15

vulnerability within each of these socio-professional categories, the Objective Measure reflects the expected gradient, with 'unskilled workers' representing the category with the highest prevalence of income-poverty (43%), followed at some distance by 'self-employed and farmers' and 'skilled-manual workers' (26% and 25%). The very high rate of income-poverty among the 'self-employed' might be related to the flipside of the previously mentioned aspect of the Swiss legal framework for old age provisions: because self-employed individuals are exempted from the mandatory contribution to occupational benefit plans (2nd pillar), those who were not able to save enough money during their professionally active lives tend to have a meager income in old age.

Moving on to the Self-Assessed Measure, the share of vulnerability among the 'unskilled' is much lower than for the Objective Measure, but still at 27%, followed by 'skilled manual' with 24%. These two socio-professional categories record also the highest share of vulnerability according to the Perceived Measure (30% and 29%). The vulnerability rates among the higher-ranking SPCs, 'top executives and academic professions' and 'intermediary professions', are noticeably higher for the subjective measures compared to the monetary measure (Table 15.1).

For all three measures, the association with former SPC is statistically significant, Cramér's V indicating a moderately strong effect size. The Objective Measure has by far the strongest association with former socio-professional status ($V = 0.29$, $\chi^2\ (5) = 138.4$; $p < 0.001$).

The distribution of socio-professional categories by vulnerability type (excluding majority type AAA) is shown in Appendix 15.1. We will first consider the most frequent SPC within each vulnerability type: the largest proportion of type BBB is made up of former 'skilled manual' workers (34.6%); this SPC is also ranking as the most important category within types BBA, BAB, ABB and BAA. This socio-professional group is clearly overrepresented in comparison to its relative size within the total sample (17%). For types ABA and AAB, 'top executives' are the dominating socio-professional group.

Table 15.1 Sample distribution by socio-professional status and measures of economic vulnerability

	Sample distribution		obj_ev	sa_ev	perc_ev
Socio-professional Category (SPC)	n	%	% vulnerable within SPC		
Top executives/academic prof.	583	36.1	6.5	8.2	12.7
Self-employed/farmers	181	11.2	26.0	16.6	21.0
Intermed.Prof.	244	15.1	5.7	9.0	13.9
Skilled non-manual	252	15.6	14.3	15.1	15.1
Skilled manual	276	17.1	25.4	24.3	29.4
Unskilled	77	4.8	42.9	27.3	29.9
Total	*1613*	*100.0*	*14.8*	*14.2*	*17.8*
Cramér's V			0.29	0.19	0.17
χ^2 (5)			138.4	58.3	48.1
p			< 0.001	< 0.001	0.001

Table 15.2 Percentage distribution of income within each socio-professional category

	Monthly household income					
Socio-professional category	< 2400	2400–3600	3600–4800	4800–6000	> 6000	Total
1. Top executive/academic	6.5	11.5	27.3	24.4	30.4	*100*
2. Self-empl./farmers	26.0	19.3	27.1	12.2	15.5	*100*
3. Intermed.Prof.	5.7	20.9	41.0	23.4	9.0	*100*
4. Non-manual	14.3	19.4	36.5	20.2	9.5	*100*
5. Skilled manual	25.4	33.7	35.9	4.0	1.1	*100*
6. Unskilled	42.9	31.2	23.4	1.3	1.3	*100*
Total	*14.8*	*19.8*	*32.1*	*17.6*	*15.8*	*100*

With regard to the distribution of vulnerability types within socio-professional categories, the mode is most frequently found in type BAA, namely for the following categories: liberal, non-manual, skilled manual and unskilled workers. The association of SPC and the economic Vulnerability Typology (AAA included) is statistically significant ($p < 0.001$) and of a moderate effect size (Cramér's $V = 0.15$).

At this stage, it is helpful to consider the income levels that are typical for each socio-professional category in order to be able to extract as much information as possible in view of our hypotheses. Table 15.2 shows the percentage distribution of income for each socio-professional category.[1] Let us recall that both variables are located at the household level, income having been rendered equivalent based on household size and, in the case of married couples, the highest SPC having been extended to both spouses.

It becomes obvious that the professional categories only partially correspond to a ranking in income: in terms of monthly income, top executive are clearly the wealthiest professional category, followed by intermediary professions, then non-manual workers and, on rank 4, the formerly self-employed and farmers. As expected, skilled manual and unskilled workers conclude the ranking. Consequently, we rediscover the pattern of those SPC that record the greatest shares among type BAA: the self-employed, non-manual, skilled manual and unskilled workers. These individuals seem to be unaware of their objective vulnerability and therefore they neither experience economic strain nor are they easily stressed because of it. In order to test whether the discrepancy between objective and subjective measurement angles of this particular constellation is more frequent among respondents who have had a small income during their lives, we will have to make a more nuanced analysis later.

[1] For facilitating the interpretation the most frequent income class within each SPC is underlined with a double line, the second most frequent income class is underlined with a simple line.

15.2 Health

In the context of our research question, the variable health is of interest because a poor health status represents a potential increase in financial needs, all other things remaining equal. The expected sense of the correlation is between a poor health status and economic vulnerability, while the locus of interest is the Self-Assessed Measure: a poor health status is thought to increase the likelihood of experiencing economic strain because more financial resources may be required to ensure the same level of economic quality of life. Among our sample population, the self-assessed general health status is reported as ‘poor’ or ‘very poor’ for 6.5% and ‘satisfactory’ for 32% of the respondents. Among the group who assessed their health status as ‘very bad’ or ‘bad’, there is close to double the share of objectively vulnerable respondents (22%) as compared to those who say their health is ‘very good’ or ‘good’ (12%) (Table 15.3).

For the subjective angle, the health gradient is still more pronounced: among the group who is self-assessing or perceiving themselves as economically vulnerable, one third reports ‘poor’ or ‘very poor’ health, compared to a mere 10% (perceived) or 15% (self-assessed) among those who are in ‘good’ or ‘very good’ health. Those who consider their health as ‘satisfactory’ are closer to those in ‘good’ health as far as the Perceived and Self-Assessed Measures are concerned and closer to ‘bad’ health in terms of the Objective Measure. With cross-sectional data it is not possible to exclude the possibility of reverse causation: for those who are objectively vulnerable, for example, health issues could be the result of a life spent in poverty, or it could be part of the reasons why a person ended up sliding into poverty. However, considering the response pattern across measurement angles for ‘bad’ health, it is plausible that the high share of respondents who suffer from economic strain and stress at least partly reflects the burden of additional expenditures (Pavalko and Caputo 2013).

In terms of effect size, the association between health and the Objective and the Perceived Measure is weak with Cramér’s V = 0.10 and V = 0.11, respectively. The effect size for the Self-Assessed Measure is moderate with V = 0.17. For all three measures of economic vulnerability, the association is statistically highly significant

Table 15.3 Sample distribution by self-assessed health status and measures of economic vulnerability

	Sample distribution		obj_ev	sa_ev	perc_ev
Self-assessed health status	n	%	% vulnerable within health		
(very) bad	106	6.5	21.7	31.1	31.1
Satisfactory	520	31.9	19.0	18.5	20.0
(very) good	992	60.8	12.0	10.2	15.2
Total	*1618*	*100.0*	*14.8*	*14.2*	*17.8*
Cramér’s V			0.10	0.17	0.11
χ^2 (2)			17.5	58.3	48.1
p			< 0.001	< 0.001	0.001

at $p < 0.001$. When considering the distribution by vulnerability type (Fig. 15.1), it becomes obvious that it is again members of type ABB who tend to be most vulnerable in terms of rating their health status as being 'bad' or 'very bad' (18%).

The next highest ranks for poor and very poor health are occupied by groups of the generic type XBX: type BBA (15%), BBB (13%) and ABA (11%). The contrast between types XBX and BAA is striking: among respondents who are vulnerable exclusively by an objective standard (= BAA), only 6% are assessing their health as 'bad' or 'very bad'. Considering the strong empirical support for an economic health gradient (Adler et al. 1994), this finding raises the question whether this group of respondents tends to generally assess *any* situation more optimistically than other groups, possibly as a result of the psychological mechanisms referred to as 'downward adaptation'. This speculation seems all the more reasonable recalling that type BAA occurs more frequently in higher age groups, whose members should on average be more affected by physical frailty. Indeed, in the area of health measures, evidence has been found that the probability of false reporting is significantly income graded (Johnston et al. 2009): there is an increased risk for respondents in lower income classes to be unaware of the real state of their physical condition and report overly positive on their health status. Suspecting an excessive optimism among group BAA, we decided to examine this question based on alternative health-indicators.

Since there were no strictly speaking 'objective' health variables (such as biomarkers) in our survey questionnaire, we have to rely on other self-reported measures, but ones that (in contrast to the survey question on 'the general state of health') provide specific criterion of assessment. From among a series of ten questions on physical complaints we choose those three complaints that are most

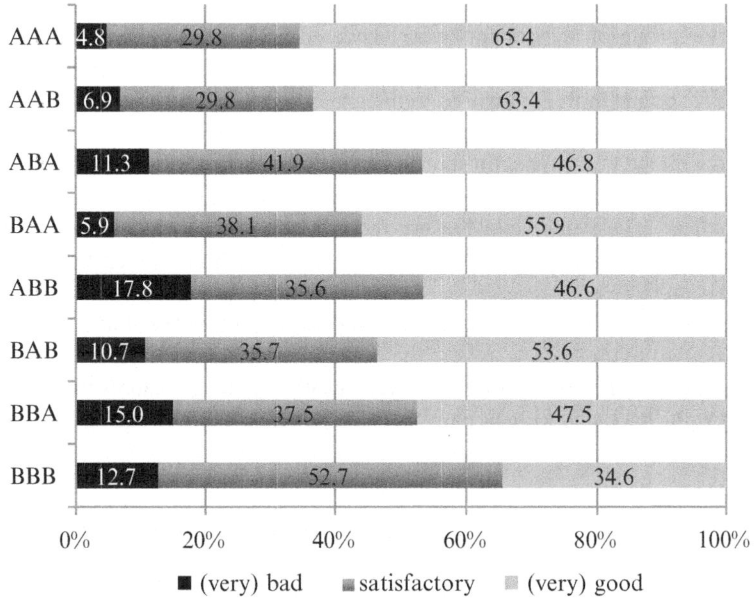

Fig. 15.1 Distribution of self-assessed general health status, by vulnerability type

prevalent in our sample population. The wording of the questions reads as follows: 'Over the course of the last 4 weeks, have you suffered from…

- pain, cramps, trembling or swelling in your *lower members* (feet, legs, knees…)
- pain, cramps, trembling or swelling in your *upper members* (hands, arms, shoulders…)
- pain in the *back or kidneys*?'

As fourth variable we included the Activities of Daily Living (ADL[2]) scale (Katz 1963). For the question at hand we consider this variable as relevant because it allows distinguishing three levels of functional health (independent, in difficulties, handicapped) based on a list of every day activities and the criterion, whether the person is able to perform these activities autonomously. Due to the small number of respondents who reported being 'in difficulties' and 'handicapped', these two categories were collapsed, resulting in a binary indicator for poor functional health.

The two competing hypotheses we will confront by means of these four variables is whether (a) respondents of type BAA tend to *respond too positively* to the question on self-assessed health or whether (b) respondents of type BAA are indeed *very similar to type AAA* in terms of their health condition, indicating that objective economic vulnerability does not – at least not by itself, without the effect of economic strain and stress – increase health risks. We will not be able to verify these hypotheses in a methodologically robust way at this stage, but we can find out, which one is more plausible by comparing health rates among vulnerability types.

Let's take a look at the prevalence of the four health troubles within our sample and their distribution among each group of economic vulnerability (Appendix 15.2). Statistically, the relationship between the four health-variables is only significant for the two subjective measures but not for the monetary measure: for all three types of pain, the group of respondents who answered having 'a lot of pain or troubles' also had a considerably higher rate of subjective economic vulnerability (between 10 and 20% points more) compared to the group who had no pain. The relationship between the three types of pain and the Self-Assessed Measure is similar to the one observed with the general health status in terms of statistical significance (high) and effect size (moderate); the relationship with ADL shows a small though highly significant effect. The Perceived Measure records only a weak relationship with the four health problems and only 'pain in upper members' proves statistically significant at $p < 0.001$.

The prevalence by vulnerability type is shown for pain in lower and upper members in the two figures below, where the column charts are ordered from highest to lowest share of the response category 'yes, a lot' of pain. The analysis of the relationship with 'back and kidney pain' and 'activities of daily living' is not pursued

[2] We would like to remind that we had considered using the, a more 'objective' health measure, but had found its association with our dependent variables to be relatively weak (see introduction to the variable 'health', p.89). In the context of this specific question, the ADL scale seems appropriate because it provides a clear and objectively observable criterion as basis for self-assessment.

any further since both variables turned out not to have any statistically significant relationship with two of the tree measures (obj_ev and perc_ev).

What stands out immediately from looking at the bars charting the prevalence of 'pain in lower members' (Fig. 15.2) and 'pain in upper members' (Fig. 15.3) is the extremely similar proportions of responses of type BAA and type AAA: respondents of these two groups are no different with regard to pain in feet, legs, hands and arms.

It is noteworthy that for all three complaints, the three vulnerability types with the highest share of respondents affirming 'a lot of pain' are invariably types BBB, ABB and ABA (see ranking in Table 15.4), which are all of the overarching type XBX: all of these groups of respondents share in common that they report difficulties in making ends meet. The difference of the highest ranks to ranks 4 and lower is considerable, amounting to double the prevalence in terms of percentage share (e.g. prevalence of 'a lot of pain' in upper members is 17% for BBB and 7% for BAA).

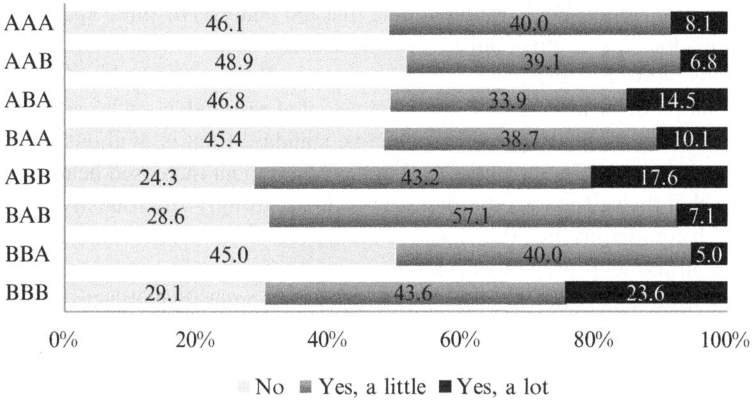

Fig. 15.2 Prevalence of pain in lower members (feet, legs, etc.), by vulnerability type

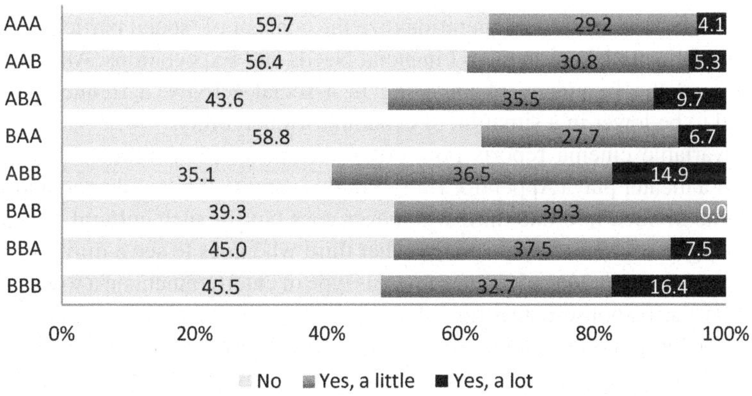

Fig. 15.3 Prevalence of pain in upper members (hands, arms, etc.), by vulnerability type

Table 15.4 The three highest ranking vulnerability types in terms of the frequency of response category 'a lot of pain/troubles'

Type/health symptoms:	Lower members	Upper members	Back or kidney
BBB	1.	1.	3.
ABB	2.	2.	2.
ABA	3.	3.	1.

As far as the statistical significance of the association with the Vulnerability Typology is concerned, we find a significant relationship for three variables: for pain in lower members, the effect size as measured by Cramér's V is weak to moderate (V = 0.11, $X^2(14) = 39.01$, $p < 0.001$); for pain in upper members the effect is slightly stronger (V = 0.13, $X^2(14) = 49.193$, $p < 0.001$) and for pain in back or kidneys the effect is again weak though statistically significant (V = 0.9, $X^2(14) = 27.638$, $p = 0.016$).

In response to the question that triggered this excursus, the risk of false response by members of group BAA, we conclude that the analysis of three specific health complaints does not confirm our suspicion. Instead, based on our results we find it more plausible to assume that members of group BAA are very similar to group AAA in terms of health. In fact, this suggests that respondents who are economically vulnerable exclusively by the objective standard (that is, without the experience of economic strain and stress) are not exposed to an increased health risk. We conclude that the self-assessed dimension is more strongly and robustly associated with health complaints than the objective dimension of economic vulnerability. This finding is inline with previous research stating that subjective measures of socioeconomic standing are more predictive of health outcomes than objective indicators (Rios and Zautra 2011).

15.3 Social Participation

Three variables were used to operationalize the concept of 'social participation', an element of the model component Financial Needs and Expectations. All three variables record the frequency of engaging in a social activity, a frequency that is expected to be lower in a situation of economic vulnerability.

The variable cinema reports how often, on average, respondents go to see a movie or a theater play (Appendix 15.3). For this activity the sample population can roughly be divided into one third who never goes out for such cultural activities, a third who goes out once a year and another third who goes to see a movie or a play at least once a month. Only 5% consume this type of entertainment on a weekly basis.

Within the response categories, the proportions of vulnerable respondents are similar for the three measures: the rate of vulnerability by the perceived angle is above the sample total of 18% among those who 'never' go out (23%) and those who go out 'at least once a year' (20%), and it is clearly above average for the self-assessed angle and the objective angle among those who 'never' go out (23% for objectively vulnerable people compared to the sample total of 15%, and 22% for the self-assessed vulnerable compared to the sample total of 14%).

Kendall's tau-b and Kruskal's Gamma were calculated to assess the strength of the relationship and they both indicated a statistically significant negative relationship with all three measures ($p < 0.001$). Given the high number of tied observations in the present contingency table, Gamma is more appropriate: effect sizes are moderate, the strongest one being recorded for the association between cinema and obj_ev ($\gamma = -0.36$), followed by sa_ev ($\gamma = -0.31$) and lastly perc_ev ($\gamma = -0.28$).

The variable trip is based on two survey questions about the frequency of taking a trip or a short holiday of at least 1 day (Appendix 15.4). Most retirees in our sample take a short vacation or trip at least once a year (48%), or at least once a month (32%).

Among those respondents who never take a trip, self-assessed vulnerability is much more prevalent at 30%, which is more than twice the rate found in the overall sample (16%). On the other side of the spectrum, among respondents who take a trip 'at least once a week' only 4% are vulnerable according to the self-assessed angle. The vulnerability rates for respondents who 'never' travel is of a comparable size for the objective angle and the perceived angle (25% and 28%, respectively); both are above sample proportions by some 10% points.

As expected from looking at the percentage rates, the effect size measured by the gamma coefficient is strongest for the Self-Assessed Measure ($\gamma = -0.45$), the Objective Measure taking the second ($\gamma = -0.32$) and the Perceived Measure the third rank ($\gamma = -0.26$). These effect sizes are of a moderate strength and statistically highly significant ($p < 0.001$).

The variable restaurant gives insight into the habits of going out for a meal or something to drink, an activity that costs money while being indicative of a person's participation in social life (Appendix 15.5). For a minority (13%) of the Swiss population aged 65–84 this is a daily habit, while the majority (40%) goes out on a weekly basis. 30% go to a restaurant or a coffee shop at least once a month and 18% say to go out 'at least once a year' or 'never'.

The distribution of 'vulnerable' versus 'non-vulnerable' within each response category of restaurant indicates, that the Objective Measure has the strongest effect on the frequency of going out: Even more than for cinema and trip, there is a strong concentration of the frequency of going to a restaurant 'never' or only 'once a year' among those who are below the poverty line. Self-assessed and perceived economic vulnerability also seems to have an impact. The effect size measures confirm the pattern that is discernable from looking at the contingency table but this time, the measures of association are only significant for the Objective and the Self-Assessed Measures ($p < 0.001$). Gamma indicates a moderate effect between the frequency of going to a restaurant and objective vulnerability ($\gamma = -0.29$) and a barely moderate effect for self-assessed vulnerability ($\gamma = -0.20$). No noticeable effect size is registered for the relationship between the variable restaurant and economic stress.

We proceed to assessing the differences in social participation by vulnerability type. Figures 15.4 and 15.5 depict the frequency distribution of seeing a movie/

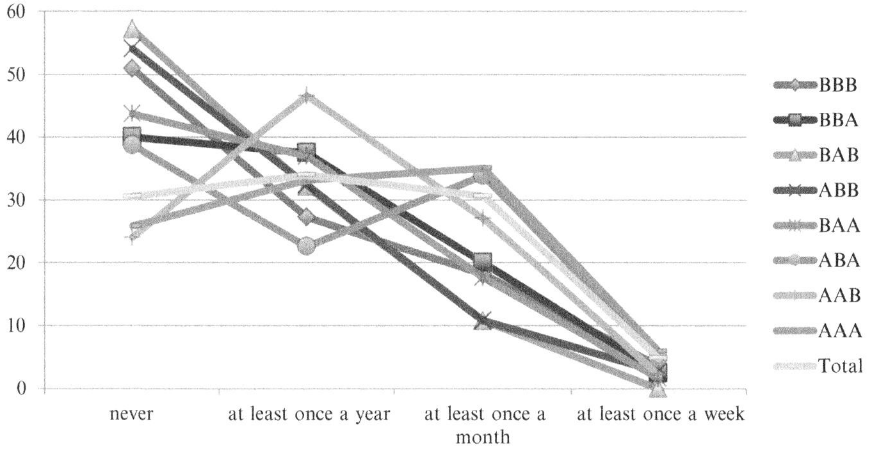

Fig. 15.4 Frequency distribution of going to the movie theater/seeing a theater play, by vulnerability type

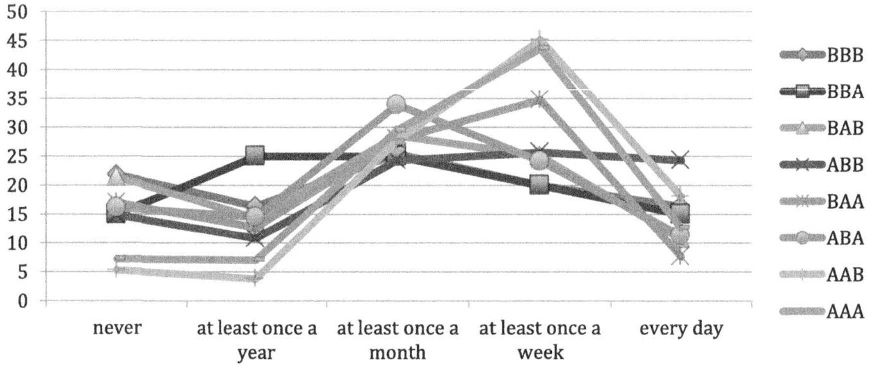

Fig. 15.5 Frequency distribution of going to a restaurant or coffee shop, by vulnerability type

theatre play or going to a restaurant/coffee shop. The representation of a percentage distribution by a line graph is unusual, however, we find that it paints an easily readable 'picture' that is helpful for taking stock of the general trend: while the frequency distribution of each type remains visible, the stacking of the lines allows to see which response categories (e.g. within the variable restaurant) are most frequent and whether response patterns are similar across vulnerability types. The most homogeneous response pattern is found for the frequency of taking a trip: across all types, most respondents travel at least once a year (figure not shown).

The frequency distribution of the variable cinema reflect two trends that distinguish types AAB and AAA from all other types: for most types, responses are most concentrated in 'never', while among these two types, most respondents go to see a movie or a play at least once a year if not once a month. It is noteworthy that among these two types, respondents of AAB (who are worried about current finances), clearly distinguishes themselves from the non-vulnerable type AAA by going out less frequently.

Type ABA also reveals an unusual frequency pattern with two response items of similar size, 'never' at 39% and 'at least once a month' at 34%, which is in fact equivalent to the response rates for the same categories of type AAA. This could be seen as another clue for a pattern we already discovered when analyzing the variable 'wealth', namely that there are indeed two distinct groups within this type, one that resembles non-vulnerable type AAA, and one with a tight budget and little or no savings for a rainy day.

The habit of eating out or going to a coffee shop (figure below) is unequally distributed across vulnerability types. Going to a restaurant 'at least once a week' is the most frequent response for generic types XAX (AAA, AAB and BAA). Here, type BAA is especially interesting: this group only slightly diverges from groups 'above the poverty line' with regard to this social activity. Overall we see that for half of the vulnerability types, respondents tend to eat out at least once a week. Types ABA, BAB and BBA have their peak in 'at least once a month'.

In order to examine our hypothesis concerning the effect of economic vulnerability on social participation, it is interesting to consider the ranking of the vulnerability types with the lowest rates of participation in social activities (Table 15.5). BBB, BAB and ABB share most of the highest ranks among themselves. They share in common that they are worried about their financial situation (generic type XXB), plus one other dimension of vulnerability (sa_ev *or* obj_ev). BAA is the exception to this pattern with as many as 17% who never go out to eat. This stands in contrast to the relatively high frequency (35%) of individuals who go out once a week. These results indicate that group BAA is relatively heterogeneous, with some who have gotten used to a more modest life and others, whose social participation behavior is similar to type AAA.

Going out to a restaurant or to a coffee shop being such a prevalent activity, adding up the lowest two frequencies (respondents who say they go to a restaurant 'never' or 'at least once a year') can shed further light on those groups who tend to have a low level of social participation: BBA (40%), BBB (38%) and BAB (36%) are the types where it is most common to never or very rarely eat out; in contrast, these responses are only recorded in 14% of respondents for type AAA and 9% for AAB.

On the other side of the spectrum, it is noteworthy that among type ABB, as many as 25% (highest frequency among all types) go to a coffee shop or a restaurant on a daily basis. This finding needs to be integrated with the previously reported

Table 15.5 Ranking of the three vulnerability types with the highest frequency of respondents who 'never' participate in the social activity

Rank	Restaurant	Cinema	Trip
1.	BBB (22%)	BAB (57%)	ABB (41%)
2.	BAB (21%)	ABB (54%)	BBB (38%)
3.	BAA (17%)	BBB (51%)	BAB (32%)

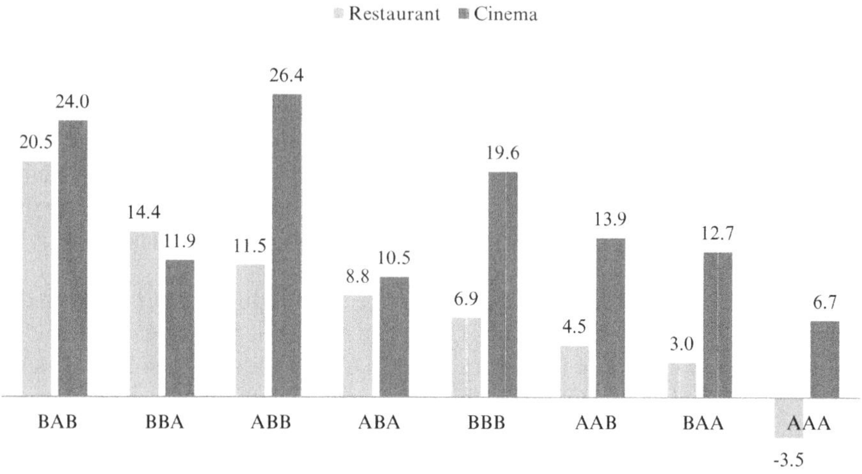

Fig. 15.6 Percentage points above sample proportion for going *less* frequently to a restaurant or to a movie than when aged 45 years, by vulnerability type

evidence that group ABB also stands out as having a below average endowment of socioeconomic resources (e.g. lower than average educational attainment levels, higher than average rates of financial support and supplementary benefits, low rate of home ownership and wealth, etc.). At the same we need to keep in mind that type ABB by definition encompasses a large range of income categories (every category except less than 2400) thus the heterogeneity of available resources in this group is great. For this reason ABB also represents those pensioners who are sensibly affected by the drop of income at retirement and who are worried about not being able to continue the same lifestyle as before.

In order to verify this hypothesis, which taps into the phenomenon of 'expensive taste', we consider two additional variables that ask respondents to compare the current frequency of doing a given activity with how frequently they used to do the same activity at age 45. The underlying survey question is of a likert-type, with four response options ('did not do it at 45', 'yes, less often than now', 'yes, more or less the same', 'yes, more often than now'). We posit that by looking at past habits of social participations we will see that members of group ABB are economically vulnerable by subjective standards because of the perceived 'price' they had to pay by substantially reducing their customary levels of social participation.[3]

Figure 15.6 shows how much, for each vulnerability type, the reduced frequency (less often than at age 45) of going to a restaurant or to the movies deviates from the sample average. The sample average of respondents saying they are less frequently

[3] Of course there are also other reasons why individuals may have reduced the frequency of certain activities, for example, because of changes in daily routines after retirement or as a result of health problems. However, these reasons should affect other vulnerability types in a similar manner and therefore should not significantly distort our results.

going out is 33% for restaurant and 43% for cinema; in group ABB, 45% have reduced the frequency of going out to eat and as many as 60% are not going as often to the movies as they used to.

Two other types stand out from the sample average: congruently with the ranking shown in Table 15.5, types BAB and BBB also record very high shares of respondents who report a decrease in the frequency of engaging in these social activities. In combination with previous findings, we can begin to sketch a profile for (the theoretically unlikely) type BAB: respondents of this type seem to have made considerable adjustments towards a more modest life style, not least by reducing their social activities, to a point where they are now making ends meet easily, though not without the experience of economic stress. The previous finding that this group has above average scores of neuroticism reinforces this impression. BBA, which featured highest among those who never or rarely go to a restaurant, records the second to highest share (14.4%) of respondents who report a reduction in the frequency of going out to eat compared to when they were 45 years old. This suggests that for members of this group, reducing the money spent on social participation may be a functional coping strategy that effectively reduces economic stress.

The analysis of three variables of social participation provides a complex picture yet one that confirms the relevance of the model component Financial Needs and Expectations for understanding subjective economic vulnerability. Among the three measures, the Objective Measure is most strongly correlated with the frequency of going to the movies and to a restaurant, while the Self-Assessed Measure is most strongly correlated with the frequency of taking a trip. With regard to the Vulnerability Typology, several incongruences between measurement angles are reflected in the different habits of current and/or past habits of social participation.

Appendices

Appendix 15.1: Percentage Distribution of Vulnerability Types by Socio-professional Category, Without Type AAA

	Vulnerability type							
Socio-professional Category	BBB	BBA	BAB	ABB	BAA	ABA	AAB	Total
1. Top executive/academ.	12.7	12.8	17.9	23.6	18.1	31.7	34.1	*23.7*
2. Self-empl./farmers	23.6	20.5	14.3	5.6	19.0	8.3	12.1	*14.3*
3. Intermed.Prof.	7.3	2.6	3.6	11.1	6.9	15.0	15.9	*10.4*
4. Non-manual	9.1	20.5	10.7	16.7	17.2	21.7	13.6	*15.7*
5. Skilled manual	34.6	33.3	28.6	34.7	25.9	16.7	22.0	*26.7*
6. Unskilled	12.7	10.3	25.0	8.3	12.9	6.7	2.3	*9.2*
Total	*100*	*100*	*100*	*100*	*100*	*100*	*100*	*100*

(continued)

	Vulnerability type							
Socio-professional Category	BBB	BBA	BAB	ABB	BAA	ABA	AAB	Total
1. Top executive/academ.	5.9	4.2	4.2	14.3	17.7	16.0	37.8	*100*
2. Self-empl./farmers	18.1	11.1	5.6	5.6	30.6	6.9	22.2	*100*
3. Intermed.Prof.	7.7	1.9	1.9	15.4	15.4	17.3	40.4	*100*
4. Non-manual	6.3	10.1	3.8	15.2	25.3	16.5	22.8	*100*
5. Skilled manual	14.2	9.7	6.0	18.7	22.4	7.5	21.6	*100*
6. Unskilled	15.2	8.7	15.2	13.0	32.6	8.7	6.5	*100*
Total	*11.0*	*7.8*	*5.6*	*14.3*	*23.1*	*12.0*	*26.3*	*100*

Appendix 15.2: Sample Distribution by Health Symptoms and Measures of Economic Vulnerability

Variable	Category	Sample distribution		obj_ev	sa_ev	perc_ev
		n	%	% vulnerable within health problem		
Pain in lower members	**No**	725	44.5	13.2	11.2	14.9
	Yes, a little	654	40.1	15.6	14.2	19.0
	Yes, a lot	151	9.3	19.2	24.5	24.5
	Missing	101	6.2	14.9	19.8	21.8
	Total	*1631*	*100.0*			
	Cramér's V			0.05	0.11	0.08
	χ^2 (2)			4.04	19.00	9.46
	p			0.133	<0.001	0.009
Pain in upper members	**No**	919	56.4	13.5	10.5	15.0
	Yes, a little	495	30.4	15.6	16.6	19.6
	Yes, a lot	90	5.5	22.2	32.2	30.0
	Missing	127	7.8	16.5	18.9	22.8
	Total	*1631*	*100.0*			
	Cramér's V			0.06	0.16	0.10
	χ^2 (2)			0.07	37.61	15.22
	p			0.067	<0.002	<0.001
Pain in back or kidney	**No**	716	43.9	14.8	10.8	16.5
	Yes, a little	657	40.3	15.1	15.5	18.1
	Yes, a lot	158	9.7	15.2	23.4	23.4
	Missing	100	6.1	13.0	15.0	17.0
	Total	*1631*	*100.0*			
	Cramér's V			0.00	0.11	0.05
	χ^2 (2)			0.03	18.95	4.28
	p			0.958	<0.001	0.131

(continued)

Variable	Category	Sample distribution		obj_ev	sa_ev	perc_ev
		n	%	% vulnerable within health problem		
ADL	**Independent**	1534	94.8	14.8	13.5	17.3
	Difficulties/ handicap	84	5.2	14.3	26.8	27.4
	Total	*1618*	*100.0*			
	Phi coefficient			0.00	0.08	0.06
	χ^2 (1)			0.02	11.30	5.56
	p			0.897	0.001	0.018

Appendix 15.3: Frequency of Going to the Movies/Theater Play, by Measure of Economic Vulnerability

	Sample distribution		obj_ev	sa_ev	perc_ev
Cinema	n	%	% vulnerable within cinema		
Never	498	30.7	22.5	21.9	23.5
At least once a year	551	33.9	15.1	12.4	20.0
At least once a month	496	30.5	8.5	9.5	11.5
At least once a week	79	4.9	6.3	10.1	8.9
Total	*1624*	*100.0*	*14.8*	*14.2*	*17.8*
Gamma			−0.36	−0.31	−0.28
z			7.2	5.6	5.7
p			<0.001	<0.001	<0.001
Kendall's tau-b			−0.15	−0.13	−0.12
z			6.9	6.6	5.6
p			<0.001	<0.001	<0.001

Appendix 15.4: Frequency of Taking a Trip, by Measure of Economic Vulnerability

	Sample distribution		obj_ev	sa_ev	perc_ev
Trip	n	%	% vulnerable within trip		
Never	254	15.7	25.2	30.4	27.6
At least once a year	775	47.8	15.5	14.0	18.5
At least once a month	512	31.6	9.6	8.4	13.5
At least once a week	82	5.1	9.8	3.7	11.0
Total	*1623*	*100*	*14.8*	*14.2*	*17.8*

(continued)

	Sample distribution		obj_ev	sa_ev	perc_ev
Trip	n	%	% vulnerable within trip		
Gamma			−0.32	−0.45	−0.26
z			5.9	8.8	5.0
p			<0.001	<0.001	<0.001
Kendall's tau-b			−0.13	−0.19	−0.11
z			5.7	8.1	4.9
p			<0.001	<0.001	<0.001

Appendix 15.5: Frequency of Eating out or Going to a Coffee Shop, by Measure of Economic Vulnerability

	Sample distribution		obj_ev	sa_ev	perc_ev
Restaurant	n	%	% vulnerable within restaurant		
Never	152	9.4	<u>29.0</u>	<u>25.7</u>	<u>23.7</u>
At least once a year	138	8.5	<u>27.5</u>	<u>26.1</u>	18.8
At least once a month	469	28.9	13.9	13.5	16.6
At least once a week	648	39.9	10.3	8.2	15.0
Every day	218	13.4	12.4	18.5	24.8
Total	*1625*	*100.0*	*14.8*	*14.2*	*17.8*
Gamma			−0.29	−0.20	0.00
z			5.6	3.6	0.02
p			<0.001	<0.001	0.492
Kendall's tau-b			−0.13	−0.09	0.00
z			5.4	3.5	0.02
p			<0.001	<0.001	0.492

References

Adler, N. E., Boyce, T., Chesney, M. A., Cohen, S., Folkman, S., Kahn, R. L., & Leonard, S. (1994). Socioeconomic status and health: The challenge of the gradient. *American Psychologist, 49*(1), 15–24.

Johnston, D. W., Propper, C., & Shields, M. A. (2009). Comparing subjective and objective measures of health: Evidence from hypertension for the income/health gradient. *Journal of Health Economics, 28*(3), 540–552.

Katz, S. F. A. (1963). Studies of illness in the aged: The index of adl: A standardized measure of biological and psychosocial function. *JAMA, 185*(12), 914–919.

Pavalko, E. K., & Caputo, J. (2013). Social inequality and health across the life course. *American Behavioral Scientist, 57*(8), 1040–1056. 0002764213487344.

Rios, R., & Zautra, A. J. (2011). Socioeconomic disparities in pain: The role of economic hardship and daily financial worry. *Health Psychology, 30*(1), 58–66.

Chapter 16
Psychosocial Consequences and Symptoms

Our theoretical model posits that, in addition to the direct relationship between economic strain (sa_ev) on economic stress (perc_ev), this effect is moderated by *Psychosocial Symptoms and Consequences*: it is thought that, at the same level of difficulties in making ends meet, the experience of role strains and a diminished self-concept is going to increase the probability of higher levels of economic stress. The variable block Psychosocial Symptoms and Consequences encompasses *social isolation*, such as a low frequency of contact with friends and family via phone calls or visits, a frequent feeling of *loneliness*, and symptoms of a *sense of diminishment,* expressed in negative statements about one's self, and lastly, the perceived level of *mastery* or sense of control over life. Just like previous variables based on likert-type response scales, these variables were treated as continuous variables in order to facilitate interpretation and comparison. All variables are coded in such a way that the lowest category represents the worst outcome or the most severe symptom.

16.1 Social Isolation

This chapter presents the average frequencies of contact with friends and family among our sample population and the association of these variables with the three measures of economic vulnerability. Only the association between the Perceived Measure and calling family (see Table 16.1) and the association between the Objective Measure and visit friends (see Table 16.2) are significant at $p < 0.05$; the other tables are shown in the Appendix (16.1 and 16.2).

We conclude that the frequency of contact with loved ones varies very little between vulnerable and non-vulnerable groups. Contrary to the general hypothesis stating that economic vulnerability is accompanied by social isolation, groups of respondents who perceive themselves as vulnerable and groups whose income is below the poverty line tend to make or receive *more* phone calls than the non-vulnerable share of the population: for example, among those who receive phone

J. Henke, *Revisiting Economic Vulnerability in Old Age*, Life Course Research and Social Policies 11, https://doi.org/10.1007/978-3-030-36323-9_16

Table 16.1 Frequency of calling family, by measure of economic vulnerability

Call family	Sample distribution		obj_ev	sa_ev	perc_ev
	n	%	% vulnerable within call family		
No or hardly ever	53	3.3	9.4	13.5	18.9
At least once a year	83	5.2	15.7	16.9	13.3
At least once a month	347	21.7	17.0	15.9	19.3
At least once a week	759	47.4	12.1	11.4	15.0
(almost) every day	359	22.4	18.4	18.4	24.2
Total	1601	100.0	*14.8*	*14.2*	*17.8*
Gamma			0.04	0.04	0.10
z			0.7	0.6	1.8
p			0.242	0.267	0.034
Kendall's tau-b			0.02	0.02	0.04
z			0.7	0.6	1.9
p			0.242	0.267	0.322

Table 16.2 Frequency of visiting friends, by measure of economic vulnerability

Visit friends	Sample distribution		obj_ev	sa_ev	perc_ev
	n	%	% vulnerable within visit friends		
No or hardly ever	136	8.4	14.6	15.3	13.2
At least once a year	369	22.8	23.3	16.2	20.1
At least once a month	691	42.7	35.4	38.4	35.4
At least once a week	354	21.9	22.1	23.1	25.0
(almost) every day	68	4.2	4.6	7.0	6.3
Total	1618	100.0	*14.8*	*14.2*	*17.8*
Gamma			−0.09	0.02	0.02
z			1.68	0.28	0.31
p			0.05	0.39	0.38
Kendall's tau-b			−0.04	0.01	0.01
z			1.64	0.28	0.31
p			0.05	0.39	0.38

calls almost every day, 24% are worried about their financial situation (compared to the overall sample proportion of 18%). While it cannot be verified here, it might be that their state of emotional stress prompts them to solicit frequent moral support from friends and family.

As far as the relationship with the Vulnerability Typology is concerned, differences in social isolation between types are also small (Table 16.3). A Kruskal-Wallis H test confirms that differences in frequency of contact with friends and family are statistically insignificant, with the exception of the variable 'visiting friends',[1] where group-differences proved significant. Since these four variables have previously been shown to follow a normal distribution, summary statistics are helpful for

[1] The Kruskal-Wallis H test for visiting friends is $\chi^2(7) = 17.307$, p = 0.016.

Table 16.3 Deviation from the mean of frequency of contact (call/visit) with friends and family

Frequency of contact (1–5)	Mean	AAA	AAB	ABA	ABB	BAA	BAB	BBA	BBB
Call friends	3.5	0.01	−0.06	0.05	−0.11	−0.21	−0.25	−0.17	0.03
Call family	3.8	−0.01	0.06	−0.01	0.02	−0.02	0.31	−0.08	0.18
Visit friends	2.9	0.02	0.10	0.07	−0.06	−0.12	−0.51	−0.23	0.19
Visit family	3.3	−0.04	0.16	−0.15	0.03	0.09	−0.19	−0.09	−0.14

getting a general idea about the size of the difference between groups. Table 16.3 shows the deviation from the mean in the frequency of contact for each vulnerability type, though, because of the results of the Kruskal-Wallis Test, only the variable visiting friends will be considered further. Type BAB stands out as diverging considerably from the mean, a finding that is in line with previous results. Comparing type BAB with BAA allows to verify whether two individuals who are in very similar situations may experience different levels of stress, depending on additional indirect sources of stress that have to do with the self-concept and with role strains. Groups BAB and BAA sharing identical values for the objective and the self-assessed angle, we would expect the incongruence in the Perceived Measure to be confirmed by group BAB recording a lower frequency of visits from friends.

For the examination of statistically significant differences in the distribution of calling family and visiting friends, a non-parametric Wilkoxon rank-sum test is calculated. Taking into consideration the median of groups BAB and BAA, the test is significant for the variable visiting friends ($z = 2.03$, $p = 0.042$). The Mann-Whitney effect size statistic indicates that for 38% of the pairs, the frequency of visits with friends is higher for BAB than for BAA, meaning that the average frequency of contact is significantly higher for type BAA than for type BAB. Integrating this finding with the findings about social participation (going out to restaurants and movie theaters), we are able to confirm the previously outlined profile for type BAB, stating that this group of people tends to benefit less from social contacts than other types.

In summary, the findings presented in this chapter suggest that for our sample population, social isolation as measured by low frequency of contact with friends and family is not as strongly associated with economic vulnerability as was expected. For this reason, only the variable visiting friends will be retained for further analysis.

16.2 Loneliness

Table 16.4 shows the frequency of the feeling of loneliness in our sample population and its association with measures of economic vulnerability. Among the age group 65–84, 8% say they feel 'always' or 'often' lonely, 37% feel 'rarely' and 55% feel 'never' lonely. Among those who feel lonely always or often, economic vulnerability is a more frequent experience across the three measures. Though the association

Table 16.4 Frequency of feeling lonely, by measure of economic vulnerability

	Sample distribution		obj_ev	sa_ev	perc_ev
Lonely	n	%	% vulnerable within lonely		
Always/often	33	8.2	27.8	28.6	27.8
Rarely	597	36.7	11.1	12.1	17.4
Never	895	55.1	15.4	13.5	16.7
Total	*1525*	*100*	*14.8*	*14.2*	*17.8*
Gamma			−0.03	−0.13	−0.12
z			0.4	2.0	2.0
p			0.345	0.023	0.022
Kendall's tau-b			−0.01	−0.05	−0.05
z			0.4	1.9	1.9
p			0.345	0.281	0.026

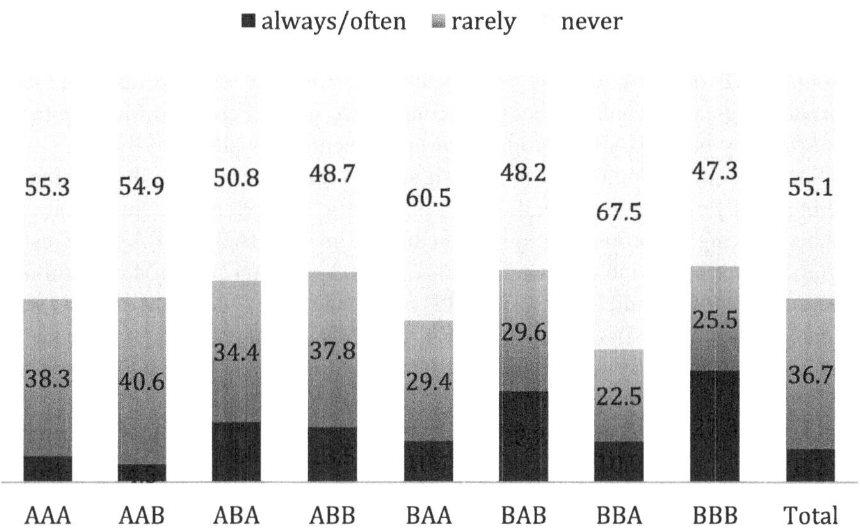

Fig. 16.1 Distribution of frequency of feelings of loneliness, by vulnerability type

is statistically significant for the Self-Assessed and the Perceived Measure ($p < 0.05$), the effect size is very small with gamma = −0.13 and – 0.12.

A Kruskal-Wallis H test was calculated to assess the association of loneliness with the vulnerability type, and it just failed to be significant: $\chi2\ (7)\ =\ 13.107$, $p = 0.070$. The distribution of frequency by vulnerability type (Fig. 16.1) shows that feeling lonely 'always' or 'often' is by far most frequent among types BBB and BAB (27% and 22%, respectively). As the next highest in rank we identify types ABA and ABB (15% and 14%), confirming the previously made observation that the Self-Assessed Measure is the most relevant dimension of economic vulnerability for explaining an increased risk of feeling lonely.

16.3 Sense of Diminishment

A diminished sense of self can be a long-term consequence of economic vulnerability, in particular, of economic stress (Pearlin et al. 1981). The concept was operationalized with the self-worth scale and two other survey questions capturing respondent's levels of self-confidence and the worry about not receiving sufficient recognition for what they do. Table 16.5 shows the scores of the self-worth scale and the association with the three measures, measured with coefficients suitable for ordinal-level data since the scale does not follow a normal distribution. The association is not significant for the Objective Measure, while small and barely significant for the Self-Assessed (gamma = −0.08) and the Perceived Measure (gamma = −0.11).

In order to assess, whether certain vulnerability types distinguish themselves significantly from each other in terms of the median value of the self-worth score, a Kruskal-Wallis H test was again calculated: the omnibus test indicated no significant differences in the median $\chi2\ (7)\ = 10.2$, $p = 0.177$.

Turning to the next variable operationalizing the concept 'sense of diminishment', shows the frequency of respondents who worry about not being recognized for what they do. 4% are 'strongly' worried, 8% are 'somewhat' worried and 17% say that lack of recognition worries them 'a little'. Among respondents who are 'strongly' worried about lacking recognition, the vulnerability rates are systematically higher: 29% for the Objective Measure, 39% for the Self-Assessed Measure and 55% for the Perceived Measure. Among those who are 'somewhat' worried,

Table 16.5 Frequency of scores and measures of association of the self-worth scale

Self-worth	Sample distribution		
Score	n	%	Cum. %
1	134	8.4	8.4
2	93	5.8	14.3
3	95	6.0	20.2
4	138	8.7	28.9
5	193	12.1	41.0
6	211	13.3	54.2
7	190	11.9	66.2
8	224	14.1	80.2
9	214	13.4	93.7
10	101	6.3	100.0
Total	*1593*	*100*	
	obj_ev	**sa_ev**	**perc_ev**
Gamma	0.00	−0.08	−0.11
z		1.6	2.6
p		0.055	0.049
Kendall's tau-b	0.00	−0.03	−0.06
z		1.6	2.6
p		0.055	0.005

Table 16.6 Worry about lack of recognition, by measure of economic vulnerability

	Sample distribution		obj_ev	sa_ev	perc_ev
Recognition	n	%	% vulnerable within recognition		
Strongly	55	3.6	29.1	38.9	54.6
Somewhat	122	7.9	17.2	21.3	41.0
a little	267	17.3	14.6	15.4	25.5
No worry/does not concern	1104	71.3	13.9	12.6	12.4
Total	*1548*	*100*	*14.8*	*14.2*	*17.8*
Gamma			−0.14	−0.27	−0.54
z			2.0	4.4	12.8
p			0.023	<0.001	<0.001
Kendall's tau-b			−0.05	−0.10	−0.25
z			1.9	3.9	9.4
p			0.037	<0.001	<0.001

only the Perceived Measure remains high at 41%.[2] The association with the other two indicators is also significant and negative though less strong: for the Objective Measure, the effect is weak (gamma = −0.14, p = 0.023) and for the Self-Assessed Measure, the effect is of a moderate size (gamma = −0.27, p < 0.001) (Table 16.6).

Figure 16.2 shows the distribution of worry about not getting sufficient recognition within each vulnerability type, in the order of the highest frequency of being 'strongly' worried. Type BBB takes the lead, closely followed by types ABB and AAB. Type BAB is most concerned by this worry when measured in terms of 'any concern at all': two thirds of group BAB are somehow fearful about not getting enough recognition, corresponding to previous findings that this group is characterized by having fewer social contacts and a tendency for being neurotic.

Turning to the last variable within the concept of 'diminishment of self', this item assesses the frequency of feeling self-confident (Table 16.7). Among our sample population, 7% feel 'never' or 'rarely' self-confident, 46% feel 'often' self-confident and the remaining 48 say that this feeling prevails 'always'.

The effect size is very small and not significant for neither of the indicators of economic vulnerability. This might also have to do with the small number of respondents in category 'never/rarely'. For this reason, because cell frequencies would be too low, the association with the Vulnerability Typology was not examined.

[2] Part of why the association with the Perceived Measure is so strong (gamma = −0.54) may have to do with the location of this question within the survey questionnaire: the item 'worry about not getting sufficient recognition' was asked at the same time as the question on 'worry about not having enough money to pay current expenses' (perc_ev).

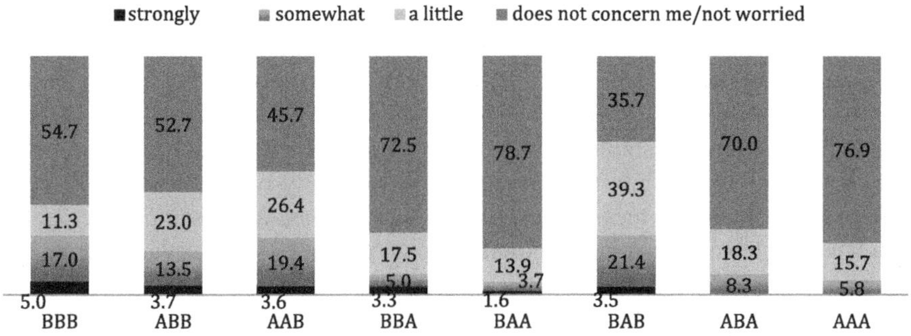

Fig. 16.2 Frequency distribution of respondents who are worried about not receiving sufficient recognition for what they do

Table 16.7 Frequency of feeling self-confident, by measure of economic vulnerability

Self-confidence	Sample distribution		obj_ev	sa_ev	perc_ev
	n	%	% vulnerable within self-confidence		
Never/rarely	109	6.8	23.9	17.4	21.1
Often	734	45.6	12.4	13.3	18.0
Always	767	47.6	15.7	14.4	17.2
Total	*1610*	*100.0*	*14.8*	*14.2*	*17.8*
Gamma			0.01	0.00	−0.05
z			0.18	–	0.78
p			0.42		0.21
Kendall's tau-b			0.00	0.00	−0.02
z			0.18	–	0.80
p			0.42		0.21

Table 16.8 Bi- and univariate statistics of the sense of control

Mastery (1–10)	obj_ev		sa_ev		perc_ev	
	Point Biserial Correlation	P > |t|	Point Biserial Correlation	P > |t|	Point Biserial Correlation	P > |t|
	−0.08	<0.001	−0.16	<0.001	−0.14	<0.001

Table 16.9 Deviation from the mean score of mastery, by vulnerability type

Mastery (1–10)	AAA	AAB	ABA	ABB	BAA	BAB	BBA	BBB
Deviation from the mean	−6.6	−6.6	−6.6	−6.6	−6.6	−6.6	−6.6	−6.6

16.4 Mastery

The composite scale 'mastery' captures the subjective sense of being in control of life's circumstances. Table 16.8 shows the point biserial correlations for the relationship between interval and binomial variables, revealing that the sense of mastery is systematically lower among the three economically vulnerable groups. The association with all three measures is negative and highly significant, the strongest effect being recorded for the Self-Assessed Measure ($r_{pb} = -0.16$).

The mastery scale following a normal distribution, a summary statistic such as the deviation from the sample mean provides a general idea of whether there are any differences in levels of mastery among the vulnerability types (Table 16.9). Type BBB stands out with a mean score that is 1.7 points below the sample average of 6.6. The next lowest mean score is found among type ABB with a deviation from the mean of −1.2.

The Kruskal-Wallis H test indicates that there is indeed a highly significant difference in the median of at least two types within the typology ($\chi 2\ (7) = 53.5$, $p < 0.001$). The question we are interested in here is whether differences in groups who share the same vulnerability profile in terms of Objective and Self-Assessed Measures but differ in terms of the Perceived Measure may be distinct in regard to their sense of being in control of life. We therefore proceed to testing whether the distribution of group BBB differs significantly from group BBA. The Wilkoxon rank sum test is significant ($z = 2.30$, $p = 0.021$) and the effect size is considerable: for 36% of the pairs, the median score of mastery is higher for type BBB than for type BBA, meaning that in the majority of cases, respondents of type BBB are feeling less in control than respondents found in type BBA. This result confirms that individuals who are experiencing economic stress are also those who are negatively affected in their self-concept, which includes a sense of mastery.

In summary, in this chapter we found that among our sample population, economic vulnerability as measured by Objective, Self-assessed and Perceived Measures is rarely and only weakly associated with disadvantageous psychosocial symptoms, indicative of role strains or a poor self-concept. An exception of this tendency represents the finding that on average, members of economically vulnerable groups have a lower sense of mastery over their lives and are more frequently worried about a lack of recognition.

Appendices

Appendix 16.1: Frequency of Calling Friends, by Measure of Economic Vulnerability

Call friends	Sample distribution		obj_ev	sa_ev	perc_ev
	n	%	% vulnerable within callfriends		
No or hardly ever	80.00	4.98	26.25	22.78	26.25
At least once a year	147.00	9.14	16.33	15.65	17.69
At least once a month	500.00	31.09	13.40	11.45	18.40

(continued)

Call friends	Sample distribution		obj_ev	sa_ev	perc_ev
	n	%	% vulnerable within callfriends		
At least once a week	695.00	43.22	13.96	13.15	15.11
(almost) every day	186.00	11.57	13.44	18.82	22.04
Total	1608.00	100.00	*14.80*	*14.20*	*17.80*
Gamma			−0.08	0.02	−0.04
z			1.5	0.3	−0.8
p			0.067	0.378	0.215
Kendall's tau-b			−0.04	0.01	−0.02
z			1.5	0.3	0.8
p			0.069	0.378	0.218

Appendix 16.2: Frequency of Visiting Family, by Measure of Economic Vulnerability

Visit family	Sample distribution		obj_ev	sa_ev	perc_ev
	n	%	% vulnerable within visitfamily		
No or hardly ever	77	4.8	16.9	22.4	23.4
At least once a year	306	19.1	15.0	16.2	16.7
At least once a month	510	31.9	14.7	11.6	15.9
At least once a week	519	32.4	13.9	13.7	18.9
(almost) every day	189	11.8	17.5	16.5	22.2
Total	1601	100.0	*14.8*	*14.2*	*17.8*
Gamma			0.00	−0.03	0.06
z				0.7	1.2
p				0.254	0.125
Kendall's tau-b			0.00	−0.01	0.03
z				0.6	1.4
p				0.268	0.087

Reference

Pearlin, L. I., Menaghan, E. G., Lieberman, M. A., & Mullan, J. T. (1981). The stress process. *Journal of Health and Social Behavior, 22*(4), 337–356.

Chapter 17
Conclusion Part III

The bivariate analysis presented in this part has provided the necessary groundwork to proceed to the more complex methods of analysis. Most important factors that previous research has identified to be at the root of economic vulnerability in old age have been confirmed by our data. Moreover, some preliminary elements of response to our research questions have been presented and we now have the basis to draw final conclusions on which variables to retain for multivariate analysis. Our major research questions pertain to the difference between the three measurement angles of economic vulnerability and aim at drawing out the complementary information that can be gained by combining these three measures. Thus, the review of our main findings will follow the logic of the sequence of the indicators, from objective to subjective, before addressing the contribution of a perspective that integrates all three angles, the Vulnerability Typology.

When judging by the magnitude of the effect size,[1] i.e. the difference between the effect of a covariate on one measure and the effect that same covariate has on another measure, we find that there are notable differences between, on the one hand, the Objective and the Self-Assessed Measure (research question I) and on the other hand, between the Self-Assessed and the Perceived Measure (research question II).

With regard to the relationship between the Objective and the Self-Assessed Measure, our investigation began with a regression analysis, confirming the very strong correlation between the objective monetary poverty line and the self-assessed difficulties in making ends meet: respondents who are vulnerable by the standard of

[1]The following interpretation of the magnitude of the differences in effect size is based on the following convention, which allows discerning the general tendency more easily: a difference of effect size of 0.05 is considered small, a difference of up to 0.10 is considered moderate, a difference larger than 0.10 is considered important (see Appendix 17.1 for an overview of effect size by measure). Since different measurement levels call for different effect size measures (i.e. Cramér's V, Phi, Gamma, Point Biserial Correlation), the effect sizes cannot directly be compared across all covariates. It is safe, however, to compare effect sizes across measures.

J. Henke, *Revisiting Economic Vulnerability in Old Age*, Life Course Research and Social Policies 11, https://doi.org/10.1007/978-3-030-36323-9_17

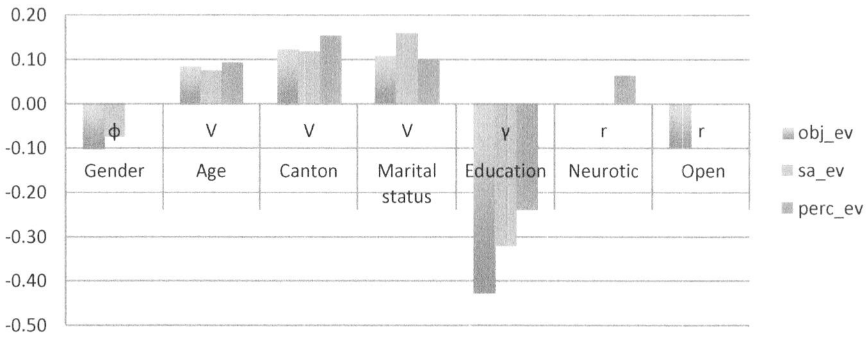

Fig. 17.1 Effect size of covariate set Background Characteristics on measures of economic vulnerability

any one of these measures are 5 times more at risk of also being vulnerable by the other measure as compared to non-vulnerable respondents.

Among the examined *Background Characteristics* (Fig. 17.1), the overall effect size is strongest for education, followed by marital status and canton. The effect of education proved to be the single most important effect on all three measures though the magnitude of the effect was greatest for the Objective Measure: low income in old age is known to be associated with low levels of education, mediated by low professional status, and possibly, more frequent spells of inactivity before the legal age of retirement (Gabriel et al. 2015); its lesser association with self-assessed and perceived economic vulnerability is consequently to be understood within the same framework.

Marital status reveals to have the strongest effect on the Self-Assessed Measure, representing the multifaceted way in which the family situation affects economic quality of life beyond the standard of a monetary poverty line. The higher risk of income poverty among individuals living alone is well documented and is thought to be due to the lack of economies of sale. Divorce and separation, too, are among the most frequent risk factors for various forms of economic vulnerability, as this family situation often ends up combining higher living costs with a decrease in income and wealth.

At the bivariate level, canton, sex and age do not account for much variance. Despite their weak effects, these three variables will be retained for multivariate analysis because they represent stratifying variables of our sample data.

Personality traits play a minor role in explaining the difference between vulnerable and non-vulnerable groups: the Perceived Measure is slightly influenced by a neurotic personality, while the character trait of 'openness to new experiences' is negatively correlated with the Objective Measure. Due to their weak association, the variables on personality traits will not be retained for further analysis.

As was to be expected, from among the sets of variables pertaining to *Economic Resources* (Fig. 17.2), wealth had the strongest association with any one of the three measures of economic vulnerability. Besides the well-documented relationship

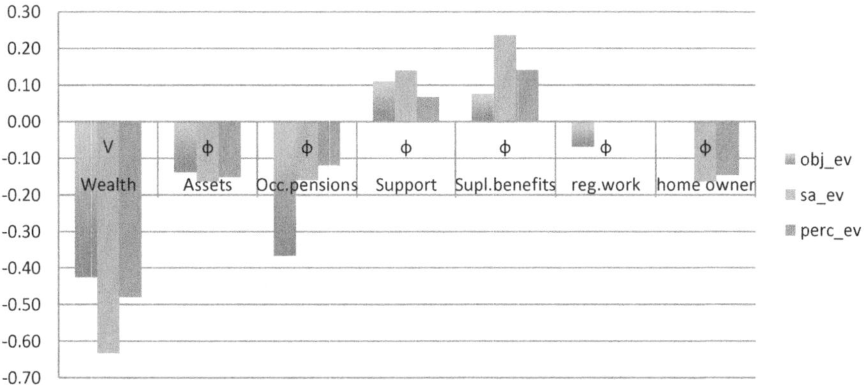

Fig. 17.2 Effect size of covariate set Economic Resources on measures of economic vulnerability

between wealth and monthly income in old age, it is interesting to note that the Self-Assessed and the Perceived Measure are even more strongly associated with wealth. The difference in the effect size of wealth between the Objective and the Self-Assessed Measure is considerable (0.21), highlighting the sensitivity of the subjective measure for capturing economic resources other than income. The effect size for supplementary benefits is higher for the Self-Assessed Measure than for the Objective Measure. This variable will be retained in the subsequent analysis because it is indicative of 'post-transfer' economic vulnerability, thus illuminating shortcomings of the highly standardized Swiss pension scheme. The variable home ownership only records a small effect on the two subjective measures; still, the absence of any association with income poverty is a noteworthy result. Since the main interest of this variable lies specifically in the comparison of the effect between the Objective and the Self-Assessed Measure, the influence of home ownership on economic vulnerability will be investigated further.

The effect of having an occupational pension, too, reveals very strong differences between the vulnerable groups, this type of regular income having a strong association with the Objective Measure but much less with the Self-Assessed and the Perceived Measure. As mentioned elsewhere, this source of income has consistently been identified as a protective factor against income poverty. The variable itself has little explanatory power beyond the description of the income composition and will not be retained for multivariate analysis. The same is true for the variables 'support from Canton, NGOs and family', and 'regular or irregular work' (the latter is not shown in the figure above because the effect size was not statistically significant); none of these three variables will therefore be retained.

The variables in the model component *Financial Needs and Expectations* (Fig. 17.3) represent the diverse factors that, combined with Economic Resources, influence the self-assessed adequacy of income (sa_ev). The roots of the effect of the last socio-professional category is likely related to the causes that also underlie the effect of education; the differences in effect size between the angles is comparable

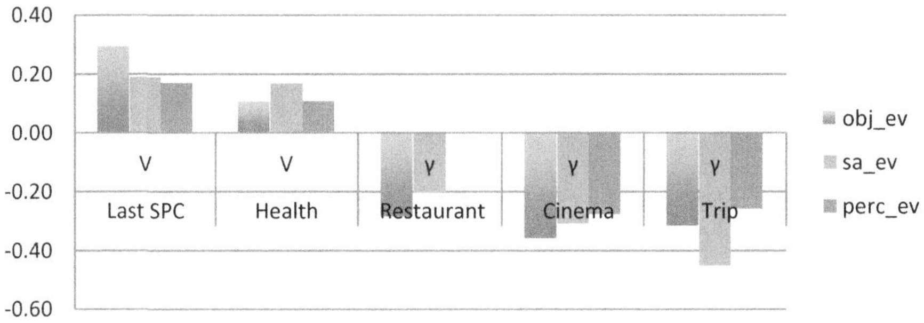

Fig. 17.3 Effect size of covariate set Financial Needs and Expectations on measures of economic vulnerability

for the two variables. Subsequent analysis will help disentangle between the effect of economic status and the way social status may shape what is considered an adequate level of quality of life.

The effect of health on self-assessed economic vulnerability reminds us of experiences reported by Pro Senectute, that expenditures for chronic ill-health constitute a frequent risk factor among their clients. Among the three variables operationalizing the concept of social participation, the frequency of going to a restaurant/coffee shop and the frequency of going to see a movie/theater play is more strongly associated with income poverty than with self-assessed economic vulnerability. However, the average frequency of taking a trip, representing a more significant expenditure, is considerably lower among respondents who are facing difficulties in making ends meet than for those who live on less than CHF 2400 per month.

The second set of hypotheses focuses on the relationship between the Self-Assessed and the Perceived Measure of economic vulnerability, among them the question of the difference between the Self-Assessed and the Perceived Measure of economic vulnerability with regard to their association with *Psychosocial Symptoms and Consequences* of economic vulnerability. These psychological and social conditions are expected to be associated more strongly with economic stress (perc_ev) than with economic strain (sa_ev), because they may reflect failing or insufficient coping mechanisms.

From among the variables representing Psychosocial Symptoms and Consequences only one variable, the worry about lack of recognition, recorded an effect of at least moderate size on the Self-Assessed Measure; it did show a strong effect on the Perceived Measure of economic vulnerability. Among the variable set operationalizing social isolation, merely the frequency of visiting friends was weakly associated with any of the three measures of economic vulnerability and, contrary to expectation, this statistically significant association turned out to be with the Objective Measure. Neither the feeling of loneliness nor low levels of self-confidence had any significant effect on economic stress. Self-worth had a weak negative association with the Perceived Measure, while mastery affected both, Self-Assessed and Perceived Measures (Fig. 17.4).

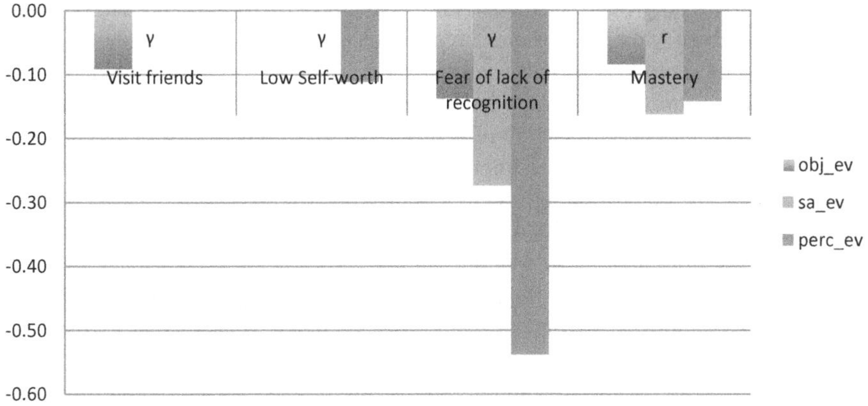

Fig. 17.4 Effect size of covariate set Psychosocial Consequences and Symptoms on measures of economic vulnerability

A cautious attempt at interpreting these findings at the bivariate level would prompt us to say that psychosocial consequences and symptoms of economic vulnerability are primarily revealed in the area of diminishment of self (operationalized by the variables 'worry about lack of recognition', 'self-worth' and 'mastery'). These findings are congruent with our theory-based predictions grounded in Pearlin's stress model as they show that pensioners who experience economic stress are more likely to suffer from an erosion of their self-concept. As far as the social life is concerned, limited economic means do not have a significant impact on the frequency of contact with loved ones, resulting in social isolation; the influence here is more visible in the choice of the place of meeting (at home rather than in a restaurant).

At this stage, the many factors associated with economic vulnerability are confluent, reinforcing and attenuating each other. It would therefore be too early to attempt to describe the various profiles of vulnerable groups that emerge at the intersection of the three measures, corresponding to the eight variations of the Vulnerability Typology. For the sake of providing a rough idea of what can be gleaned from the bivariate analysis so far, we will simply consider the comparison of two types before the backdrop of the hypotheses stated in our theoretical model.

According to our theoretical model, a comparison between group BBA and BBB would yield that members of both groups share similar characteristics in terms of their socioeconomic background. Since both groups recognize their situation as an adversity, we expect to find their overall levels of economic resources to be rather low, though the comparable serenity manifested among members of group BBA (absence of economic stress) might indicate a superior endowment. We certainly expect a notable difference in the association with the variable block Psychosocial Consequences and Symptoms as it is this area where members of type BBB are likely to struggle more than members of type BBA.

In terms of Background Characteristics, while type BBA is composed of about an equal share of women and men, type BBA features almost double as many women as men. People who have only reached primary level education and those who have gone through a divorce are strongly overrepresented among type BBB and, moreover, they tend to be more neurotic. Both groups record below average shares of home ownership and an important proportion of individuals with no wealth or less than CHF 60,000 though members of type BBB have even less economic resources than those of type BBA. As far as Financial Needs and Expectations are concerned, the dominant social standing in both types is the professional group of skilled manual workers. Both groups have an above average share of respondents who suffer from poor health and an important share never go out to eat. Indeed, the main differences between the groups crystalize in the area of economic stress, associated with a sense of diminishment of self: group BBB records three times the share of individuals who fear not being recognized sufficiently for what they do, and two and a half times the share of individuals who feel lonely (very) often as compared to group BBA. Lastly, the average sense of mastery for group BBB is considerably lower than for type BBA.

This comparison between two detailed profiles of vulnerability types, which are alike on two measures but diverge with regard to their experience of economic stress (perc_ev) suggests that the combination of three measures allows shedding some light on differentials in people's ability to deal with economic strain. If judged merely by their monthly income these people would have been classified as being 'equally' vulnerable even though their actual experience of economic quality of life is very different. Thus, this first round of analysis at the binary level is encouraging with regard to the potential gain of combining objective and subjective indicators for obtaining a more nuanced picture of the heterogeneity of experiences of economic vulnerability that are common among older adults in the Swiss context.

The outline of the next three parts will follow our overarching research questions: Part IV will focus on the relation between the Objective and the Self-Assessed Measure of economic vulnerability and in Part V, the relevance of distinguishing between two measures of subjective economic vulnerability will be put to the test some more.

Appendix

Appendix 17.1: Overview of Effect Size of All Covariates on Measures of Economic Vulnerability

Variable set	Variable	Effect size measure	obj_ev	sa_ev	perc_ev
Background	Sex	Phi	–0.10	–0.08	ns
	Age	Cramér's V	0.09	0.08	0.09
	Canton	Cramér's V	0.12	0.12	0.15
	Marital status	Cramér's V	0.11	0.16	0.10
	Education	Gamma	–0.43	–0.32	–0.24
	Personality:				
	neurotic	Point bis.corr.	ns	ns	0.06
	open	Point bis.corr.	–0.10	ns	ns
	extraverted	Point bis.corr.	ns	ns	ns
	agreeable	Point bis.corr.	ns	ns	ns
	conscientious	Gamma	ns	ns	ns
Economic resources	Wealth	Cramér's V	–0.43	–0.63	–0.48
	Assets	Phi	–0.14	–0.16	–0.15
	Occ.pensions	Phi	–0.37	–0.16	–0.12
	Support	Phi	0.11	0.14	0.07
	Supl.benefits	Phi	0.08	0.24	0.14
	Irreg.work	Phi	ns	ns	ns
	Reg.work	Phi	–0.07	ns	ns
	Home owner	Phi	ns	–0.16	–0.15
Needs and expectations	Last SPC	Cramér's V	0.29	0.19	0.17
	Health	Cramér's V	0.10	0.17	0.11
	Social participation: Freq. of going to.				
	restaurant	Gamma	–0.29	–0.20	0.00
	cinema	Gamma	–0.36	–0.31	–0.28
	take a trip	Gamma	–0.32	–0.45	–0.26
Psycho-social consequences and symptoms	Social isolation: Freq. of contact.				
	calling family	Gamma	ns	ns	ns
	calling friends	Gamma	ns	ns	ns
	visiting family	Gamma	ns	ns	ns
	visiting friends	Gamma	–0.09	ns	ns
	Loneliness	Gamma	ns	ns	ns
	Self-worth	Gamma	ns	ns	–0.11
	Self-Confidence	Gamma	ns	ns	ns
	Recognition	Gamma	–0.14	–0.27	–0.54
	Mastery	Point bis.corr.	–0.08	–0.16	–0.14

Reference

Gabriel, R., Oris, M., Studer, M., & Baeriswil, M. (2015). The persistence of social stratification? A life course perspective on poverty in old-age in Switzerland. *Swiss Journal of Sociology, 41*(3), 465–487.

Part IV
The Self-Assessed Measure of Economic Vulnerability

This part focuses on the analysis of the Self-Assessed Measure of economic vulnerability and its distinctiveness from the Objective Measure. After having examined the relationship between each potential explanatory variable and the outcome measures individually we now proceed to estimating a series of multivariate models, beginning by binary logistic models before fine-tuning our analysis with structural equation models.

Our theoretical model (extract shown below) presents the Self-Assessed Measure as being predicted by Economic Resources and Financial Needs and Expectations, the effect of resource-variables partly being mediated by the covariates of Financial Needs and Expectations. The Objective Measure – operationalized as monthly household income below CHF 2400 – is strongly correlated with, but distinct from, the variable complex Economic Resources. In our analysis we are interested in the predictive power of both, financial resources generally speaking as well as the specific effect of the Objective Measure on the Self-Assessed Measure. Integrating all resources that may influence the 'difficulties in making ends meet' is necessary for understanding how this self-assessment comes about and for understanding the relative share that objective material resources play, respectively, to what degree Financial Needs and Expectations contribute to the experience of economic strain. The specific relationship between the Objective Measure and Self-Assessed Measure merits a careful analysis in its own right because of the important place the Objective Measure continues to hold in policy-relevant data analysis in form of the absolute poverty line. For this reason, particular attention will be given to comparing these two measures with regard to the effects of different sets of explanatory variables that predict economic vulnerability for the population groups identified by each of these two measures.

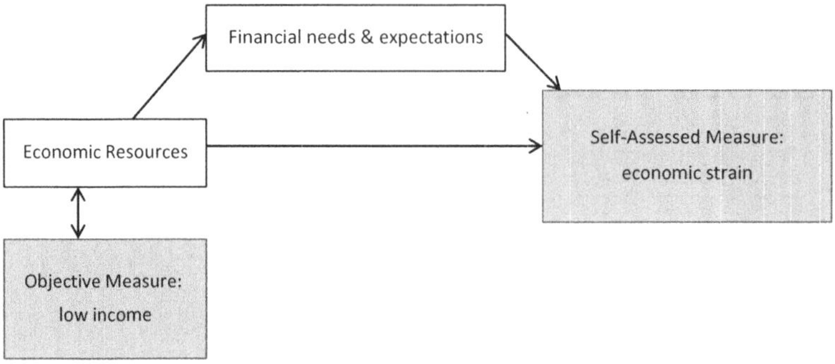

Theoretical model of the relationship between the Objective and the Self-Assessed Measure

The different statistical methods used in this and the next parts are meant to complement each other. The main advantage of regression analysis lies in its ability to improve the accuracy of our understanding of the relationship between each explanatory variable and the dependent variables by controlling for the possible influences of other potential explanatory variables. In the context of this volume, we apply binary logistic regression models primarily for conducting exploratory analysis and final variable selection, though the results will of course provide us with elements of response to some of our hypotheses. Among the strengths of regression analysis is its capacity to model nominal variables and variables that are not normally distributed; both cannot be analyzed by structural equation models (SEM). SEM on the other hand permits a more rigorous investigation of our research questions because it allows aligning the statistical model more precisely with our theoretical framework. This statistical technique is clearly confirmatory in nature and both model development and modification must be grounded in theory. With regard to our research question on the relationship between the Objective and the Self-Assessed Measure, SEM will allow to simultaneously test several of our hypotheses pertaining to the Self-Assessed Measure, how it is predicted by Economic Resources and whether variables related to Financial Needs and Expectations play a mediating role. Chap. 8 provides a detailed description of the statistical techniques used in this part. For each regression model fitted in this part, the exact same sample was used by beforehand eliminating any cases with missing values on any of the independent variables. All models were estimated in two versions, one using weights and one without weights, in order to assess whether the model without weights represents the data well. Since the coefficients of the two versions were close to identical, the models without weights were given preference and are presented here. The results are reported using odds ratios[1] for a more straightforward interpretation of the effect size. Since the objective of using regression models is not primarily to find the best

[1] *Odds* are the probability of occurrence of an event (i.e. of economic vulnerability) divided by the probability of non-occurrence. Odds Ratios refer to the *ratio* of the *odds* of occurrence of the event in group 1 as compared to group 2.

fitting model but to compare two outcome measures of economic vulnerability in an exploratory manner, post-estimation tests and the question of multi-colinearity are not going to be discussed in much depth.

This part is organized in chapters according to the hypotheses (stated in Part II Chap. 7). Chapter 18 begins by estimating an initial regression model in order to assess the difference between the two population groups that are vulnerable (a) according to the Self-Assessed Measure and (b) according to the Objective Measure with regard to Background Characteristics. In Chap. 19, the difference and similarity between the two population groups are further examined with regard to types and size of Economic Resources, a question that will be deepened in Chap. 20. Adding variables of Financial Needs and Expectations to the model on Economic Resources completes the exploratory part in Chap. 21. Chapter 22 introduces the SEM model for testing the general theoretical model linking the Self-Assessed Measure to Economic Resources via the mediating role of Financial Needs and Expectations. Chapter 23 focuses on fitting the most parsimonious model representing the relationship between the Self-Assessed and the Objective Measure. Lastly, Chap. 24 puts the Vulnerability Typology (based on the combination of the two measurement angles) to the empirical test, using ordinal regression models: firstly, by assessing its capacity for identifying the most vulnerable groups according to low levels of wealth and secondly, by checking whether the phenomena of 'expensive taste' could undermine the reliability of the typology. Throughout this part, there will be results that suggest that psychosocial factors could yield elements of response; these elements will only be discussed in Part V when the analysis of our theoretical model will take into account the Perceived Measure and the corresponding psychosocial covariates.

Chapter 18
Regressing Background Characteristics on the Self-Assessed and the Objective Measure of Economic Vulnerability

The first step in our plan of analysis consists in estimating a base model of Background Characteristics, yielding a general profile of the two population groups that are vulnerable according to the Objective Measure and the Self-Assessed Measure. Considering the similarities and possible differences between the groups will allow to respond to question Ii) 'Is there a significant difference between the two (overlapping) population groups that are vulnerable (a) according to the Self-Assessed Measure and (b) according to the Objective Measure with regard to their Background Characteristics?' The covariates analyzed in this chapter have been identified as pivotal in positioning individuals within the social strata. They are expected to be associated with the ability to build up financial resources over the life course as well as to be foundational with regard to the endogenous reference system that underlies the self-assessment of economic strain.

Table 18.1 shows the odds ratios of a logistic regression performed with the Self-Assessed Measure (difficulties in making ends meet) and the Objective Measure (equivalized household income less than CHF 2400 per month) as outcome variables and the covariate set Background Characteristics (sex, age, canton, marital status, education). After removing all cases that had missing values on any of the predictors used in this and subsequent nested models, the estimation sample comprises 1465 cases. The number of individuals classified as economically vulnerable is $n_{sa\_ev} = 216$ (14.3%) and $n_{obj\_ev} = 225$ (15.9%). The overlap of the two vulnerable groups is 5.8%.

Starting with the contextual variable canton, we see that residency in the canton Ticino doubles the risk of economic vulnerability by a monetary and a self-assessed standard. Respondents living in the canton Valais, too, bear a higher risk of income poverty though they do not experience the economic strain more frequently than inhabitants of other cantons. Although only two (respectively one for sa_ev) of four categories of the variable canton were significant predictors, the variable canton as a whole does have an effect on self-assessed and objective economic vulnerability

J. Henke, *Revisiting Economic Vulnerability in Old Age*, Life Course Research and Social Policies 11, https://doi.org/10.1007/978-3-030-36323-9_18

Table 18.1 Odds ratios and confidence intervals of Background Characteristics on the Self-Assessed and the Objective Measure

	Model 1: Background Characteristics					
	sa_ev			ob_ev		
Variable (Ref.)	OR	[95% CI]		OR	[95% CI]	
Gender (male)	0.81	0.58	1.12	0.67*	0.49	0.94
Age	0.95	0.82	1.10	1.07	0.93	1.24
Canton (Basel)						
Genève	1.65*	1.01	2.70	0.98	0.58	1.67
Valais	1.62	0.98	2.68	1.93**	1.18	3.15
Bern	0.92	0.55	1.54	1.06	0.64	1.74
Ticino	2.24**	1.37	3.68	2.36**	1.45	3.86
Marital status (married)						
Single	0.67	0.32	1.38	1.31	0.73	2.34
Divorced/sep.	3.45***	2.30	5.16	2.30**	1.49	3.55
Widowed	1.55*	1.04	2.33	1.17	0.78	1.77
Educational attainment (primary school)						
Lower secondary	0.86	0.47	1.59	0.50*	0.27	0.93
Prof. Training	0.45*	0.28	0.71	0.38***	0.25	0.59
Highschool	0.32**	0.19	0.56	0.21***	0.12	0.36
University of Appl. Sciences	0.33**	0.19	0.60	0.25***	0.14	0.43
University	0.24***	0.13	0.45	0.11***	0.05	0.23
McFadden's R^2		*0.086*			*0.088*	

Key: *** < 0.001; ** < 0.01; * < 0.05

in this particular model.[1] Based solely on socio-demographic characteristics it becomes obvious that elderlies living in the two alpine cantons are more frequently affected by economic strain than those living in Basel, Berne or Geneva.

At the individual- and family level, the state of being divorced is a very strong predictor of economic vulnerability for both measures, though the predictive power is stronger for the Self-Assessed Measure: divorced and separated individuals have a 3.5 times higher risk of having difficulties in making ends meet than those who are married, while their comparative risk of finding themselves below the poverty line is 2.25 times higher. Widowhood, too, affects the subjective appraisal of one's economic situation: it increases the odds of self-assessed vulnerability by 60%, after controlling for sex. The absence of any significant effect of marital status on the Objective Measure is in line with previous research (Gabriel et al. 2015), suggesting that the protective mechanisms introduced with the tenth reform of the pension scheme in 1997 are having the desired effect. In Switzerland, the reduction of the poverty risk in the case of widowhood has been observed in particular for pensioners

[1]A maximum likelihood test was performed to test the hypothesis that the coefficients for all the indicator variables are zero (H_0: $\beta_{Geneve}=\beta_{Valais}=\beta_{Bern}=\beta_{Ticino}= 0$.) The test results were $LRX^2(4) = 30.0$, $p < 0.001$ for sa_ev; and $LRX^2(5) = 68.1$, $p < 0.001$ for obj_ev, thus we rejected the null hypothesis.

Table 18.2 Probability of being economically vulnerable based on an ideal type of Background Characteristics

Ideal type	Probability of being economically vulnerable [95% CI] obj_ev	sa_ev
Female, living in Ticino, divorced, primary education	0.67 [0.54, 0.81]	0.59 [0.44, 0.75]

who are younger than 85 years of age and who tend to benefit from occupational pensions (Wanner and Fall 2012).

Educational attainment records the single strongest predictive power, especially on the Objective Measure: having a university degree increases the odds of *not* falling into income poverty by a factor of 9,[2] the odds of finding it easy or very easy to make ends meet at the end of the month are increased by a factor of 4, compared to the reference group of individuals who have only attained the level of primary school. For graduates of high school and the university of applied sciences, the protection against economic vulnerability is similar, decreasing the risk by a factor close to 3 for the Self-Assessed Measure and 4.5 for the Objective Measure. Interestingly, having attended secondary school cuts the poverty risk into half, but does not affect self-assessed vulnerability. It remains to be seen whether the effect of educational attainment stands alone or whether it serves, in this model, as a proxy for economic resources.

From comparing the odds ratios of these two regression models we conclude that the risk profile with regard to Background Characteristics is very similar for the population group that is vulnerable in terms of low income and the one that is vulnerable in terms of their self-assessed economic vulnerability. Moreover, the confidence intervals of the two regression models overlap to a large degree, indicating that there is indeed no statistically significant difference in the odds ratios of covariates when regressed on the two measures. The McFadden's R^2 reveals that, all covariates of Background Characteristics taken together, they predict the Objective Measure and the Self-Assessed Measure with close to the same accuracy, though explaining only a relatively small share of the variance in the original covariance matrix ($R^2 = 0.09$).

Table 18.2 shows the predicted probability for a hypothetical 'ideal type' of an economically vulnerable person, holding all other variables at their mean. If we consider a situation where the most important risks of exposure for old age poverty are united, the predicted probability show that it is slightly more likely to be vulnerable by the Objective Measure (0.67) than by the Self-Assessed Measure (0.59). However, here too the large overlap of the confidence intervals indicates that there is no significant difference between the two probabilities.

Based on these findings, we cannot conclude that there is a significant difference in terms of Background Characteristics between the two groups identified by the Objective Measure and the Self-Assessed Measure of economic vulnerability.

[2] Odds ratios being multiplicative, it is in some cases easier to interpret the inverse of the negative effects.

References

Gabriel, R., Oris, M., Studer, M., & Baeriswil, M. (2015). The persistence of social stratification? A life course perspective on poverty in old-age in Switzerland. *Swiss Journal of Sociology, 41*(3), 465–487.

Wanner, P., & Fall, S. (2012). *La situation économique des veuves et des veufs (Bd. 5)*. Berne: Office fédéral des assurances sociales.

Chapter 19
Regressing Economic Resources on the Self-Assessed and the Objective Measure of Economic Vulnerability

The next step in our analysis consists of adding a set of Economic Resources to the model component Background Characteristics. According to theory as well as from empirical research we know that the Objective and the Self-Assessed Measures differ in terms of the spectrum of what is taken into account for the determination of a minimum standard of economic quality of life. Our general hypothesis posits that the Self-Assessed Measure is more sensitive to variations in Economic Resources compared to the Objective Measure. Included in the regression model are all variables of Economic Resources that have proven to be statistically significant[1] at the binary level of analysis: supplementary benefits, home ownership and wealth.

Among the three variables operationalizing Economic Resources, wealth stands out as the most important predictor (Table 19.1): the odds of having difficulties in making ends meet are over 13 times higher for respondents who have hardly any savings or no assets at all, compared to the reference group[2] with 'more than CHF 500,000'. For respondents who are income poor, the difference in odds between those who are best-off financially and those who have no wealth is substantially lower but still high with a factor of 6. Already the endowment with some capital – less than CHF 60,000 – has the effect of reducing the risk of self-assessing oneself as economically vulnerable (OR = 4.8). For the Objective Measure, the odds of being income poor are four times higher for the middle-range wealth categories '<60,000' and '60,000–150,000' compared to the reference category. The small overlap between the confidence intervals (for 'no/almost no wealth' the intervals are [7.27, 25.06] for sa_ev and for obj_ev [3.33, 11.30]) confirms the statistical as well

[1] We deliberately did not include income as an explanatory variable, even though it would obviously be a very important predictor of self-assessed economic vulnerability, because our primary aim is to be able to compare the models between the two outcomes, and the Objective Measure being a binary indicator based on the ordinal variable income, introducing the same variable as explanatory variable is not an option.

[2] The majority of the sample population (31%) is found in the wealth category of ,more than CHF 500,000'.

J. Henke, *Revisiting Economic Vulnerability in Old Age*, Life Course Research and Social Policies 11, https://doi.org/10.1007/978-3-030-36323-9_19

Table 19.1 Odds ratios and confidence intervals of Background Characteristics and Economic Resources on the Self-Assessed and the Objective Measure

	Model 2: Background Characteristics + Economic Resources					
	Self-assessed measure			Objective measure		
Variable (Ref.)	OR	[95% CI]		OR	[95% CI]	
Gender (male)	0.84*	0.58	1.21	0.70*	0.50	0.99
Age	0.95	0.81	1.12	1.09	0.94	1.26
Canton (Basel)						
Genève	1.35	0.79	2.32	0.90	0.52	1.56
Valais	2.19*	1.25	3.84	1.96*	1.17	3.30
Bern	0.92	0.52	1.63	1.05	0.63	1.76
Ticino	2.03*	1.17	3.53	2.11**	1.27	3.51
Marital status (married)						
Single	0.44*	0.20	0.96	1.02	0.55	1.91
Divorced/sep.	1.65*	1.02	2.66	1.69*	1.05	2.72
Widowed	1.13	0.71	1.77	1.02	0.66	1.55
Educational attainment (primary school)						
Lower secondary	1.05	0.53	2.07	0.56	0.29	1.06
Prof. Training	0.54*	0.33	0.89	0.44***	0.28	0.69
Highschool	0.48*	0.26	0.88	0.29***	0.16	0.50
University of Appl. Sciences	0.47*	0.24	0.90	0.31***	0.17	0.55
University	0.55	0.27	1.14	0.20***	0.09	0.43
Supplementary benefits	2.46***	1.43	4.23	1.09	0.61	1.94
Home owner	0.93	0.62	1.40	1.49*	1.02	2.18
Wealth (>500,000)						
(almost) nothing	13.49***	7.27	25.06	6.14***	3.33	11.30
<60,000	4.78***	2.59	8.83	4.24***	2.37	7.60
60,000–150,000	2.34**	1.26	4.37	4.30***	2.49	7.42
150,000–500,000	1.34	0.73	2.47	1.78*	1.04	3.05
McFadden's R^2	*0.208*			*0.127*		

Key: *** < 0.001; ** < 0.01; * < 0.05

as practical significance of the magnitude of the difference in effect size between the two outcome measures: among older adults in Switzerland, low levels of wealth are clearly more strongly associated with the self-assessed 'difficulties in making ends meet' than with 'income poverty'.

The association between home ownership and indicators of economic vulnerability remains surprisingly weak and insignificant, in the case of the Self-Assessed Measure, after controlling for economic resources. In this country, where the rate of home ownership among older adults is generally low compared to other wealthy European countries (Kolb et al. 2013), one possible explanation for this disjunction has been the 'high cost of owning caused by the taxation of imputed rent and capital gains, and the lack of financial incentives to save for a down payment' (Werczberger 1997, p. 350).

For respondents who count on supplementary benefits, the odds higher of self-assessing themselves as economically vulnerable are 2.5 times higher. The lack of predictive power of supplementary benefits for the objective outcome was to be expected since the eligibility threshold is close to the poverty line.[3]

Comparing Table 19.1 to Table 18.1 shows that the effect of wealth weakens the influence of several of the background variables: the predictive power of educational attainment decreases substantially, lower secondary- and university-level education even cease to be significant for self-assessed economic vulnerability. At constant levels of wealth, living in canton Valais increases the risk of self-assessing oneself as economically vulnerable (OR = 2.2).[4] After controlling for wealth, the effect of marital status 'divorced/separated' decreases substantially for both outcomes. The adverse effect of being a widow/widower is also accounted for by Economic Resources. This finding resonates with a recent study based on SHARE data[5]: comparing 12 European countries, it was found that the adverse effects of widowhood on financial wealth holdings are close to zero and statistically insignificant for Switzerland (Bíró 2013).

Adding the covariate set Economic Resources clearly improves the model fit of model 2 for both dependent variables, but much more so for the Self-Assessed Measure, for which the likelihood-ratio index goes up to $R^2 = 0.21$ (compared to 0.13 for the equivalent model predicting the Objective Measure). In terms of the differences in the magnitude of effects of covariates on either the Objective or the Self-Assessed Measure, this preliminary assessment of model 2 shows clear differences in the size of odds ratios. Two more analyses will prove helpful for substantiating this conclusion: a look at the predicted probabilities of ideal types and the comparison of the relative magnitude of change caused by the variable wealth in each of the two dependent variables.

Table 19.2 shows the predicted probabilities for an average pensioner and three ideal types of vulnerable respondents: all of them are found in the lowest category of wealth (little or no savings) and are either single, widowed or divorced/separated – states that previous research has identified as exposure factors to an increased risk of economic vulnerability. All other variables are kept at their mean.

For the average pensioner, the predicted probability of being economically vulnerable is close to the same by either the Objective or the Self-Assessed Measure (0.11 and 0.10). For individuals who have never gotten married, disposing of no assets (as is the case for all subsequent types), the probability of economic vulnerability is twice as high compared with the average pensioner; there is no difference in probability between the two measures (0.23 and 0.22). Respondents whose mari-

[3] As the eligibility criteria for receiving supplementary benefits is based on wealth, too, we proceeded to estimating model 2 without wealth: the variable supplementary benefits still did not predict the objective outcome variable.

[4] These findings will be discussed later when we will consider the effects of the variable block Needs and Expectations.

[5] The Survey of Health, Ageing and Retirement in Europe (SHARE) is a multidisciplinary and cross-national panel database of micro data on individuals aged 50 and older.

Table 19.2 Probability of being economically vulnerable based on ideal types with regard to low wealth and marital status

	Probability of being economically vulnerable [95% CI]	
Ideal types	Objective Measure	Self-Assessed Measure
Average pensioner	0.11 [0.09, 0.13]	0.10 [0.07, 0.11]
Lowest wealth category, single	0.23 [0.11, 0.35]	0.22 [0.08, 0.35]
Lowest wealth category, widowed	0.24 [0.15, 0.33]	0.42 [0.30, 0.54]
Lowest wealth category, divorced or separated	0.33 [0.21, 0.45]	0.50 [0.37, 0.63]

tal status is 'widowed' have a barely noticeable higher risk of objective vulnerability compared to single individuals (0.24) but their risk of self-assessing themselves as vulnerable doubles. Widowhood therefore can be considered a risk factor if it converges with other risk factors. In a thorough study of the financial situation of widowers and widows based on tax data, Philippe Wanner points to a possible explanation: 'If widowhood is accompanied by other factors of precarization, it can lead to particularly difficult financial situations for some groups of widows and widowers. This is the case of widows aged 64 and older who should theoretically be entitled to an old age pensions but who continue to benefit from a survivors' pension.[6] This group is made up of women who have a low old age provision and whose survivors' pension is higher than the old age pension, which they could claim. As a reminder, there is a cap on the survivor's pension at the equivalent amount of 80% of the old age pension, which is why this group is mainly comprised of women who have not paid into the first pillar throughout their entire lives (Wanner & Fall 2012, p. 95). Thus, our results confirm the existence of sub-groups among widows and widowers who are more vulnerable than average pensioners, which is especially reflected in their subjective experience. For individuals with the status 'divorced' or 'separated', the predicted probability of objective vulnerability is clearly higher (0.33) compared to single individuals (0.23), while the probability of self-assessed vulnerability is more than double (0.50 compared to 0.22) representing a highly significant result.

These simulations reveal a substantial difference in the ability to cope with objectively low economic resources among individuals with different risk profiles. Our results confirm a gradient of objective vulnerability found in previous income-based research, ranging from (decreasing order) married to widowed to single to divorced (Wanner & Fall 2012, p. 16). Looking at it merely on a material level, the

[6] Survivors' insurance, old age pension and invalidity insurance jointly make up the first pillar of the Swiss social security system with pensions intended to cover basic living costs. When they not do so, there are supplementary benefits to top-up income to the required level.

Table 19.3 Change in probability of being vulnerable for each increment of discrete change in wealth

	Change in probability [95% CI]	
Change in category of wealth	Objective Measure	Self-Assessed Measure
1.-2. From 'no wealth/almost nothing' to '<60,000'	−0.06 [−0.14, 0.03]	−0.20 [−0.29, −0.10]
2.-3. From '<60,000' to '60,000–150,000'	0.00 [−0.07, 0.07]	−0.08 [−0.15, −0.02]
3.-4. From '60,000–150,000' to '150,000–500,000'	−0.10 [−0.16, 0.04]	−0.05 [0.01, 0.10]

analyzed combinations of the explanatory variables 'marital status' and 'wealth' have relatively similar outcomes for 'divorced/separated' and 'single'. Yet individuals whose current status is 'divorced/separated' are much more likely to suffer from their lack of finances in form of economic strain. The forthcoming analysis of Financial Needs and Expectations will shed further light on other factors that may increase the probability of experiencing economic strain.

In order to know whether the observed differences in this second model are significant, we compare by how much an increment of discrete change in the predictor wealth affects the probability of the two outcomes of economic vulnerability, while holding all other variables constant. Table 19.3 shows that for the Objective Measure, none of the changes from a lower to a higher wealth category are significant while each discrete change in wealth provokes a significant change in the probability of being vulnerable by the Self-Assessed Measure. The most important change occurs when moving from category 'no wealth/almost nothing' to 'less than 60,000' where the probability of self-assessed economic vulnerability is reduced by 0.20.

Figure 19.1 visualizes the predicted probabilities of being economically vulnerable by the Self-Assessed and the Objective Measure and the corresponding confidence intervals according to wealth categories.[7] The two graphs show that for both outcome measures, the confidence interval is smaller for the higher categories of wealth, however, the probability of being economically vulnerable at increasing levels of wealth starts at a higher level and declines more sharply for the Self-Assessed Measure than for the Objective Measure.

This finding is noteworthy with regard to the discussion on the *relevance* of an indicator for measuring economic quality of life: as far as the ability to reflect low levels of economic resources beyond income is concerned, the Self-Assessed Measure clearly outperforms the Objective Measure. In Part V we will take a closer look at the psychological interpretation of the subjective experience of economic vulnerability.

[7] 1 = no wealth or almost nothing, 2 = less than 60,000, 3 = 60,000–150,000, 4 = 150,000–500,000, 5 = more than 500,000.

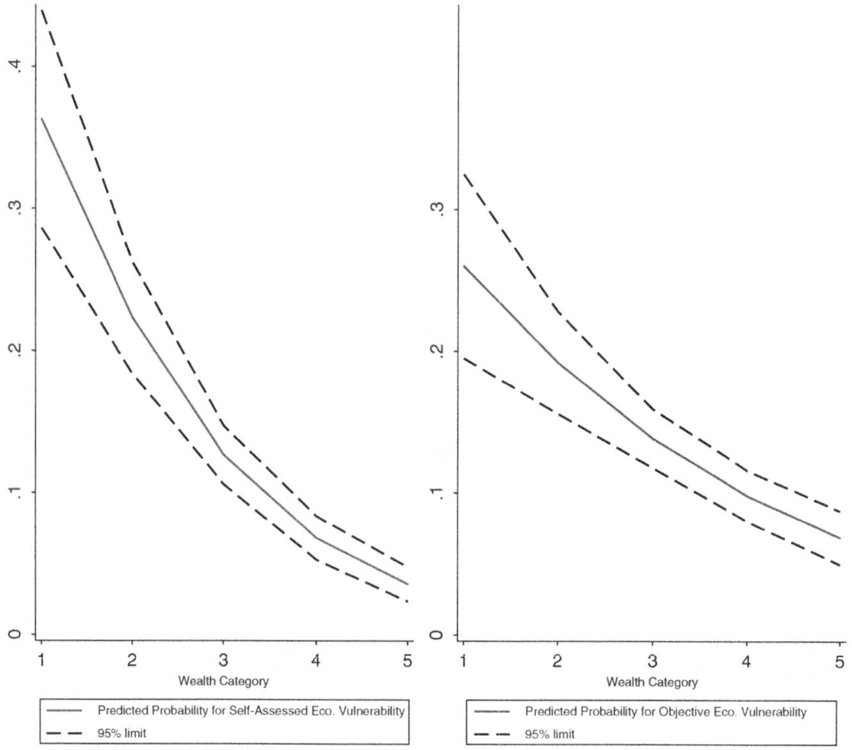

Fig. 19.1 Comparison of predicted probability for the Self-Assessed and the Objective Measure according to wealth category, with confidence intervals

In view of these findings, we respond positively to hypothesis (III) by affirming that there is a significant difference between the two population groups that are vulnerable according to the Self-Assessed Measure and according to the Objective Measure with regard to the types and size of economic resources, specifically with regard to levels of wealth.

References

Bíró, A. (2013). Adverse effects of widowhood in Europe. *Advances in Life Course Research, 18*(1), 68–82.

Kolb, K., Blossfeld, H.-P., & Skopek, N. (2013). The two dimensions of housing inequality in Europe. Are high home ownership rates an Indicator of low housing values? *Comparative Population Studies – Zeitschrift für Bevölkerungswissenschaft, 38*(4).

Wanner, P., & Fall, S. (2012). *La situation économique des veuves et des veufs (Bd. 5)*. Berne: Office fédéral des assurances sociales.

Werczberger, E. (1997). Home ownership and rent control in Switzerland. *Housing Studies, 12*(3), 337–353.

Chapter 20
Exploring the Relationship Between Economic Resources and the Self-Assessed Measure of Economic Vulnerability

Understanding the relationship between the Self-Assessed Measure of economic vulnerability and economic resources is particularly relevant with regard to the hypothesis that the subjective difficulties in making ends meet may capture differences in the ability to cope with low monthly income. Wealth consumption has previously been mentioned as one of the coping strategies that moderates the level of economic strain an individual experiences at a given level of income. The question is in what way monthly household income and household wealth interact in their effect on self-assessed economic vulnerability. In the following, we will estimate multiple models in order to test competing concepts for the relationship between the three variables. The models were performed using maximum likelihood estimation[1] since all variables follow a normal distribution. First, we will consider a model that incorporates both income and wealth as predictors of the Self-Assessed Measure (Fig. 20.1). In this model, both types of economic resources have an identical and highly significant effect on the Self-Assessed Measure (standardized $\beta = 0.25$, $\rho < 0.001$) while being even more strongly correlated amongst each other (0.44, $\rho < 0.001$).

According to our theoretical framework, however, it is probable that the effect of wealth mitigates the effect of low income on self-assessed economic vulnerability so that the relationship between income and the Self-Assessed Measure is stronger when levels of wealth are lower, a relationship that is referred to as moderation.[2]

[1] The Stata command for this SEM option is > *method(ml).*

[2] In contrast to a mediator variable that explains the relationship between two variables, a moderator variable is one that influences the strength of a relationship between a predictor and a dependent variable. This influence is also referred to as a statistical 'interaction effect' (Baron and Kenny 1986). According to Holmbeck (1997), a *mediational model* attempts to respond to the following questions: how (by what means) does an effect occur or what accounts for the impact of A (predictor) on C (dependent variable). A mediational model is therefore a causal model, whereby it is hypothesized that A 'causes' B and that B then 'causes' C. In contrast, a *moderational model* examines the following: when (under what conditions) does the effect occur or under what conditions of B (moderator variable) is A (predictor) significantly associated with C (dependent variable) (Holmbeck 1997).

J. Henke, *Revisiting Economic Vulnerability in Old Age*, Life Course Research and Social Policies 11, https://doi.org/10.1007/978-3-030-36323-9_20

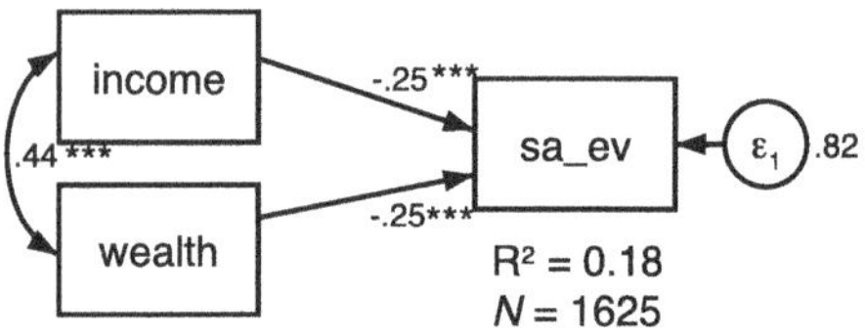

Fig. 20.1 Joint effects of monthly household income and wealth on the Self-Assessed Measure (standardized coefficients; *** p < 0.001)

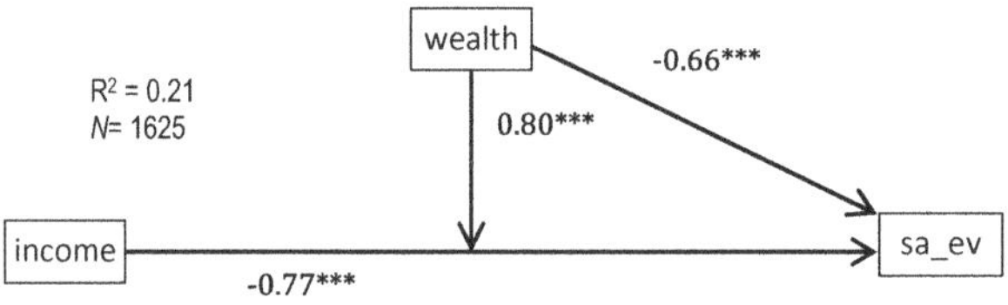

Fig. 20.2 A path model: the relationship between income and the Self-Assessed Measure economic vulnerability as moderated by wealth (standardized coefficients; *** p < 0.001)

Figure 20.2[3] shows how the magnitude of wealth moderates the influence of income on self-assessed economic vulnerability. Even when controlling for Background Characteristics (age, sex, canton, education, and marital status), the strength and nature of the relationship between income, wealth and the Self-Assessed Measure remains robust.[4]

This moderation model provides a better representation of the observed data: it yields a slightly better model fit[5] with $R^2 = 0.21$ and notably higher coefficients for income, where the standardized coefficient is $\beta = -0.77$ ($\rho < 0.001$) and for wealth, where it is $\beta = -0.66$ ($\rho < 0.001$); the interaction coefficient between the two is $\beta = 0.80$ ($\rho < 0.001$).

Figure 20.3 visually represents the predicted values of self-assessed economic vulnerability as moderated by wealth. We see that, as income categories grow higher, lower wealth categories have a sharper drop-off in their estimated difficulties in making ends meet than higher wealth categories. For example, for wealth category '(almost) nothing' represented by hallow diamonds, the effect of income on the Self-Assessed Measure declines more sharply than for any of the higher wealth categories. On the other hand we see that for lower income, the mediating effect of wealth is stronger. For example, respondents who are in the 2nd to lowest income category (CHF 2400–3600) but who have '(almost) nothing' in terms of wealth (1 in Fig. 20.3) have close to the same probability of self-assessing themselves as economically vulnerable as those who are in the lowest income category and in wealth category 'less than 60'000' (2 in Fig. 20.3). In contrast, respondents who are well dotted with wealth (150,000 or more) are much less likely to have

[3] This figure was drawn freehand because the Stata SEM-builder is unable to graphically represent this type of model.

[4] The standardized beta coefficients vary by a maximum of 0.03.

[5] Since the model is just identified, it is not possible to calculate any other fit statistics.

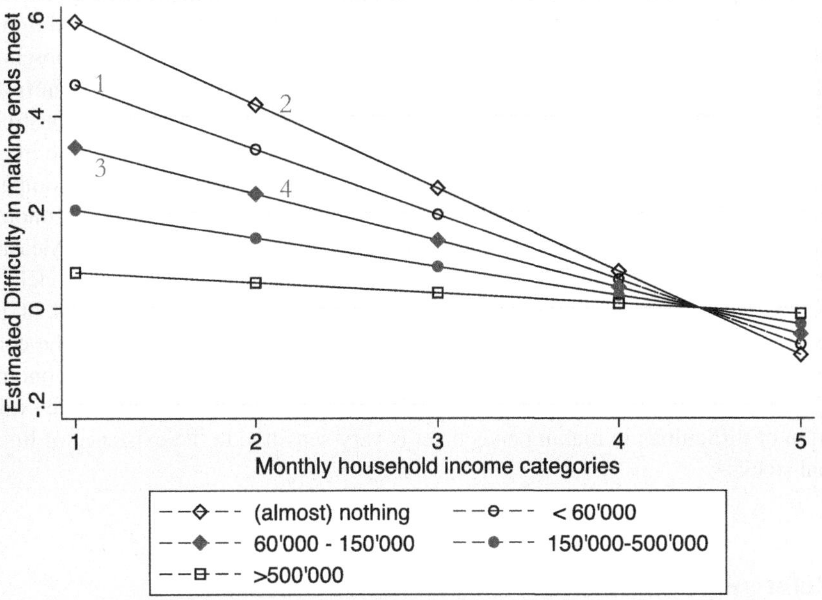

Fig. 20.3 Predicted values between income and the Self-Assessed Measure (difficulties in making ends meet) as moderated by wealth

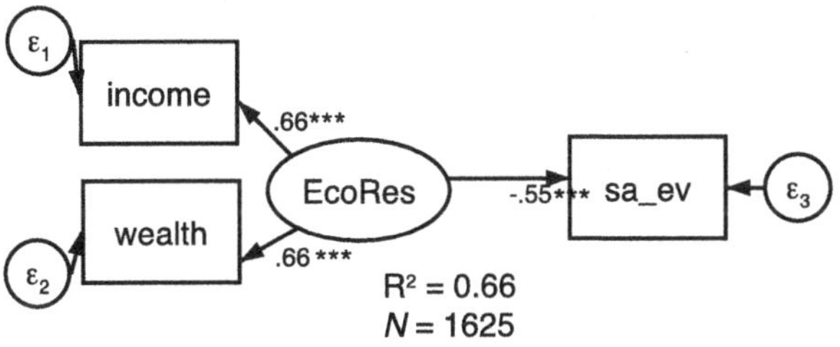

Fig. 20.4 A latent construct Economic Resources: income and wealth as latent variables predicting the Self-Assessed Measure of economic vulnerability (standardized coefficients; *** $p < 0.001$)

difficulties in making ends meet, even if they are in lower income categories (3 and 4 in Fig. 20.3). This analysis shows that income and wealth interact in a non-linear manner in their association with the Self-Assessed Measure of economic vulnerability.

One more model, shown in Fig. 20.4, was estimated to investigate the nature of the relationship between self-assessed economic vulnerability and Economic Resources, now operationalized as a latent construct EcoRes, with income and wealth as reflective indicators. The rational for conceptualizing Economic Resources as latent construct lies in the idea that a person who is endowed with a high level of resources can be expected to have a higher income and more wealth – in addition to enjoying a generally higher level of economic quality of life that cannot be observed in our data. Thus, the variables income and wealth are considered to be manifestations of a broader, less tangible reality that is ideally captured by a latent construct.

This model is able to explain a substantially higher share of the variance in our data ($R^2 = 0.66$) and does better at predicting sa_ev (R^2 sa_ev = 0.30). The comparison of the latent model and the path model reveal that operationalizing Economic Resources as the joint effect of wealth and income on the Self-Assessed Measure comes even closer to the observed covariance matrix than the mediation model.

The conclusion emerging from these findings corroborate the results of the regression analysis while allowing for a more nuanced understanding of the combined effect of Economic Resources on the Self-Assessed Measure of economic vulnerability. The findings presented here further confirm that the subjective evaluation of difficulties in making ends meet is very sensitive to the existence of financial stocks.

References

Baron, R. M., & Kenny, D. A. (1986). The moderator–mediator variable distinction in social psychological research: Conceptual, strategic, and statistical considerations. *Journal of Personality and Social Psychology, 51*(6), 1173–1182.

Holmbeck, G. N. (1997). Toward terminological, conceptual, and statistical clarity in the study of mediators and moderators: Examples from the child-clinical and pediatric psychology literatures. *Journal of Consulting and Clinical Psychology, 65*(4), 599–610.

Chapter 21
Regressing Financial Needs and Expectations on the Self-Assessed and the Objective Measure

Adding the variable set Financial Needs and Expectations will allow testing hypothesis I iv) positing that there are significant differences between the two vulnerable groups defined by the Objective and the Self-Assessed Measure of economic vulnerability. According to our theoretical model, the Self-Assessed Measure entails a juxtaposition of monthly income with the individual's needs in terms of finances. Therefore, our general hypothesis states that the Self-Assessed Measure is relatively more sensitive to variations in Financial Needs and Expectations compared to the Objective Measure. The nature of the association with the two dependent variables may vary considerably among covariates: according to the reviewed literature, it is possible that the Self-Assessed Measure partly captures the phenomenon of 'expensive taste' as expressed in respondents self-assessing themselves as economically vulnerable despite enjoying an above average living standard. If this is the case, we would find that the pre-retirement socio-professional category (SPC) mediates the effect of Economic Resources on the Self-Assessed Measure in such a way that, at the same level of wealth, a higher ranking SPC[1] increases the odds of respondents self-assessing themselves as economically vulnerable. As for the predictive power of SPC on the Objective Measure, SPC is expected to have little effect when controlling for wealth.

The variable general health status is expected to have more predictive power on the Self-Assessed Measure than on the Objective Measure when controlling for Economic Resources because poor health is likely to capture a situation where more financial assets are required to attain the same level of economic quality of life as enjoyed by a healthy person. However, since this effect is only probable for respondents who are in the middle-income categories, it is possible that the significance of the effect is blurred in a model estimated simultaneously on all income categories.

[1] When speaking of ‚ranks' here we must keep in mind that the variable socio-professional category does not provide a scale with equal distance intervals; however, in a two-by-two comparison it is most of the time possible to determine which one is higher in a hierarchy of social status.

J. Henke, *Revisiting Economic Vulnerability in Old Age*, Life Course Research and Social Policies 11, https://doi.org/10.1007/978-3-030-36323-9_21

As for the variables operationalizing the concept of social participation, the frequency of going to a restaurant/coffee shop, seeing a movie/theater play or taking at trip, the frequency of engaging in social activities were originally expected to be more predictive of the Self-Assessed than of the Objective Measure at any given level of wealth. However, the bivariate analysis revealed a more ambiguous picture, showing that the association of the two measures of economic vulnerability with the frequency of social participation varies considerably among different activities.

On a technical note it should be mentioned that all three covariates are going to be treated as interval variables after having tested that the successive intervals within each variable are equally spaced, guaranteeing that there is no loss of information when using this format in regression models. In order to gain a better understanding of the effect of the covariates operationalizing Financial Needs and Expectations on each dependent variable, we first fit a model with only the Background Characteristics (model 3) before adding the variables on Economic Resources (model 4). This step-by-step procedure helps disentangling the effects of SPC and wealth, which are correlated (Cramér's V = 0.19, $p < 0.001$) and the effects of SPC and educational attainment (Cramér's V = 0.27, $p < 0.001$). As these three covariates operationalize very distinct model components they cannot substitute one another even though they are correlated.

The odds ratios shown in Table 21.1 reveal that adding the variable set Financial Needs and Expectations to Background Characteristics (model 3) accounts for the increased risk of self-assessed economic vulnerability of respondents living in the canton of Valais. The effect of educational attainment, too, looses its significant effect on the Self-Assessed Measure. This means that, once social position is controlled for, education becomes an unreliable predictor of the difficulties in making ends meet.[2] For the Objective Measure, the effect of education is weakened as well, but there remains a lower likelihood of being income poor for the categories professional training, high school or university compared to primary level education. However, after performing a likelihood ratio test we find that all coefficients of educational attainment levels taken together equals zero ($LRX^2 = 8.09$, $df = 5$, $p = 0.15$). Thus, when controlling for the variable last socio-professional status, educational attainment has no longer any statistically significant effect on either measure of economic vulnerability.

An omnibus test reveals that socio-professional categories taken on the whole is predictive of the objective outcome ($LRX^2 = 59.94$, $df = 5$, $p < 0.001$) and also has a significant effect on the self-assessed outcome ($LRX^2 = 19.05$, $df = 5$, $p = 0.002$). It is indeed only the category top executives and academic professions for which there is weak evidence of reducing self-assessed economic vulnerability (OR 0.63, not significant). It is noteworthy that the confidence intervals for this indicator of SPC cover identical ranges for the Self-Assessed as for the Objective Measure ([0.37, 01.08], [0.35, 1.09]).

[2] The variables SPC and education are highly correlated, which is not surprising, since the variable SPC is based directly on a categorization that takes into account educational attainment, see Fig. 9.3. The rationale behind including both is that we expect the last pre-retirement SPC to shape expectations with regard to what is considered an adequate level standard.

Table 21.1 Odds ratios and confidence intervals of Background Characteristics and Financial Needs and Expectations on the Self-Assessed and the Objective Measure of economic vulnerability

	Model 3: Background Characteristics + Needs and Expectations					
	Self-assessed measure			Objective measure		
Variable (Ref.)	OR	[95% CI]		OR	[95% CI]	
Gender (male)	0.84	0.59	1.20	0.69*	0.48	0.99
Age	0.83*	0.71	0.96	0.96	0.83	1.12
Canton (Basel)						
Genève	1.68*	1.00	2.82	1.11	0.63	1.96
Valais	1.45	0.86	2.47	1.94*	1.14	3.29
Bern	0.92	0.54	1.59	1.04	0.61	1.76
Ticino	1.74*	1.03	2.96	2.32**	1.36	3.95
Marital status (married)						
Single	0.65*	0.30	1.40	1.51	0.80	2.86
Divorced/sep.	3.28***	2.13	5.05	2.21**	1.39	3.51
Widowed	1.47	0.96	2.26	1.09	0.70	1.69
Educational attainment (primary school)						
Lower secondary	1.01	0.52	1.96	0.55	0.28	1.08
Prof. training	0.65	0.40	1.06	0.57*	0.36	0.92
High school	0.58	0.31	1.06	0.42**	0.23	0.78
University of appl. sciences	0.68	0.36	1.31	0.56	0.30	1.06
University	0.59	0.28	1.26	0.35*	0.15	0.80
SPC (skilled non-manual)						
Top executive/academic	0.63	0.37	1.08	0.62	0.35	1.09
Self-employed/farmers	1.20	0.66	2.16	2.44**	1.40	4.27
Intermed. prof.	0.61	0.33	1.14	0.43*	0.22	0.87
Skilled manual	1.50	0.91	2.46	1.74*	1.05	2.88
Unskilled	1.16	0.57	2.37	2.92**	1.49	5.71
Health (good/very good)						
Bad/very bad	2.19**	1.29	3.74	1.47	0.82	2.64
Satisfactory	1.42*	1.01	2.01	1.43*	1.02	2.02
Restaurant	0.86*	0.75	0.98	0.75***	0.65	0.85
Cinema	0.87	0.70	1.08	0.83	0.67	1.03
Trip	0.57***	0.45	0.73	0.85	0.68	1.07
McFadden's R^2	*0.146*			*0.172*		

Key: *** < 0.001; ** < 0.01; * < 0.05

As for the Objective Measure, the effect of a singular indicator of SPC is most pronounced for pensioners whose pre-retirement status was unskilled workers: holding all variables constant, their odds of being income poor is 3 times larger than for the reference group skilled non-manual workers. Respondents whose last SPC was self-employed/farmers, too, have an increased risk of income poverty (OR 2.4). In contrast, the odds of former top executives and academic professions and intermediary professionals *not* being objectively vulnerable are by a factor of 1.7 and 2.4 higher than for skilled non-manual workers (OR 0.6 and OR 0.4).[3]

For the variable health, the trend observed for the bivariate associations with the outcome measures is confirmed after controlling for Background Characteristics and social status: a bad or very bad general health status doubles the odds of respondents self-assessing themselves as economically vulnerable compared to respondents in good or very good health (OR 2.2, $p < 0.01$). In contrast, when conducting an omnibus test on the significance of the effect of all indicators within the variable health on objective economic vulnerability, there is no evidence of a significant effect ($LRX^2 = 4.61$, $df = 5$, $p = 0.09$). The expected sensitivity of the Self-Assessed Measure to the potentially higher financial needs as a result of poor health is not confirmed. However, since the size of the effect may be non-linear across income categories, it will be worthwhile to take a closer look at it in a later section.

Among the variables representing social participation, the pattern observed for the bivariate statistics is maintained: for each increment in the frequency of going out to eat, the odds of being objectively vulnerable increase by a highly significant factor of 1.3; for seeing a movie the odds increase by a weakly significant factor of 1.2. The frequency of taking a trip is again strongly associated with the positive assessment of one's economic situation: increasing the frequency of this social activity raises the odds of finding it easy to make ends meet by a factor of 1.8.

In order to assess to what degree covariates related to social class may have an influence on social participation we refitted model 3 without SPC and education (results not shown). The coefficients remained close to constant: for the self-assessed outcome, the effect was still not significant for restaurant (OR 0.89, $p = 0.078$) and cinema (OR 0.80, $p = 0.027$) and the highly significant effect of trip remained unchanged (OR 0.56, $p < 0.001$). For the objective outcome, the highly significant effect of restaurant (OR 0.78, $p < 0.001$) and cinema (OR 0.65, $p < 0.001$) stayed close to the same, and there was no significant effect of trip on income poverty. The robustness of these results indicates that social participation as operationalized by these three covariates does not vary significantly by social class and is more the result of individual preferences and, as we will shortly investigate more closely, by access to financial means.

Indeed, when adding Economic Resources to Financial Needs and Expectations (model 4, Table 21.2) SPC entirely looses its effect on the self-assessed outcome ($LRX^2 = 9.97$, $df = 5$, $p = 0.076$) while the predictive effect on the Objective Measure remains unchanged. This finding speaks of the different essence of what is being captured by the two covariates, confirming that, with respect to our outcome measures, they indeed measure something distinct. This finding does therefore not lend support to the concern that the Self-Assessed Measure of economic vulnerability is susceptible to be influenced by the phenomenon of 'expensive taste'. The difference in risk of being objectively vulnerable for skilled manual workers when compared to the reference group skilled non-manual workers is entirely accounted for by

[3] Odds ratios being multiplicative, it is easier to interpret the inverse of the negative effects.

Table 21.2 Odds ratios and confidence intervals of Background Characteristics, Financial Needs and Expectations and Economic Resources on the Self-Assessed and the Objective Measure of Economic Vulnerability

	Model 4: Background Characteristics + Needs and Expectations + Economic Resources					
	Self-assessed measure			Objective measure		
Variable (Ref.)	OR	[95% CI]		OR	[95% CI]	
Gender (male)	0.94	0.64	1.39	0.70	0.48	1.01
Age	0.86	0.72	1.01	0.98	0.84	1.15
Canton (Basel)						
Genève	1.37	0.78	2.40	1.06	0.59	1.90
Valais	1.99*	1.10	3.59	1.914*	1.09	3.36
Bern	0.95	0.52	1.72	1.07	0.62	1.85
Ticino	1.66	0.93	2.97	2.16**	1.24	3.75
Marital status (married)						
Single	0.50	0.22	1.14	1.24	0.63	2.42
Divorced/sep.	1.78*	1.08	2.95	1.81*	1.09	3.01
Widowed	1.12	0.70	1.80	0.97	0.61	1.53
Educational attainment (primary school)						
Lower secondary	1.17	0.57	2.40	0.58	0.29	1.16
Prof. training	0.63	0.37	1.08	0.62	0.39	1.01
High school	0.61	0.31	1.20	0.51	0.27	0.97
University of appl. sciences	0.63	0.31	1.30	0.65*	0.33	1.23
University	0.73	0.32	1.69	0.52	0.22	1.26
Supplementary benefits	2.22**	1.25	3.94	0.95	0.51	1.78
Home owner	0.97	0.64	1.48	1.54*	1.03	2.31
Wealth (>500,000)						
(almost) nothing	11.64***	6.07	22.32	4.64***	2.41	8.92
< 60,000	4.65***	2.45	8.83	3.81***	2.04	7.12
60,000–150,000	2.17*	1.14	4.14	4.35***	2.44	7.77
150,000–500,000	1.33	0.70	2.51	1.84*	1.04	3.27
SPC (skilled non-manual)						
Top executive/academic	0.86	0.48	1.54	0.64	0.36	1.15
Self-employed/farmers	1.70	0.89	3.26	2.89***	1.60	5.21
Intermed. prof.	0.79	0.40	1.53	0.42*	0.21	0.86
Skilled manual	1.35	0.78	2.31	1.58**	0.95	2.64
Unskilled	0.83	0.38	1.83	2.64**	1.32	5.31
Health (good/very good)						
Bad/very bad	1.76	0.98	3.16	1.45	0.80	2.63
Satisfactory	1.21	0.83	1.77	1.42	0.99	2.03
Restaurant	0.86	0.75	1.00	0.75***	0.66	0.87
Cinema	1.01	0.80	1.28	0.84	0.67	1.05
Trip	0.59***	0.46	0.77	0.90	0.71	1.14
McFadden's R^2	*0.241*			*0.198*		

Key: *** < 0.001; ** < 0.01; * < 0.05

wealth. In contrast, controlling for wealth, the risk for unskilled workers to be found objectively vulnerable remains stable while the risk for self-employed/farmers increases (from OR 2.4 to OR 2.9) and becomes highly significant.[4]

In other words, the more precise the definition of Economic Resources, the more important becomes this socio-professional category as a predictor of income poverty. The contrast between the two outcome measures being most pronounced for respondents who used to be self-employed/farmers and unskilled workers, these categories will be of particular interest for further analysis based on the Vulnerability Typology.

Taking into account Economic Resources causes the variable health to further loose significance. This means that at a given level of wealth, a poor general health status no longer predicts the probability of self-assessed economic vulnerability. This finding suggests that the association of general health status with Economic Resources overrides the above-described expected effect that poor health could have on economic strain by increasing the need for additional financial resources.

For the variables operationalizing social participation, the ambiguity found in the previous analysis is not reversed: the effects of restaurant on the Objective Measure, and of trip on the Self-Assessed Measure are maintained even after controlling for levels of wealth. Clearly, for individuals who live on less than CHF 2400 a month, the level of their savings does not affect their habit of eating out. Going to a restaurant or a coffee shop is a less expensive activity than taking a trip but one that is normally undertaken at a higher frequency and more indicative of everyday life. As few people would be willing to touch on savings on a regular basis for consumption habits, it makes sense that monthly income is more relevant for this activity than how much is on the bank account. However, taking into account wealth does explain differences in how often people go to the movies or see a theater play. Also, respondents who have difficulties in making ends meet are less likely to take a trip often (OR 1.7) – independent of how much money they have put aside. The variable taking a trip of at least 1 day encompasses trips of varied duration and destination and consequently of varied cost. According to our findings, this occasional expenditure is more easily subject to cancellation when respondents are finding it difficult to make ends meet than regular and less expensive social enjoyments such as going to a restaurant.

Reviewing the results found in models 3 and 4 we can affirm that there are significant differences between the two vulnerable population groups according to their Financial Needs and Expectations. However, contrary to our expectations, the sensitivity of the Self-Assessed Measure by reflecting increased needs for additional finances has not been conformed. Rather, the variable wealth outperformed any other covariate in predicting the self-assessed difficulties in making ends meet. This strong effect erases any differences in financial needs due to health problems

[4]As mentioned previously, a possible explanation for this is a structural risk linked to the fact that self-employed individuals are exempt from the otherwise mandatory occupational pension plan.

or life style expectations or so called 'expensive taste'. The exception to this is the variable trip, which represents an infrequent expenditure of considerable size, which is spent significantly less frequently by individuals who self-assess themselves as economically vulnerable. In contrast, the predictive power of any other variables operationalizing Financial Needs and Expectations is generally stronger for the objective outcome measure: even after controlling for wealth, former socio-professional status retains its significant effect on the Objective Measure. Moreover, monthly income below CHF 2400 is predicted by less frequent visits to restaurants and coffee shops. Lastly, the Pseudo R^2 must be mentioned: according to Mc Fadden's R^2 the overall fit of model 4 is superior for the Self-Assessed as compared to Objective Measure (0.241 and 0.198), a result that again must be attributed to the predominant role of wealth in explaining the self-assessed indicator.

Chapter 22
A Structural Equation Model for Self-Assessed Economic Vulnerability

We will now proceed to a confirmatory type of analysis using structural equation modeling (SEM) in order to examine our research question Iv): Is there statistical evidence that Financial Needs and Expectations play a mediating role between Economic Resources and the Self-Assessed Measure? Among various approaches that exist in moving from theory to a structural equation model, the one proposed here has been referred to as 'model generating' (Schumacker and Lomax 2004, p.73): if the initially estimated theoretical model fails to fit the data well, we will consult modification indices with the objective of finding a model that is rooted in theory and that fits the data well. Three general types of criteria will guide this specification search: global measures of model fit, the statistical significance of parameter estimates and the magnitude (and expected direction) of parameter estimates. The question of model fit being less straightforward for SEM compared to other multivariate techniques, several different measures of fit will be reported and discussed.

So far we have mainly been interested in comparing the Objective and the Self-Assessed Measures of economic vulnerability with regard to different variable sets we had previously identified in our theoretical model as possibly relevant in view of understanding and comparing what the two measures were exactly capturing. For this reason, an important variable for explaining the Self-Assessed Measure was omitted: income.[1] In this chapter, we are turning our attention exclusively to the Self-Assessed Measure with the objective of finding the model that best explains why individuals self-assess themselves as economically vulnerable, this time taking into account the categorical variable income. The other variables that have proven statistical significance in the regression models are wealth, for the model component Economic Resources, and health and social participation (trip, cinema, restaurant) for the variable block Financial Needs and Expectations. The last

[1] As previously explicated, it was not possible to include income into the regression models because this variable constitutes the basis on which the Objective Measure is built.

J. Henke, *Revisiting Economic Vulnerability in Old Age*, Life Course Research and Social Policies 11, https://doi.org/10.1007/978-3-030-36323-9_22

Table 22.1 Correlation coefficients between observed variables used in the model

Observed Variables	1	2	3	4	5	6	7
1. sa_ev	1.0	–	–	–	–	–	–
2. Income	−0.4	1.0	–	–	–	–	–
3. Wealth	−0.4	0.4	1.0	–	–	–	–
4. Restaurant	−0.1	0.2	0.1	1.0	–	–	–
5. Trip	−0.2	0.2	0.2	0.2	1.0	–	–
6. Cinema	−0.1	0.3	0.2	0.2	0.3	1.0	
7. Health	−0.2	0.1	0.2	0.1	0.2	0.2	1.0

socio-professional category, though significant in the regression models, will not be used because it is a nominal-level variable that is not suitable for SEM-techniques. The predictive power of Background Characteristics having been very limited in those regression models that included wealth, we did not deem it necessary to account for age, sex, marital status and education here.[2] Table 22.1 shows a correlation table of all variables utilized in the SEM models.

We chose to use the default method using maximum likelihood estimation and listwise deletion.[3] This method is considered robust even in the case of slight violations of normality (Acock 2013). However, because several of our variables diverge considerably from the assumption of normality (trip, cinema, health), the standard errors and confidence intervals of model components containing these variables were examined separately. Specifically, the latent measure Social Participation (SocPart, with the reflective indicators trip, cinema and restaurant) was estimated with a bootstrap method. Both, standard errors and confidence intervals were identical, with the exception of a single deviation of 0.01 for the variable cinema (standard error = 0.18 when estimated with maximum likelihood method, and 0.19 when estimated with bootstrap procedure). We thus concluded that it is safe to proceed with the default procedure. Figure 22.1 depicts an adaptation of the general theoretical model introduced in Part I. Rectangular boxes are observed variables and elliptical shapes are latent variables. An arrow indicates a hypothesized path and ε symbolizes an error term.

Since this chapter focuses on explaining the Self-Assessed Measure, the model uses the ordinal version of the variable income, with five categories instead of the binary version of the Objective Measure. In addition to the direct influence of monthly income, the Self-Assessed Measure is predicted by wealth. The variable health is expected to mediate the effect of low Economic Resources (EcoRes) on the Self-Assessed Measure: the lower the health status, the stronger the effect of low

[2]The general assumption that Background Characteristics are not relevant in models that integrate the variable wealth was tested for the model shown in Fig. 20.2. The standardized beta coefficients of the model controlling for age, sex, marital status and education only varied by maximum 0.03 from the original model, confirming the choice of the original, more parsimonious model.

[3]With listwise delition, the entire record of a given respondent is excluded from analysis if any single value is missing.

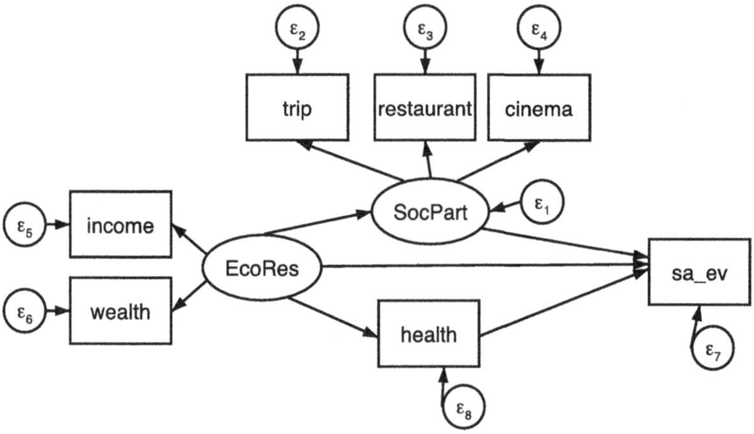

Fig. 22.1 Hypothesized structural equation model for predicting the Self-Assessed Measure of economic vulnerability (sa_ev)

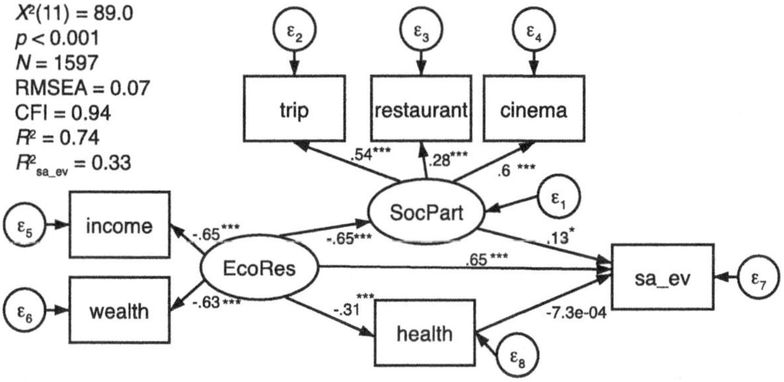

Fig. 22.2 Initial model: the relationship between Economic Resources, Financial Needs and Expectations and the Self-Assessed Measure of economic vulnerability (standardized coefficients; $*** \; p < 0.001$, $* \; p < 0.05$)

income on the Self-Assessed Measure because poor health is expected to increase the need for additional financial resources. Health is expected to be predicted by EcoRes, although reverse causation cannot be excluded: it is possible that persistently low levels of health in the past may have led to lower levels of income and wealth at the time of measurement. The latent variable social participation (SocPart) is hypothesized to mediate the effect of low levels of income on the Self-Assessed Measure: the more frequently respondents engage in social activities (reflective indicators: restaurant, trip, cinema), the stronger the effect of low levels of Economic Resources on the Self-Assessed Measure. Figure 22.2 shows the standardized[4] coefficients and fit statistics of the initial model.

[4]The standardized solution rescales all variables – observed and latent – to have a variance of 1.0.

The measurement part of the model has moderate to strong loadings: for the standardized solution, all estimates for EcoRes and SocPart are between 0.54 and 0.65, except the variable restaurant (0.28). The loadings for both latent variables are statistically highly significant at $p < 0.001$.

For the structural part of the model, we recognize the – by now well known – effect of low levels of income and wealth (EcoRes) on the Self-Assessed Measure: $\beta = 0.65$, $p < 0.001$. Low levels of EcoRes also has a moderate negative effect on health: $\beta = -0.31$, $p < 0.001$ and it has a strong negative effect on SocPart: $\beta = -0.65$, $p < 0.001$. The path coefficient from health to the Self-Assessed Measure is, however, small and not statistically significant: $\beta = 0.00$, $p = 0.978$ and the one from SocPart to the Self-Assessed Measure is rather small but significant: $\beta = -0.13$, $p = 0.047$.

The goodness-of-fit post-estimation at the equation level indicates that the model was able to explain 32.9% of the variance in the Self-Assessed Measure of economic vulnerability and 42.8% in the endogenous latent variable SocPart. At the model level, three indicators were considered: the chi-squared statistic of $\chi^2(11) = 89.0$, $p < 0.001$ tells us that this model significantly fails at fully reproducing the original covariance matrix. The root mean squared error of approximation (RMSEA[5]) of 0.07 reveals a reasonably close, though not a good fit (Browne and Cudeck 1992). The standardized root mean squared residual (SRMR[6]) of 0.04 is clearly below the cut-off point of 0.08 and the comparative fit index (CFI[7]) of 0.94 is just below the recommended threshold of 0.95 (Hu and Bentler 1999).

In order to assess possible ways for improving the initial model, we asked for modification indices, which estimate of how much we can reduce the chi-squared for our model if we free individual parameters. The modification index for the covariance of health and SocPart stands out as being high (52.4). It would conceptually make sense for health to predict SocPart, reflecting the idea that – at any given level of EcoRes – a higher general health status results in more frequent social participation. Also, we decide to remove the reflective indicator restaurant because of its minor contribution to the latent variable SocPart. The new model has more difficulties in converging[8] and the output indicates that the log-likelihood function is essentially flat in a particular part of the model.[9] Upon closer inspection it seems that the latent variable EcoRes is under-identified because the two items income and wealth have a close to identical loading. Constraining the unstandardized path estimates of

[5] The RMSEA is a goodness-of-fit indicator that penalizes the model for unnecessary added complexity by measuring how much error there is for each degree of freedom. It is recommended that the RMSEA be less than 0.05 for a good fit and less than 0.10 for an acceptable fit.

[6] The SRMR reports how close the model comes to reproducing each correlation, on average.

[7] The CFI compares the estimated model with a null model, indicating how much better the estimated model fits the data.

[8] It took 15 instead of the previous 6 iterations.

[9] Stata indicates this situation with the report ‚not concave'.

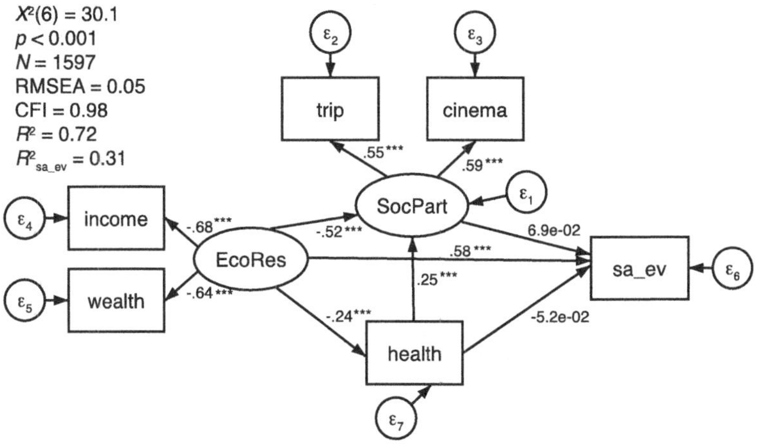

Fig. 22.3 Improved model: the relationship between Economic Resources, Financial Needs and Expectations and the Self-Assessed Measure (standardized coefficients; *** p < 0.001)

income and wealth to load equally on EcoRes resolves the problem.[10] Figure 22.3 shows the improved model.

The newly added path from health to SocPart is highly significant with $\beta = 0.25$, $p < 0.001$. The decreased effect of EcoRes on SocPart ($\beta = -0.52$, $p < 0.001$) suggests that part of the effect of EcoRes on SocPart is in fact mediated by health.[11] More importantly for our research question is the erased effect of SocPart on the Self-Assessed Measure: after accounting for an effect of health on social participation, the frequency of engaging in social activities ceases to be significant for explaining self-assessed economic vulnerability. Table 22.2 shows the standardized and unstandardized coefficients with p-values.

At the equation level, the model does slightly less well at explaining the variance in the Self-Assessed Measure of economic vulnerability (31%) and SocPart (40%). At the model level, however, the goodness-of-fit statistics indicate a net improvement: though the chi-squared statistic of $\chi^2(10) = 29.6$, $p < 0.001$ still fails at fully accounting for the original covariance matrix, the RMSEA of 0.05 is good and the SRMR of 0.02 is very good. Lastly, a CFI of 0.98 confirms that this model fits the data very well.

Diverging results of fit indices are not unusual in SEM (Hooper 2008). The poor model fit according to χ^2 may have several possible explanations: according to Hu and Bentler, a significant goodness-of-fit of the χ^2 value may be the result of 'model misspecification, power of the test or violation of some technical assumptions underlying the estimation method' (Hu and Bentler 1995, p.78). Previous research

[10] The number of iteration decreased to 9.

[11] Since this particular effect is not the object of our research, we refrain from discussing in detail the direct and indirect effects of economic resources on social participation, which is, however, important for social policy interventions. (See Baeriswyl, Marie, 2016).

Table 22.2 Standardized and unstandardized coefficients for the final model

Outcome	Standardized coefficients	Unstandardized coefficients	P-value
Structural part			
sa_ev			
Health - > sa_ev	−0.05	−0.03	0.076
SocPart - > sa_ev	0.07	0.06	0.240
EcoRes - > sa_ev	0.58	1 (fixed)	p < 0.001
Health			
EcoRes - > health	−0.24	−0.74	p < 0.001
SocPart			
Health - > SocPart	0.25	0.18	p < 0.001
EcoRes - > SocPart	−0.52	−1.11	p < 0.001
Measurement part			
SocPart			
SocPart <– trip	0.55	1 (fixed)	p < 0.001
SocPart <– cinema	0.59	1.23	p < 0.001
EcoRes			
EcoRes <– income	−0.68	−4.28	p < 0.001
EcoRes <– wealth	−0.64	−4.28	p < 0.001

has found that large sample sizes very often lead to a significant χ^2 even if the model is correctly specified (Bentler and Bonett 1980; Hooper 2008). Since our sample size is relatively large, a small difference between the sample covariance matrix and the fitted model could have provoked the rejection of the specified model. Another, more severe reason could be the violation of the multivariate normality assumption (McIntosh 2007). We had previously examined this risk for the latent construct SocPart by estimating it with the bootstrap procedure and had come to the conclusion that it was not a problem (p.180). An estimation of the entire model with a maximum likelihood technique that does not assume normality[12] confirms that our variables do not deviate from normality to the extent that lead to biased estimates because the parameter estimates proved to be identical to those yielded by the maximum likelihood method. We conclude that the significant χ^2 is not the result of violating the normality assumption. At last, it cannot be excluded that the less than satisfactory result of the chi-squared statistic is due to the fact that parts of the model are miss-specified: the lack of statistical significance of several of the path estimates (the ones leading from SocPart and health to the Self-Assessed Measure) is likely to penalize the overall model fit. This indication must be taken all the more seriously as the chi-squared statistic is the only test of significance that assesses the theoretical model (Schumacker and Lomax 2004).

Despite the satisfactory fit of several test statistics, the main conclusion we must draw from these results with regard to our research question is that, based on the

[12] The Stata command for this option is method(ml) vce(robust). It uses the Huber-White sandwich estimator for the variance-covariance matrix (Acock 2013, p.15).

variables utilized in this model, we cannot affirm that Financial Needs and Expectations play a mediating role between Economic Resources and the Self-Assessed Measure of economic vulnerability: there is no evidence for a statistically significant effect of health on the difficulties in making ends meet and, though our results indicate that a higher frequency of engaging in various social activities causes people to experience more economic strain at any given level of Economic Resources, the effect is small and of little practical significance. Rather, what stands out is the predominant importance of wealth and income for predicting whether a person finds it difficult or easy to make ends meet, suggesting a very real relationship between the objective and the self-assessed angle of measuring economic vulnerability.

In the next chapter, we will narrow our focus again and pursue an adaptation of the same research question but instead of considering income and wealth (EcoRes), we will assess the potential of the Objective Measure for playing the role of a proxy indicator for EcoRes.

References

Acock, A. C. (2013). *Discovering structural equation modeling using stata*. College Station, TX: Stata Press.

Bentler, P. M., & Bonett, D. G. (1980). Significance tests and goodness of fit in the analysis of covariance structures. *Psychological Bulletin, 88*(3), 588.

Browne, M. W., & Cudeck, R. (1992). Alternative ways of assessing model fit. *Sociological Methods & Research, 21*(2), 230–258.

Hooper, D. C. (2008). Structural equation modelling: Guidelines for determining model fit. *The Electronic Journal of Business Research Methods, 6*(1), 53–60.

Hu, L.-T., & Bentler, P. M. (1995). Evaluating model fit. In R. H. Hoyle (Hrsg.), *Structural equation modeling: Concepts, issues, and applications* (S. 76–99). Thousand Oaks: Sage.

Hu, L., & Bentler, P. M. (1999). Cutoff criteria for fit indexes in covariance structure analysis: Conventional criteria versus new alternatives. *Structural Equation Modeling: A Multidisciplinary Journal, 6*(1), 1–55.

McIntosh, C. N. (2007). Rethinking fit assessment in structural equation modelling: A commentary and elaboration on Barrett (2007). *Personality and Individual Differences, 42*(5), 859–867.

Schumacker, R. E., & Lomax, R. G. (2004). *A beginner's guide to structural equation modeling*. Psychology Press.

Chapter 23
Modeling the Relationship Between the Objective and the Self-Assessed Measure

The last research question with regard to the Self-Assessed Measure concerns its relationship with the Objective Measures of economic vulnerability and specifically, whether the model component Financial Needs and Expectations helps explain the incongruence between the Objective Measure and the Self-Assessed Measure of economic vulnerability by playing a mediating role (research question I vi).

As we have seen in the literature as well as in our preliminary analysis of the overlap of the measures (Chap. 12), there is only a very limited congruence between the monetary poverty line and the self-assessed difficulties in making ends meet. The lack of coherence between these measures has theoretically been explained by the mechanisms that emanate from the concept of 'reference utility'. Conceptually, the lack of congruence of the two measures is the result of the fact that the two measures focus on two different aspects of low levels of economic quality of life, though both make reference to monthly income: the Objective Measure registers the lack of a minimum level of monthly income, while the Self-Assessed Measure is supposed to take into consideration any type of resources that is available for current use, including income, and assess whether in their totality these resources allow meeting the day-to-day needs. Empirically, our results corroborate previous research in finding that the Self-Assessed Measure also captures availability of stock in form of wealth. The focus of this chapter is to find the most parsimonious model representing the relationship between the Self-Assessed and the Objective Measure. Given the great popularity of the Objective Measure in poverty research, it is of interest to assess the extent to which the absolute monetary poverty line is able to replace a more complex construct combining income and wealth. In the (observed) case of incongruence between the two, the attempt will be to find the monetary threshold that provides the best model fit by changing the threshold of the binary income categories underlying the Objective Measure.

Figure 23.1 shows the standardized coefficients and goodness of fit statistics of the initial model with the Objective Measure set at the usual threshold of a monthly

J. Henke, *Revisiting Economic Vulnerability in Old Age*, Life Course Research and Social Policies 11, https://doi.org/10.1007/978-3-030-36323-9_23

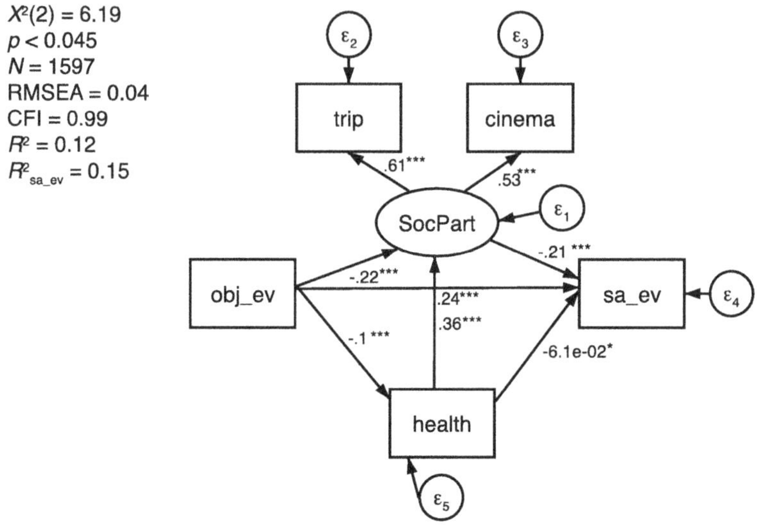

Fig. 23.1 The relationship between the Objective Measure and the Self-Assessed Measure of economic vulnerability, mediated by social participation and health. (Standardized coefficients; $***$ $p < 0.001$, $*$ $p < 0.05$)

income of less than CHF 2400. As for the previous models, a maximum likelihood estimation was used.

In comparison to the model with the latent construct EcoRes (Fig. 22.3) one of the first things that stands out is the substantially lower effect of the Objective Measure on the Self-Assessed Measure: $\beta = 0.24, p < 0.001$ (compared to $\beta = 0.58$). Not surprisingly, the monetary indicator obj_ev also performs less in predicting Financial Needs and Expectations: for health, the coefficient is $\beta = -0.1, p < 0.001$ (compared to $\beta = -0.31$) and for social participation $\beta = -0.\ 22, p < 0.001$ (compared to $\beta = -0.52$). On average, the effect sizes in the model based on the monetary poverty line (obj_ev) is cut in half compared to the model based on EcoRes.

In contrast to the model based on EcoRes, the path coefficients from Financial Needs and Expectations to the Self-Assessed Measure are statistically significant. For social participation (SocPart) we find $\beta = -0.21$, $p < 0.001$, which is almost equally strong as the direct effect of the Objective Measure. For health, the path to the Self-Assessed Measure is only weakly significant with $\beta = -0.06$, $p = 0.030$. Since there is indication for a partial mediation of obj_ev on sa_ev via SocPart and health, it is worthwhile decomposing and aggregating the effect of these path coefficients. Table 23.1 shows the direct, indirect and total effects of the structural part of the model.

The proportion of the direct effects is clearly dominant, except in the case of the effect of health on the Self-Assessed Measure: here, the indirect effect via SocPart represents close to 60% of the total effect, though only the direct effect is statistically significant. The effect of the Objective Measure on the Self-Assessed Measure is to 80% of a direct nature, summing up to $\beta = 0.30$, $p < 0.001$. The statistically

Table 23.1 Standardized direct, indirect and total effects

Outcome	Direct effect	Indirect effect	Total effect
sa_ev			
SocPart	−0.22***	–	−0.22***
health	−0.06*	−0.08	−0.14**
obj_ev	0.24***	0.06	0.30***
Health			
obj_ev	−0.10***	–	−0.10***
SocPart			
health	0.36***	–	0.36***
obj_ev	−0.22	−0.04***	−0.26***

The significance levels shown here are for the unstandardized solution. *** $p < 0.001$, ** $p < 0.01$, * $p < 0.05$

most significant mediation is at work in the path going from the Objective Measure to social participation via health. However, the effect size is very small ($\beta = -0.04$), suggesting that the negative effect of income poverty on social participation is mainly related to the constrained purchasing power, though its negative effect on health ($\beta = -0.10$) may also be the result of it being an impediment for going out frequently.

The evaluation of goodness-of-fit of this model presents us with a paradox: when looking at the post-estimation statistics for the equation level we find that the model was only able to explain 12% of the overall variance. Compared to the model integrating income and wealth (EcoRes), which was able to account for 33% in the variance of the Self-Assessed Measure, these results are very poor. As for the variance in the Self-Assessed Measure of economic vulnerability, this model only explained 15%, compared to 31% of the previous model. This low R^2 is obviously explained by the large proportion of respondents who are either vulnerable by the Objective Measure but not by the Self-Assessed, or vice versa. The question, whether the effect of Financial Needs and Expectations is different on individuals just above the poverty line will be pursued later in this chapter.

At the model level, however, all indicators coherently speak of a better fit: the chi-squared statistic comparing this model to a saturated model with no degrees of freedom of $\chi^2(2) = 6.19$, $p < 0.045$[1] indicates that the estimated solution comes closer at reproducing the original covariance matrix than EcoRes, though it just fails to be considered sufficient for rejecting the null hypothesis. The root mean squared error of approximation (RMSEA[2]) of 0.04 represents a very good value, especially

[1]A non-significant χ^2 indicates that the observed and the estimated variance-covariance matrices are similar and that the theoretical model significantly reproduces the sample variance-covariance matrix.

[2]The RMSEA is a goodness-of-fit indicator that penalizes the model for unnecessary added complexity by measuring how much error there is for each degree of freedom. It is recommended that the RMSEA be less than 0.05 for a good fit and less than 0.10 for an acceptable fit.

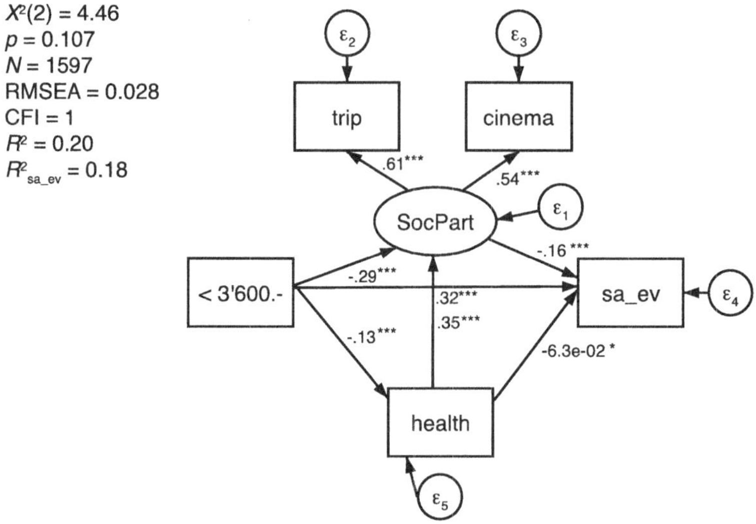

Fig. 23.2 The relationship between an Objective Measure with modified cut point (monthly income less than 3600.-) and the Self-Assessed Measure of economic vulnerability, mediated by social participation and health. (Standardized coefficients; *** $p < 0.001$, * $p < 0.05$)

considering that RMSEA tends to improve with increasing the number of variables in the model. The comparative fit index (CFI[3]) of 0.99 is also excellent.

As we have seen in Part II, there exists an important number of individuals who are above the poverty line but who are self-assessing themselves as economically vulnerable. On both, theoretical and empirical grounds, we could expect the model estimated above to have an even better fit if the threshold of the Objective Measure of economic vulnerability was changed to encompass the income category just above the poverty line: income less than CHF 3600 per month. Individuals whose income has been very low during most of their lives may have adapted their life style, in particular by reducing social participation, in order to minimalize the experience of economic strain. In contrast, individuals who have had to readjust their budgets after retirement – even though their income level was maintained above the poverty line – are likely to be more conscious of what they are missing. Figure 23.2 shows the same model as in Fig. 23.1 except for the change in the cut point of the Objective Measure, which now opposes monthly income below CHF 3600 (1) to income above CHF 3600 (0).

The most notable change compared to the previous model is the path coefficient between the Objective Measure (income below CHF 3600) and SocPart, which increases from $\beta = -0.22$ to $\beta = -0.29$. Also, the direct effect of the Objective on the Self-Assessed Measure is strengthened ($\beta = -0.32$ compared to $\beta = -0.24$). The

[3]The CFI compares the estimated model with a null model, indicating how much better the estimated model fits the data. The recommended cutoff is 0.95.

direct effect of income on health is also augmented (β = −0.13 compared to β = −0.10) when the threshold of obj_ev is set higher.

The overall model fit and the R^2 of the Self-Assessed Measure are improved significantly: the non-significant chi-squared statistic ($\chi^2(2) = 4.46$, p = 0.107) indicates that the observed and the estimated variance-covariance matrices are similar. The values of the RMSEA and the CFI, which were both very good before, are even better now and the shares of variance explained by the model and specifically for the Self-Assessed Measure have increased as well, though they clearly remain below the values obtained in the model based on EcoRes.

These findings confirm that there exist a significant number of individuals in our sample whose income is above the poverty line but who are nonetheless having difficulties in making ends meet, an experience of economic strain that is partly mediated by the frequency of their engagement in social activities. In fact, the effect of low income on social participation and health seems to be harder to bear for the low-middle income group than for the lowest income group (Seifert and Pilgrim 2009).

In order to further test this trend, a third model was estimated with an even higher threshold for the Objective Measure at income category 'less than 4800' (Appendix Fig. 23.1). The path coefficient of the Objective Measure on the Self-Assessed Measure dropped by half (β = 0.15). The effect of this dichotomization of income on health, too, showed a weakened effect (β = −0.10) while the negative direct effects of the Objective Measure on SocPart, and SocPart on the Self-Assessed Measure were slightly strengthened (β = −0.32, β = −0.21). The overall model fit deteriorated significantly; only the value of CFI remained at an acceptable level.

Integrating the results of these three models (Fig. 23.3) we see that effect of the strength of association between the objective angle of measurement and the self-assessed difficulties in making ends meet for this sample population peaks some-

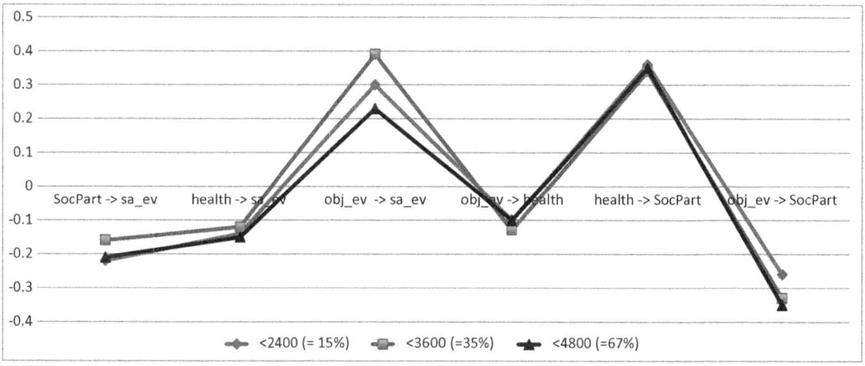

Fig. 23.3 Total effects[a] of three models based on different thresholds of the Objective Measure. (% sample population)

[a] The total effects were thought to adequately summarize the differences because the indirect effects only varied by 0.01 between the models. The indirect and direct effects are shown in Appendix Tables 23.1 and 23.2

where between a monthly income of CHF 2400 and 3600. For the three income thresholds considered, the effect of social participation is strongest for the highest income category though the difference between income 'less than 3600' and income 'less than 4800' is minimal.

This chapter started out with the goal of finding the model that best describes the relationship between the Objective and the Self-Assessed Measure. While it was known from preliminary analysis that there is an important incongruence between these two measures, in particular for respondents whose income lies below or just above the poverty line, the interest of changing the threshold for the income-based Objective Measure was primarily in the expected change of effect size with social participation and health, representing the model component Financial Needs and Expectations. The comparison of three models with Objective Measures based on different income-thresholds shows that the effect of income on social participation increases as the threshold for objective economic vulnerability increases, though the effect seems to level off for income levels higher than CHF 4800 per month, where more income does no longer contribute to increasing the frequency of social participation.

From the results in this analysis we cannot confirm that the model component Financial Needs and Expectations plays a mediating role between the Objective Measure and the Self-Assessed Measure of economic vulnerability. The proposed model reveals that the effect of the Objective Measure on the Self-Assessed Measure is more complex and that the mediation occurs in fact *among* the variables of the model block Financial Needs and Expectations: the degree to which finances (obj_ev) affect the frequency of social participation is influenced by the general health status. Again, we see the dominant effect of monetary resources on the self-assessed difficulties of making ends meet. What we can affirm based on this analysis with regard to the relationship between the Objective Measure and the Self-Assessed Measure of economic vulnerability is that the poverty line at 2400.- is an inadequate proxy for capturing low levels of economic quality of life because it fails to represent the totality of relevant resources that determine economic quality of life. We will now turn to a last step in our analysis of the Self-Assessed Measure of economic vulnerability where we will consider the added value of an outcome variable that combines the objective and the self-assessed outcomes in form of a Vulnerability Typology.

Appendix

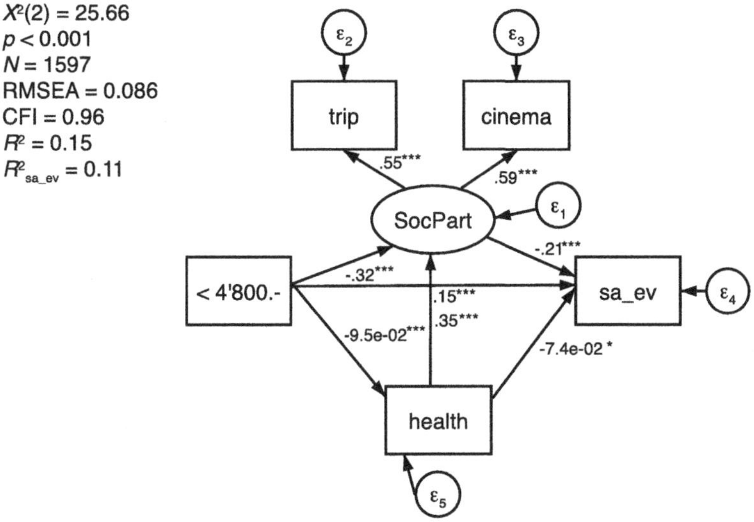

Appendix Fig. 23.1 The mediation of social participation and health relationship between a modified Objective Measure (monthly income less than 4800.-) and the Self-Assessed Measure of economic vulnerability. (Standardized coefficients; ∗∗∗ p<0.001, ∗ p<0.05)

Appendix Table 23.1 Standardized direct, indirect and total effects of the model shown in Fig. 23.2 with the Objective Measure defined as income below CHF 3'600

Outcome	Direct effect	Indirect effect	Total effect
sa_ev			
SocPart	−0.16∗∗∗	–	−0.16∗∗∗
health	−0.06∗	−0.06	−0.12∗
obj_ev	0.32∗∗∗	0.06	0.39∗∗∗
Health			
obj_ev	−0.13∗∗∗	–	−0.13∗∗∗
SocPart			
health	0.34∗∗∗	–	0.34∗∗∗
obj_ev	−0.29	−0.04∗∗∗	−0.33∗∗∗

The significance levels shown here are for the unstandardized solution. ∗∗∗ $p < 0.001$, ∗∗ $p < 0.01$, ∗ $p < 0.05$

Appendix Table 23.2 Standardized direct, indirect and total effects of the model (no figure shown) with the Objective Measure defined as income below CHF 4'800

Outcome	Direct effect	Indirect effect	Total effect
sa_ev			
SocPart	−0.21***	–	−0.21***
health	−0.07*	−0.07	−0.15*
obj_ev	0.15***	0.08	0.23***
Health			
obj_ev	−0.10***	–	−0.10***
SocPart			
health	0.35***	–	0.35***
obj_ev	−0.32	−0.03***	−0.35***

The significance levels shown here are for the unstandardized solution. *** $p < 0.001$, ** $p < 0.01$, * $p < 0.05$

Reference

Seifert, K., & Pilgram, A. (2009). *Vivre avec peu de moyens. La pauvreté des personnes âgées en Suisse*. Zürich: Edition Pro Senectute.

Chapter 24
A Typology of Economic Vulnerability Combining the Objective and the Self-Assessed Measure

So far, the focus of our analysis of the Self-Assessed Measure of economic vulnerability has been on its relationship with the objective measurement angle, specifically its dependence on material resources and in a second step, its relationship with the most commonly used objective indicator, an absolute monetary poverty line. This chapter will address another research objective of this volume, the assessment of a Vulnerability Typology consisting of a combination of three measurement angles. In the following chapter, we will take a closer look at the four types that emerge at the intersection between Objective Measure and the Self-Assessed Measure; at the end of Part V, the combination of the Self-Assessed and the Perceived Measure will be discussed.

Let us begin by restating what we concluded from comparing the Self-Assessed and the Objective Measure. Our results so far cause us to question the relevance of using an income-based cut-point for distinguishing between 'vulnerable' and 'non vulnerable' among Swiss pensioners: the range of individual experiences below and above the threshold of CHF 2400 monthly income challenge the legitimacy of this specific 'convention of equivalence between heterogeneous observations', that is, if the results are to be meaningful from a quality of life perspective. Our findings also reveal that the Self-Assessed Measure is above all predicted by the presence of (objectively measurable) economic resources and that this association is statistically and practically more significant than the association between economic resources (especially wealth) and the Objective Measure. This observations lead us to conclude that the Self-Assessed Measure, based on the survey question 'do you have difficulties in making ends meet given your monthly income' in fact provides a more accurate picture of the overall financial situation, outperforming the Objective Measure in terms of identifying the population groups that are economically vulnerable.

In this chapter, we are going to examine the added value of combining Objective and Self-Assessed Measures for identifying those individuals who experience the lowest levels of economic quality of life. The typology that emerges at the intersection of the two measurement angles allows for a more nuanced profiling of

J. Henke, *Revisiting Economic Vulnerability in Old Age*, Life Course Research and Social Policies 11, https://doi.org/10.1007/978-3-030-36323-9_24

population groups who suffer from low levels of economic quality of life; moreover, it will shed light on differentials in human agency in the context of economic vulnerability.

As recapitulated in Table 24.1, each possible combination of the Objective and the Self-Assessed Measure reflects a different chain of latent risk factors, providing insight into the theoretical probability that coping strategies have been applied and whether they have been successful. We have to keep in mind that our outcome measures, income poverty and difficulties in making ends meet, are an operationalization of *manifest* vulnerability outcomes, which means that the coping process is already underway and is captured in the outcomes. For the self-assessed measurement angle this can lead to a seemingly paradoxical situation: according to our theoretical framework, individuals may self-assess themselves as economically vulnerable even if a coping strategy has proven successful (and they have managed to get above the objective threshold of vulnerability) – simply because they experience their financial situation as strenuous. This case is best exemplified in the (high proportion of) recipients of supplementary benefits who are above the poverty line and who self-assess themselves as economically vulnerable.

Among the possible risk constellations, two (3 and 4b) lead to the exact same theoretical outcome combination of vulnerability by both a self-assessed and by an objective indicator. Let us briefly review these two risk constellations as laid out in in Part I: risk constellation 3 represents individuals who are striving to overcome a lack of resources but who are not successful because they lack coping resources or because their coping strategy is not sustainable. Respondents found in risk group 4b have chosen not to act on their self-assessed economic vulnerability – probably because they are not suffering psychologically from the scarcity of means, they esteem it as a lesser priority.

Another ambiguity in the manifest vulnerability outcomes that could not be deduced from the theoretical framework but was revealed by the empirical analysis pertains to risk constellation 4a. This outcome constellation, vulnerable by an Objective Measure but not by the Self-Assessed Measure could – as alternative to the proposed risk chain – be the result of additional resources that are not taken into account in the Objective Measure but are reflected nonetheless in the Self-Assessed Measure. In the theoretical part we explained the differential scope of the two

Table 24.1 Risk constellations underlying congruence and incongruence between Objective and Self-Assessed Measures of economic vulnerability

Figu Group of respondents	*latent vulnerability – risks*		*manifest vulnerability – outcomes*	
Risk constellation	**exposure + threat**	**coping**	**objective**	**self-assessed**
1	↑	not applicable	non-vulnerable	non-vulnerable
2	↓	↑ success	non-vulnerable	vulnerable
3	↓	↑ absent, insufficient	vulnerable	vulnerable
4 a)	↓	↓ do not realize	vulnerable	non-vulnerable
4 b)	↓	↓ do not want to mobilize coping	vulnerable	vulnerable

Table 24.2 Frequency of the combinations of the Objective and the Self-Assessed Measure

Combination of objective and self-assessed economic vulnerability	Freq.	%
BB (objective = 1/self-assessed = 1)	95	5.9
BA (objective = 1/self-assessed = 0)	147	9.1
AB (objective = 0/self-assessed = 1)	136	8.4
AA (objective = 0/self-assessed = 0)	1247	76.7
Total	1625	100.0

measures (Figure 4.1), the Self-Assessed Measure encompassing the resources as well as the needs side of the equation. The operationalization of the Self-Assessed Measure based on the survey question 'would you say that your household has difficulties in making ends meet, given your *current monthly income*', the type of resources that should have entered into the assessment has been named explicitly as monthly income, providing a strong basis for comparison with the Objective Measure. Our empirical analysis, however, clearly showed that besides income, financial stocks also had a significant effect on self-reported economic vulnerability. We therefore need to consider the possibility of stocks influencing the economic self-assessment as an alternative to risk chain 4a in explaining the incongruence between the Objective and the Self-Assessed Measure.

Having refreshed these scenarios, we now proceed to the empirical analysis of the Vulnerability Typology. For this purpose we created a variable that combines the two binary measures of economic vulnerability, yielding four possible outcomes, shown in Table 24.2. As previously explained, the matrix of manifest outcomes is abbreviated as follows: the letter A stands for 'not vulnerable' and B stands for 'vulnerable'. When comparing Tables 24.2 to 24.1 we easily identify the groups (1 = AA, 2 = AB, 3 = BB, 4a = BA, 4b = BB).

In order to test our hypothesis based on these theoretical groups we will estimate a series of ordinal regression models (ORM) that successively take variables of Economic Resources and Financial Needs and Expectations as outcomes, while controlling for Background Characteristics. The locus of interest will be the outcome probabilities[1] of each of the four vulnerability types, which are integrated in the model as an independent variable measured at the nominal level. Having removed[2] all cases with missing values on any of the dependent and independent variables used in this and subsequent nested models, we can be sure that all models are based on the exact same sample (N = 1562) and thus comparable.

[1] For the interpretation of the ordinal regression models, outcome probabilities will be discussed; a table reporting the odds ratios is shown in Appendix 24.1.

[2] Sixty-three observations or 3.9% of the sample was dropped.

24.1 Identifying the Most Vulnerable According to Low Levels of Wealth

The first hypothesis to be tested concerns the ability of the Vulnerability Typology to identify the most vulnerable group in terms of financial assets. In Table 24.1, the most vulnerable groups are represented by risk constellation group 3 and 4b, corresponding to vulnerability type BB, which makes up 6% of the sample population (Table 24.2). In a first ORM, the ordinal variable wealth is taken as dependent variable. We posit that type BB has the highest probability among the four vulnerability types to be found in the lowest wealth category, holding Background Characteristics constant. Furthermore, comparing outcome probabilities of types BA and AB, we will be able to verify whether, in the case of incongruence between the Objective and the Self-Assessed Measure, the Self-Assessed Measure performs better than the Objective Measure at capturing low levels of economic resources: we posit that the risk of having 'no or almost no wealth' is higher for members of group AB compared to group BA.

Fig. 24.1 shows the outcome probabilities of vulnerability types for the dependent variable 'wealth', controlling for age, sex, canton, education and marital status. In order to facilitate the visibility of trends, the dots (indicating the probabilities for a given vulnerability type) are connected by a line.

As Fig. 24.1 shows, the most vulnerable type in terms of low levels of wealth is type AB and not as expected type BB. The difference in the probability outcomes between the two is small and primarily concerns the probability of owning 'almost nothing or nothing' (0.33 for AB and 0.28 for BB) while there is a strong convergence for the higher classes of wealth.

The difference in wealth distribution between types BA and AB is striking: type AB records the just mentioned – highest – outcome probability of being in the lowest category of wealth. Moreover, the probability trends between the two types is not parallel but offset: running in opposite directions, type BA ascends from a low

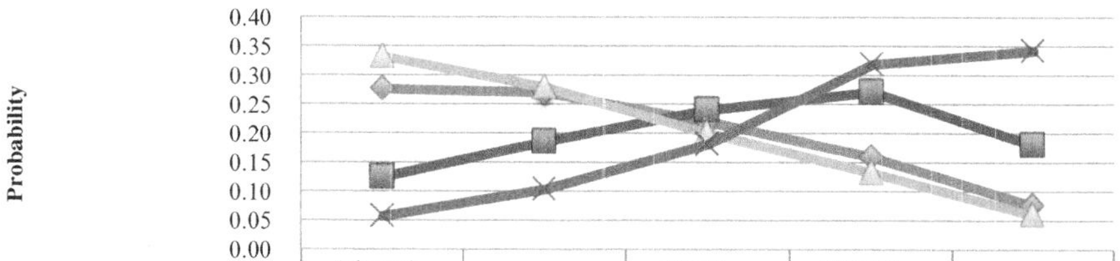

	(almost) nothing	< 60'000	60'000 - 150'000	150'000-500'000	>500'000
BB (obj = 1 / sa = 1)	0.28	0.27	0.22	0.16	0.08
BA (obj = 1 / sa = 0)	0.12	0.19	0.24	0.27	0.18
AB (obj = 0 / sa = 1)	0.33	0.28	0.20	0.13	0.06
AA (obj = 0 / sa = 0)	0.06	0.10	0.18	0.32	0.34

Fig. 24.1 Outcome probabilities of vulnerability types for the dependent variable wealth, controlling for Background Characteristics

outcome probability to be found in wealth category 'almost noting' (0.12) to the highest probability of owning between '150,000–500,000' (0.27), thus sharing more resemblance with the probability distribution of non-vulnerable type AA than with that of type AB. The magnitude of the difference in outcome probabilities between types BA and AB merits to be dwelled on for a moment. The unexpected combination of lowest income with mid-level wealth categories in the case of type AB suggests considerable intra-group heterogeneity. It prompts us to estimate a model that splits up the probability distributions by 'last professional category'. There is one professional category that we expect to be more frequent among group BA compared to group AB: individuals whose last professional status was 'self-employed' are more likely to have a lower monthly income, first of all because during the time of their self-employment they were not legally constrained to pay into an occupational pension scheme. Also, individuals who used to be self-employed are more likely to have withdrawn capital from occupational pensions in order to invest it in their business. While in the Swiss context, self-employment is known to be a risk factor for old age poverty, many of those who have been successful in their business are likely to be in category BA. Figure 24.2 shows the distribution of the last pre-retirement socio-professional category by vulnerability type.

The self-employed are strongly over-represented among type BB (22% compared to sample average of 11%). Comparing the two vulnerability types AB and BA we find an above average rate of formerly self-employed among group BA (18%), which is more than double the rate recorded among group AB (7%): individuals who used to be self-employed make up one of the socio-professional groups that is frequently classified as vulnerable by an objective standard but not by self-assessment. We proceed to estimating the same ORM as bcforc, but instead of the

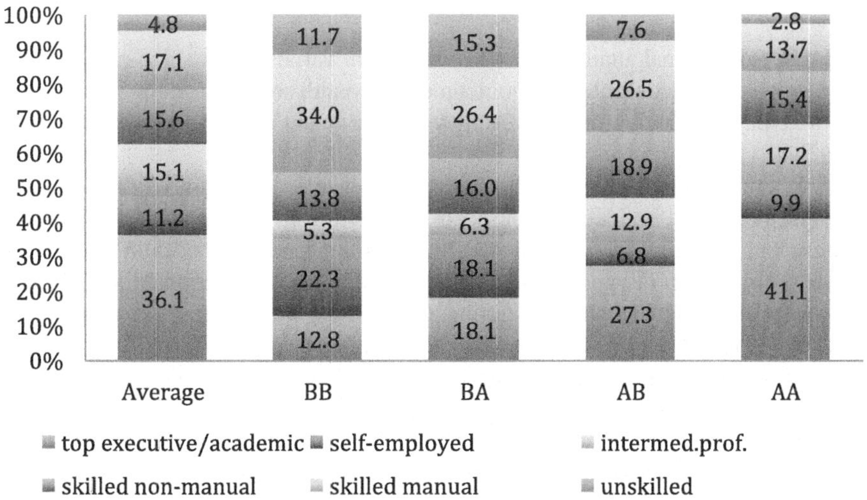

Fig. 24.2 Distribution of last pre-retirement socio-professional category (SPC) by vulnerability type

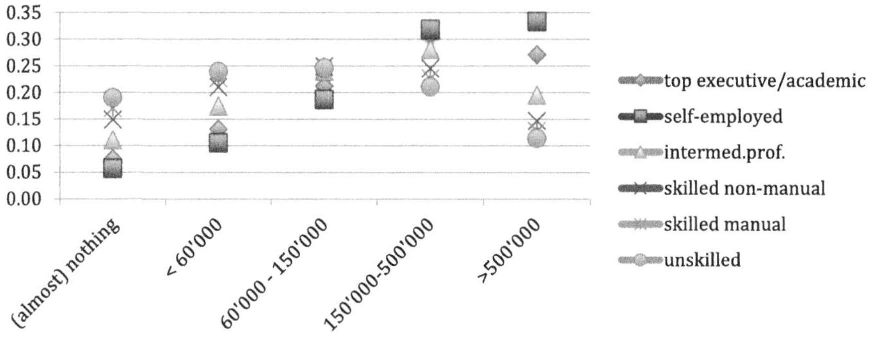

Fig. 24.3 Outcome probabilities of vulnerability type BA by socio-professional group for the dependent variable wealth, controlling for Background Characteristics

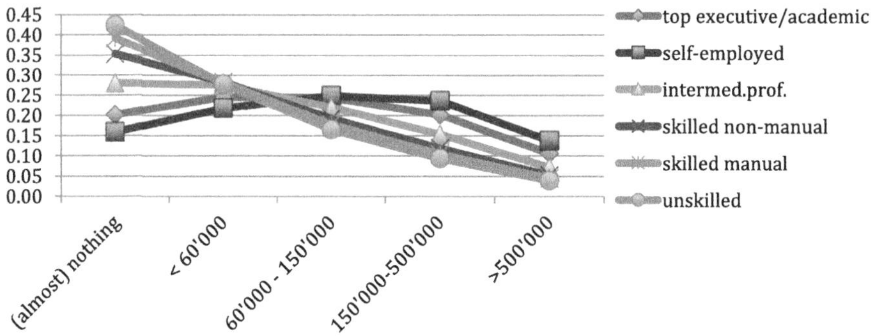

Fig. 24.4 Outcome probabilities of vulnerability type AB by socio-professional group for the dependent variable wealth, controlling for Background Characteristics

variable 'educational attainment' the independent variable 'last socio-professional category' is included. Probability-outcomes for wealth are assessed for both vulnerability types separately: for type BA, they are shown in Fig. 24.3 and for type AB in Fig. 24.4.

For both vulnerability types, the outcome probability of owning '(almost) nothing' is plotted in an identical descending order from lowest to highest socio-professional category. However, as wealth levels increase, the probability trends for the two groups run in opposite directions. Let us consider the probability distribution curve of the self-employed, starting at the lowest wealth category: among type BA, the probability of not owning any assets is 0.06, compared to 0.16 for type AB. This trend continues for wealth category below 60,000 (0.11 for BA versus 0.22 for AB). At the higher end of the wealth spectrum, the difference is greatest: the probability of formerly self-employed retirees of type BA to be in the highest wealth category (500,000 and more) is 0.33 compared to 0.14 for type AB. Together with the socio-professional group 'top executives and academics', those who were formerly self-employed and to a lesser degree the – very small fraction of – intermediate professions represent something like a 'group within the group', accounting to a large degree for the disjunction in the correlation between low income with low levels of wealth observed in vulnerability type AB.

The findings presented in this chapter highlights the benefits of measuring economic vulnerability in a way that integrates Objective and Self-Assessed Measures: in view of the outcome probabilities for wealth categories, we can claim with some degree of certainty that in terms of overall economic resources, individuals of type AB are more vulnerable than individuals of type BA. An income-based poverty line would have come to the opposite conclusion.

24.2 'Expensive Taste' or 'Downward Adaptation'?

One of the main reservations that have been brought forth against the use of self-reported economic vulnerability hast to do with the adaptation processes the may cause these measures to be influenced by either high expectations concerning an acceptable living standard (expensive taste) or on the contrary, by an adaptation to the precarious situation, reflected in a self-assessment that would be considered too optimistic from an outsider perspective (downward adaptation).

The model component Financial Needs and Expectations feature three variables based on the frequency of engaging in social activities that come with a financial cost. Our results thus far suggest that in the Swiss context, retirees' social engagement depends primarily on individual preferences and only to a small degree on the availability of financial resources. Still, there remains the question whether part of the discrepancy between the Objective and the Self-Assessed Measurement angles may be due to different aspirations concerning the level of economic quality of life that is deemed to be 'minimal'. As our findings indicate, it is likely that the heterogeneity within each group cancels out some of the effects of social participation on the two vulnerability outcomes. To take the example of the population group that is vulnerable according to the Self-Assessed Measure: it comprises both individuals who have been living a very modest life style with very little social participation as well as individuals who are experiencing financial strain precisely because they previously enjoyed a life style of frequent social and cultural activities that they are no longer able to afford. A combination of the measurement angles in form of a typology does not allow drawing a clear-cut line between these two groups but it nonetheless allows moving towards a more nuanced profiling of economically vulnerable groups.

Let us briefly remember the main hypothesis concerning social participation and economic vulnerability outcomes: as visualized in Table 24.1, members of group 2 (type AB) are thought to be committed to increasing or maintaining the current income level because of their experience of economic strain. One of the strategies they may employ is to adjust to a smaller budget by decreasing the frequency of social participation. Group 4a (type BA) theoretically represents the opposite case: individuals who have gotten used to living modestly and among whom the habit of going out to see a movie, a theater play or taking a trip is probably not very common and of a lower frequency than for the average sample population. However, our analysis of the probability-outcomes for wealth by vulnerability group revealed that about one third of this group is in fact well-off, if wealth is taken into account. This intra-group polarization must be kept in mind. Group 3 (type BB) is expected to be the most vulnerable group because of the double burden of low income and economic strain. Members of this group are expected to record the lowest frequency of social participation among the four groups.

When holding Background Characteristics constant, the difference in social participation should give us a clue as to the successfulness of coping, given that the fulfillment of social roles and conventions as well as the question of social exclusion are thought to be important factor contributing to the experience of economic strain (Kahn and Pearlin 2006). In the first ordinal regression model (ORM), the categorical variable restaurant is taken as a dependent variable. The frequency of going out to eat has been used as a measure for assessing living standards because it is considered a luxury of 'universal appeal' and thus reflects long-term strategy of managing a given household budget (Angelini et al. 2013). We expect type BB to have the highest probability among the four vulnerability types to be found in the lowest categories of social participation ('never' or at 'least once a year' in the case of going out to a restaurant). Furthermore, we expect BA to have a relatively higher probability of being in the lowest categories of social participation compared to type AB because members of type BA are expected to have embraced a modest life style already before retirement while type AB may partly experience strain due to higher life style expectations. The response pattern of type AB is expected to be most similar to type AA.

Figure 24.5 shows the outcome probabilities for each of the four vulnerability types for the dependent variable 'frequency of going out to a restaurant or a coffee shop' while controlling for Background Characteristics. Our analysis confirms that the probability distribution of 'never going out to a restaurant or a coffee shop' is the highest among group BB – members of this group are double as likely to 'never' eat in a restaurant or go to a coffee shop (0.17), or only 'once a year' (0.14), compared to non-vulnerable group AA (0.08, 0.08). In terms of this social activity, our hypothesis is confirmed that group BA behaves very similar to group BB while group AB shows a similar response pattern as the non-vulnerable group AA. In order to rule out the possibility that different levels of wealth associated with each vulnerability type might interfere, the same model was re-estimated controlling for wealth. The probability outcomes remained unaffected (results not shown). Building on the results found in the previous chapter, the outcome probabilities for type BA were estimated for each socio-professional group separately (results are shown in Appendix 24.2). The variations in outcome probabilities vary only minimally (within a range of 0.03 or less between the extremes of top-executives/academics and unskilled workers) indicating that the frequency pattern of this particular activity does not differ significantly among different social classes. The same absence of any specific trend was found for group AB (results not shown).

Compared to the variable restaurant, the variable trip reveals interesting differences in terms of the outcome probabilities of vulnerability types (Fig. 24.6).

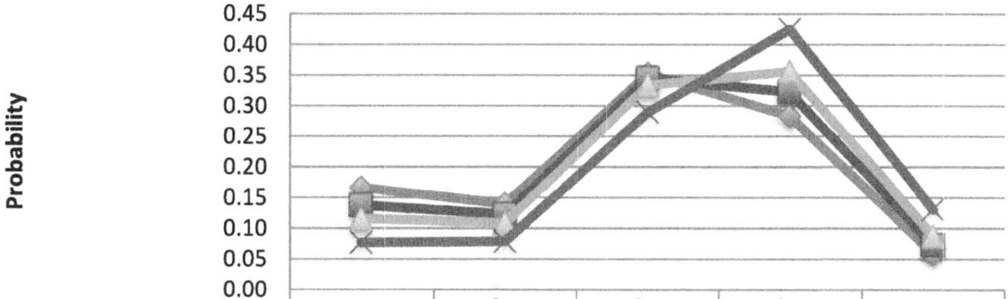

	never	at least once a year	at least once a month	at least once a week	every day
BB (obj = 1 / sa = 1)	0.17	0.14	0.35	0.28	0.06
BA (obj = 1 / sa = 0)	0.14	0.12	0.35	0.32	0.07
AB (obj = 0 / sa = 1)	0.12	0.11	0.33	0.36	0.09
AA (obj = 0 / sa = 0)	0.08	0.08	0.29	0.42	0.13

Fig. 24.5 Outcome probabilities of vulnerability types for the dependent variable 'frequency of going to a restaurant or coffee shop', controlling for Background Characteristics

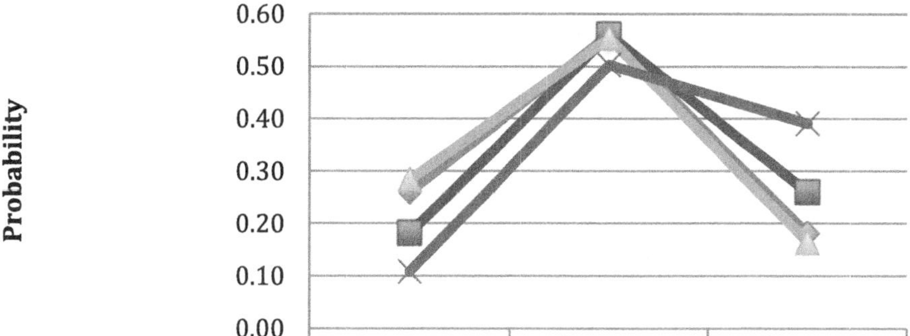

	never	at least once a year	at least once a month
BB (obj = 1 / sa = 1)	0.26	0.56	0.18
BA (obj = 1 / sa = 0)	0.18	0.56	0.26
AB (obj = 0 / sa = 1)	0.28	0.55	0.16
AA (obj = 0 / sa = 0)	0.11	0.50	0.39

Fig. 24.6 Outcome probabilities of vulnerability types for the dependent variable 'frequency of taking a trip', controlling for Background Characteristics

Generally speaking, the response pattern of group AB is basically identical to that of group BB. Members of group AB and BB record about the same likelihood of *never* taking a trip (0.28, 0.26), which is more than twice as high compared to type AA (0.11).

As before, we suspect that there are sub-groups within type BA that are characterized by different levels of socioeconomic standing. Controlling for wealth does not affect the probability outcomes (the fluctuation is within a maximum range of 0.03). However, when group BA is split up into socio-professional categories, there is a clear difference in the probability of never taking a trip between top executives/academics (0.15) and unskilled workers (0.27) the rate of the latter being equivalent to that of type BB and AB (Appendix 24.3). In contrast to going out for a meal or a drink, taking a trip seems to represent a 'luxury' that those who have limited financial means cannot afford.

Turning to the third variable operationalizing social participation, cinema, we find that the probability outcomes of the frequency of 'going out to see a movie or a theater play' (Fig. 24.7) is not any different between the three vulnerable groups: types BB, BA (0.40) and AB (0.43) are close to the same and only to be distinguished from group AA (0.24).

The bi-polar trend opposing the majority group AA to all the others suggests that for this variable, more important than degrees of economic vulnerability (be it objective or self-assessed) is a social habitus of interest and enjoyment of cultural activities that is more common among the economically well-off. Put differently: it may very well be that the threshold, above which degrees of participation in this activity would become distinguishable is higher than either of the thresholds represented by the Objective and the Self-Assessed Measure. The question whether this variable is more indicative of taste than of financial power is legitimate. On the other hand, based on our theoretical model, the scenario of a difference between

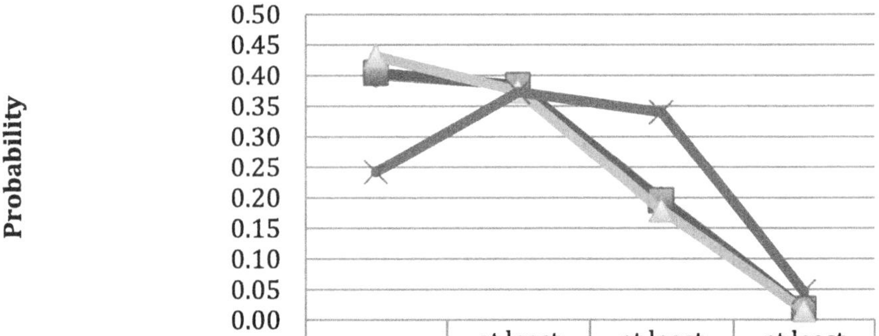

	never	at least once a year	at least once a month	at least once a week
BB (obj = 1 / sa = 1)	0.40	0.38	0.20	0.02
BA (obj = 1 / sa = 0)	0.40	0.38	0.19	0.02
AB (obj = 0 / sa = 1)	0.43	0.37	0.18	0.02
AA (obj = 0 / sa = 0)	0.24	0.37	0.34	0.05

Fig. 24.7 Outcome probabilities of vulnerability types for the dependent variable 'frequency of going out to see a movie or a theater play', controlling for Background Characteristics

type AB and type BA was to be expected and the rejection of this hypothesis allows us to refine our understanding of the Self-Assessed Measure: positing – as we did – that type BA is likely to have a higher probability of being in the category with the lowest frequency of social participation compared to type AB is tantamount to saying that the Self-Assessed Measure is prone to be biased as a result of 'expensive taste'. The underlying assumption is that income below the poverty line (BA) is a strong indicator of a low pre-retirement income and consequently members of type BA are expected to have adjusted their life style (including cultural activities) to a low budget. In contrast, type AB by definition includes a wider range of income levels and, as explained earlier, our model suggests that members of this group may voice financial discontent (difficulties in making ends meet) precisely because they no longer have the means to lavish expenditures on 'luxuries' such as cultural pleasures. The finding that the probability outcomes of type AB are *not* lower than those of type BA therefore weakens those reservations against the usage of the Self-Assessed Measure that state that it is easily distorted by 'too' high expectations and standards. Thus, the rejection of this specific hypothesis contributes to consolidating the evidence that the Self-Assessed Measure is a robust approximation of low levels of economic quality of life.

The self-assessed general health status was also used for operationalizing Financial Needs and Expectations, representing the potential need for additional financial resources to be able to maintain a given level of quality of life (Fig. 24.8). While the most vulnerable type BB is expected to have the highest probability outcome for 'poor or very poor' health, the hypothesis about which one of the two vulnerability types, AB or BA, tends to have a higher rate of 'poor health' is less obvious. In wealthy countries, subjective indicators of socioeconomic status have a stronger association with poor health, a tendency that is even more pronounced in later life (Arber et al. 2014), which speaks in favor of group AB as the most nega-

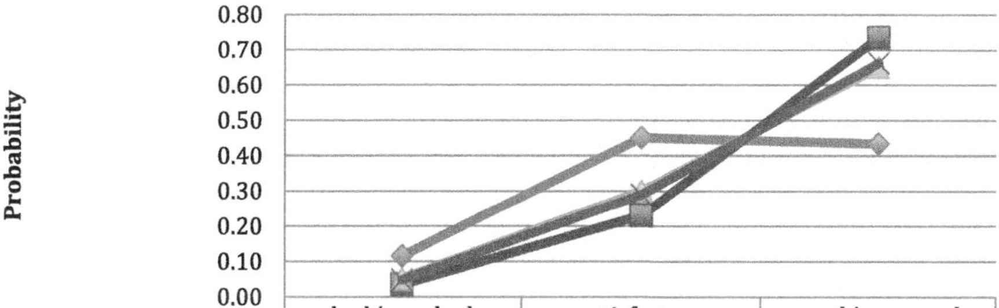

	bad/very bad	satisfactory	good/very good
	self-assessed general health status		
BB (obj = 1 / sa = 1)	0.12	0.45	0.43
BA (obj = 1 / sa = 0)	0.03	0.23	0.73
AB (obj = 0 / sa = 1)	0.05	0.30	0.65
AA (obj = 0 / sa = 0)	0.05	0.29	0.66

Fig. 24.8 Outcome probabilities of vulnerability types for the dependent variable 'self-assessed general health status', controlling for Background Characteristics

tively affected group. The theoretical risk chain model, too, suggest that group AB is likely to fare worse on average than group BA because if an income poor individual was really ill and struggling with exploding health costs, she or he would most likely self-assess themselves as 'difficulties in making ends meet', which is equivalent to category BB. On the other hand, many epidemiological studies have found low income as a marker of access to health-relevant resources across the life course (Newman and Cauley 2012), which would speak in favor or type BA as the group with a relatively higher rate of individuals with 'poor/very poor' health.

Figure 24.8 shows the results of an ORM model for the dependent variable 'self-assessed general health status', providing the outcome probabilities of all four vulnerability types while controlling for Background Characteristics.

As predicted, members of group BB have the highest probability of self-assessing their general health status as either bad and very bad (0.3), or as satisfactory (0.45), compared to any of the other groups. There is a clear difference between groups BA and AB in terms of health risk: while the average health of type AB is virtually identical to type AA, type BA fares significantly better than any of the types, even better than AA. The significantly higher risk for (very) poor health among group BB – by a factor of 4 compared to type BA – speaks a loud word in favor of combining the Objective and the Self-Assessed Measure.

Appendix

Appendix 24.1: Ordinal Regression Models with the Combination of Objective and the Self-Assessed Measure (Obj-Sa) as Independent Variable, Predicting Items of Financial Need and Expectations (Odds Ratios Are Reversed)

Outcome ->	Cinema		restaurant		trip		health	
Independent Variable (Ref.)	Rev. OR	P > |z|	Rev. OR	P > |z|	Rev. OR	P > |z|	Rev. OR	P > |z|
Sex (male)	1.55	<0.001	0.48	<0.001	0.85	0.141	0.92	0.499
Age	1.44	<0.001	1.28	<0.001	1.43	<0.001	1.37	<0.001
Canton (Basel)								
Geneva	0.83	0.214	0.94	0.655	1.80	<0.001	1.05	0.756
Valais	2.04	0.000	0.95	0.739	2.14	<0.001	1.03	0.844
Berne	1.32	0.049	1.01	0.919	0.80	0.135	0.82	0.226
Ticino	1.68	0.001	1.05	0.780	2.09	<0.001	1.09	0.624
Marital status (married)								
Single	0.86	0.446	0.52	<0.001	1.55	0.027	1.20	0.374
Divorced/sep.	0.88	0.429	0.74	0.062	1.06	0.723	1.23	0.233
Widowed	1.12	0.425	0.88	0.332	1.39	0.017	1.07	0.644

(continued)

Outcome ->	Cinema		restaurant		trip		health	
Independent Variable (Ref.)	Rev. OR	P > |z|	Rev. OR	P > |z|	Rev. OR	P > |z|	Rev. OR	P > |z|
Educational attainment (primary school)								
Lower secondary	2.92	<0.001	0.67	0.106	1.24	0.408	1.19	0.502
Prof. Training	1.84	<0.001	0.73	0.073	0.83	0.308	0.72	0.074
High school	1.22	0.198	0.74	0.120	0.65	0.033	0.52	0.002
University of Appl. Sciences	0.97	0.857	0.96	0.844	0.65	0.034	0.59	0.014
University	0.1	<0.001	0.97	0.880	0.60	0.016	0.45	<0.001
Obj-Sa (ref = AA; obj = 0/sa = 0)								
BB (obj = 1/sa = 1)	2.49	<0.001	2.30	<0.001	2.47	<0.001	2.22	<0.001
BA (obj = 1/sa = 0)	2.50	<0.001	1.84	<0.001	1.54	0.014	1.24	0.247
AB (obj = 0/sa = 1)	2.41	<0.001	1.48	0.028	2.91	<0.001	1.92	<0.001
McFadden's R^2	*0.066*		0.03		*0.071*		*0.046*	

Appendix 24.2: Outcome Probabilities of Members of Vulnerability Type BA for Being in a Given Category of 'Frequency of Going to a Restaurant or Coffee Shop', by Last Socio-professional Category

	Frequency of going to a restaurant or coffee shop				
Last pre-retirement SPC	Never	At least once a year	At least once a month	At least once a week	Every day
Top executive/academic	0.13	0.12	0.34	0.33	0.08
Self-employed	0.15	0.13	0.35	0.31	0.07
Intermed.profession	0.14	0.12	0.34	0.32	0.07
Skilled non-manual	0.13	0.12	0.34	0.33	0.08
Skilled manual	0.14	0.13	0.35	0.31	0.07
Unskilled	0.16	0.13	0.35	0.30	0.06

Appendix 24.3: Outcome Probabilities of Members of Vulnerability Type BA for Being in a Given Category of 'Frequency of Taking a Trip of at Least One Day', by Last Socio-professional Category

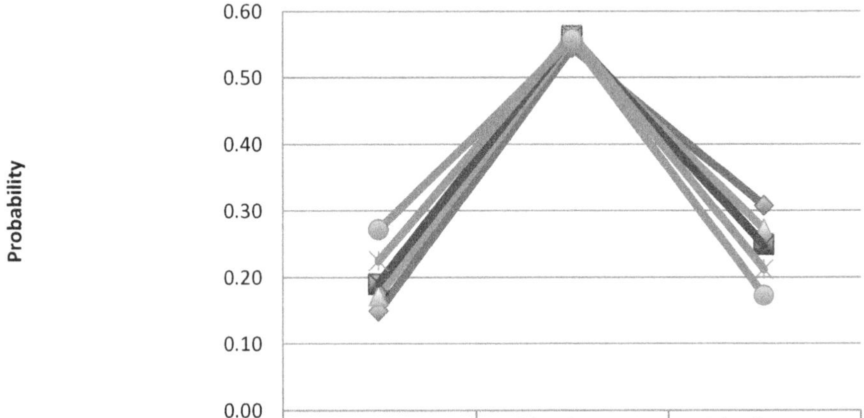

	never	at least once a year	at least once a month
top executive/academic	0.15	0.54	0.31
self-employed	0.19	0.56	0.25
intermed.prof.	0.17	0.56	0.27
skilled non-manual	0.19	0.56	0.25
skilled manual	0.22	0.56	0.21
unskilled	0.27	0.56	0.17

References

Angelini, V., Weber, G., & Brugiavini, A. (2013). Household consumption and the crisis. In A. Börsch-Supan, M. Brandt, H. Litwin, & G. Weber (Hrsg.), *Active ageing and solidarity between generations in Europe: First results from SHARE after the economic crisis* (S. 19–26). Berlin/Boston: Walter de Gruyter.

Arber, S., Fenn, K., & Meadows, R. (2014). Subjective financial well-being, income and health inequalities in mid and later life in Britain. *Social Science & Medicine, 100*, 12–20.

Kahn, J. R., & Pearlin, L. I. (2006). Financial strain over the life course and health among older adults. *Journal of Health and Social Behavior, 47*(1), 17–31.

Newman, A., & Cauley, J. A. (2012). *The epidemiology of aging*. New York: Springer.

Chapter 25
Conclusion Part IV

The systematic comparison of the two economically vulnerable population groups identified by the Objective and the Self-Assessed Measure resulted in the following key findings: while the two groups have a very similar composition in terms of age, sex, marital status and canton of origin, there exist significant differences when Economic Resources and Financial Needs and Expectations are taken into account. Vulnerability as defined by the difficulties in making ends meet (sa_ev) is more strongly associated with low levels of wealth than is the absolute monetary poverty line (obj_ev). With regard to the relationship between Economic Resources and the Self-Assessed Measure, we found that the effect of income on the difficulties in making ends meet is partly moderated by wealth: the lower the level of income, the stronger the effect of wealth on the probability of being economically vulnerable by the Self-Assessed Measure. However, in terms of reproducing the covariance matrix, the most accurate way of modeling the relationship between the objective and the self-assessed angle of measure is not a mediation model but a latent construct that combines income and wealth as reflective indicators. As far as the hypothesized sensitivity of the Self-Assessed Measure to variations in Financial Needs or Expectations is concerned, we found that the variable wealth overpowered any potential effect of health expenditures or 'expensive taste' in the form of socio-cultural habits. Only the variable 'trip of at least one day' seems to represent an activity that is undertaken significantly less frequently by individuals who self-assess themselves as economically vulnerable. In contrast, the frequency of visits to restaurants and coffee shops was negatively associated with the Objective Measure of economic vulnerability. It can only be speculated whether these habits reflect life-style adaptation to a small budget. It seems plausible that cutting the frequency of trips lends itself as a means of managing low liquidities, as it is at the same time more expensive and of a lower periodicity within a given year compared to the other two activities examined.

The strong influence of wealth on self-reports of economic vulnerability lends support to previous research: an Australian study on the relationship between income

J. Henke, *Revisiting Economic Vulnerability in Old Age*, Life Course Research and Social Policies 11, https://doi.org/10.1007/978-3-030-36323-9_25

poverty[1] and respondents' view of themselves as poor or very poor (subjective poverty) found that the judgment that someone is poor is more affected by wealth than by income (Marks 2007). A comparative research on the relationship between income and wealth in 12 European countries based on SHARE[2] data found the correlation between income and wealth among older adults to be less strong than is usually expected; in many countries, among them Switzerland, the relationship between income and wealth was found not to be linear in the older segment of the population (Skopek et al. 2012): according to this study, Switzerland belongs to the group of countries that exhibit a substantially greater inequality in terms of wealth than in terms of income. In the lowest income quartile, there is even a negative correlation between income and wealth, meaning that relatively higher levels of income are associated with relative lower levels of wealth and possibly debts. In line with these findings, a comparison of outcome probabilities of wealth levels between vulnerability type AB and BA poignantly showed that the share of individuals with no financial stocks is higher among those who are above the poverty line but who have difficulties in making ends meet than among those who are income poor but who find it easy to make ends meet. A break-down of the wealth-category-probability of vulnerability type BA by socio-professional group revealed a 'group within the group' composed by an above-average share of former self-employed individuals and a smaller-than-average share of former top executives and academics. These two groups, who together constitute a good third of the respondents within type BA, distinguish themselves from the other socio-professional groups in the same vulnerability type by an increased probability of scoring in the middle and higher wealth categories. Our findings suggest that these professional groups account to a large degree for the disjunction in the correlation between income and wealth among older adults in Switzerland.

The second guiding question of this part pertained to the role played by Financial Needs and Expectations in mediating the effect of Economic Resources on the Self-Assessed Measure, or, more specifically, mediating the effect of the Objective Measure on the Self-Assessed Measure. In both instances, the proposed structural equation models provided a very good fit to the data but clearly showed that the direct effect of Economic Resources (respectively, of the Objective Measure) on the Self-Assessed Measure is by far stronger than the indirect effect via health and items of social participation. In the context of our research, the variable self-assessed general health status seems to be more telling as an indicator of long-term socio-economic standing throughout the life course,[3] a feature that turned out to be particularly relevant in the analysis of the vulnerability types.

[1] Income poverty was defined as living in a household with an income of less than 50% of median equivalised disposable household income.

[2] Survey of Health, Ageing and Retirement in Europe. It is multidisciplinary and cross-national panel database of micro data on individuals aged 50 and older.

[3] That is, as compared to conceptualizing health as a factor that increases the need for additional financial resources, which is the second way that this variable has been considered in our analysis.

In summary, we see that the population groups identified by the Objective and the Self-Assessed Measure tend to be affected by distinct experiences of economic vulnerability. The distinctiveness of the experience of economic vulnerability stands out most clearly when the measurement angles are considered jointly: the part concluded with an application of the Vulnerability Typology, which successfully identified the most vulnerable sub-population with regard to low levels of wealth and low frequencies of social participation. In the next part, the psychosocial dimension of economic vulnerability will be the locus of our interest as we assess the difference between the Self-Assessed and the Perceived Measure.

References

Marks, G. N. (2007). Social Policy Research Paper No. 29. Income poverty, subjective poverty and financial stress. Australian Government. Department of Families, Community Services and Indigenous Affairs.

Skopek, N., Kolb, K., Buchholz, S., & Blossfeld, H.-P. (2012). Einkommensreich – vermögensarm? Die Zusammensetzung von Vermögen und die Bedeutung einzelner Vermögenskomponenten im europäischen Vergleich. *Berliner Journal für Soziologie, 22*(2), 163–187.

Part V
The Perceived Measure of Economic Vulnerability

The overarching research question of this part is dedicated to the possibility and the relevance of distinguishing between two types of subjective measures of economic vulnerability, the Perceived Measure and the Self-Assessed Measure. By 'possibility' we mean whether the observed differences in population characteristics are statistically significant; the response to the question of 'relevance' is less straightforward. We will offer elements of discussion based on a quality of life perspective at the end of this part and in the final conclusions in Part V.

The interest in attempting to distinguish between the two measures was triggered by the observation that two people who report the same level of economic strain (Self-Assessed Measure) do not necessarily suffer of it the same way when judging by the degree of stress (Perceived Measure) they report. The figure below shows an extract of our theoretical model, depicting the direct influence of economic strain on economic stress and its indirect influence via changes in the self-concept or role strains, the model component Psychosocial Symptoms and Consequences of economic vulnerability. The variable set Psychosocial Symptoms and Consequences is expected to mediate the effect of economic strain on economic stress, while exhibiting a relatively stronger association with the Perceived Measure than with the Self-Assessed Measure. Thus, the hypothesis is that inter-individual differences in the variables mastery, frequency of visits from friends, the worry about not receiving enough recognition and self-worth contribute to explaining why some individuals who are reporting difficulties in making ends meet are not worried about their lack of finances.

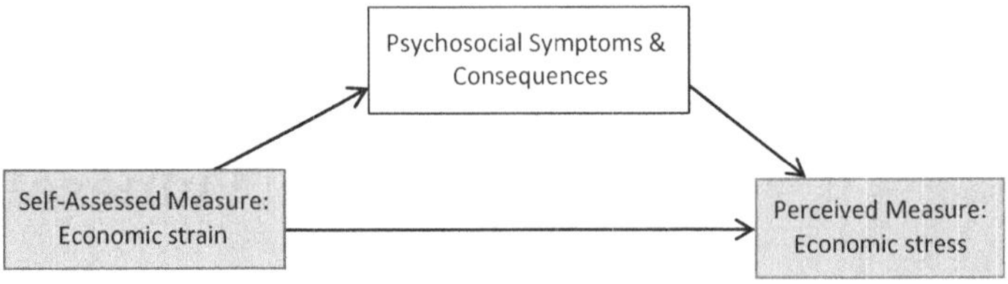

Theoretical model of the relationship between the Self-Assessed and the Perceived Measure of economic vulnerability

As outlined earlier, our theoretical model assumes that the causal pathways through the interaction between multiple risk factors occurred prior to the measurement of economic vulnerability and can therefore only be deduced indirectly. The fact that we consider the manifest outcome 'the end' of the vulnerability cycle implies that we are looking at a chronic type of economic vulnerability where the primary stressor of a diminished income has been met with coping strategies (if available and desirable), both, those aimed at changing the situation and those that merely modify the subjective meaning of the situation. Therefore, psychosocial variables such as mastery, too, reflect the success or the personal 'cost' these strategies have incurred for the individual. Thus, our objective here is not the analysis of the economic stress process but to empirically assess the potential of distinguishing between the more cognitive subjectivity of self-assessment from the more affective subjectivity of perception in view of a more nuanced profiling economic vulnerability in wealthy countries like Switzerland.

Analogous to Part IV, we will first explore the differences between the two subjective outcome measures using binary logistic regression models.[1] The technique of structural equation modeling will be employed to test those hypotheses that pertain to the relationship between the model components and lastly, ordinary regression models are estimated for estimating psychosocial probability outcomes.

The part is organized in chapters according to the hypotheses stated in Part II, Chap. 7. First, an initial regression model is estimated in order to assess the difference between the two population groups that are vulnerable according to the Self-Assessed Measure and according to the Perceived Measure with regard to their Background Characteristics. The model is then extended to include the variable sets Psychosocial Symptoms and Consequences while controlling for Economic Resources. For the testing of the hypothesis that Psychosocial Symptoms and Consequences play a mediating role between the Self-Assessed and the Perceived Measure, a structural equation model[1] will be fitted in Chap. 27. We conclude this part with a path model of the relationship between the three measurement angles, and we take a closer look at the two psychosocial variables that will turn out to be most important, mastery and recognition, this time through the lenses of the extended Vulnerability Typology based on all three measurement angles combined.

[1] Chapter 8 provides a detailed description of the statistical techniques used in this chapter.

Chapter 26
Regression Analysis

26.1 Regressing Background Characteristics on the Perceived Measure of Economic Vulnerability

The objective of this chapter is to establish whether there exist significant differences between the two groups that are economically vulnerable according to the Self-Assessed Measure and the Perceived Measure with regard to Background Characteristics. These socio-demographic variables allow locating individuals within societal systems of stratification and social arrangements that may have an effect on both, the self-assessment of economic strain and the experience of economic stress (Pearlin 1989).

Table 26.1 shows the odds ratios and corresponding confidence intervals of an initial model regressing the covariate set Background Characteristics on the Self-Assessed Measure (difficulties in making ends meet) and the Perceived Measure (worry about not having enough money for current expenses). After removing all cases with missing values on any of the predictors used in this and subsequent nested models, the estimation sample comprises 1'498 cases. The number of individuals classified as economically vulnerable is $n_{sa\_ev} = 199$ (14.4%) and $n_{perc\_ev} = 250$ (18.1%). The proportion of the population who is vulnerable according to both angles is 7.9%.

All socio-demographic variables – sex, age, canton[1] and marital status[2] – are significantly associated with perceiving oneself as economically vulnerable: women have higher odds than men (OR = 1.4), advanced age is associated with decreasing

[1] Using a maximum likelihood test, comparing the full model to a model where the indicators of the variable canton were constrained to zero. The result (LRX2 = 22.1, df = 4, p < 0.001) affirms that there is an overall effect of canton on perc_ev.

[2] We again tested the simultaneous effect of all indicators of marital status on perc_ev and rejected the hypothesis that all coefficients are simultaneously equal to zero (LRX2 = 17.3, df = 3, p < 0.001).

J. Henke, *Revisiting Economic Vulnerability in Old Age*, Life Course Research and Social Policies 11, https://doi.org/10.1007/978-3-030-36323-9_26

Table 26.1 Odds ratios and confidence intervals of Background Characteristics on the Self-Assessed and Perceived Measure of economic vulnerability

	Model 1: Background characteristics					
	sa_ev			perc_ev		
Variable (Ref.)	OR	[95% CI]		OR	[95% CI]	
Gender (male)	0.81	0.58	1.12	1.44*	1.06	1.95
Age	0.95	0.82	1.10	0.82**	0.72	0.94
Canton (Basel)						
Genève	1.65*	1.01	2.70	1.61*	1.03	2.52
Valais	1.62	0.98	2.68	1.42	0.89	2.25
Bern	0.92	0.55	1.54	1.04	0.65	1.64
Ticino	2.24**	1.37	3.68	2.59***	1.66	4.05
Marital status (married)						
Single	0.67	0.32	1.38	0.76	0.41	1.41
Divorced/sep.	3.45***	2.30	5.16	2.36****	1.59	3.50
Widowed	1.55*	1.04	2.33	1.49*	1.02	2.19
Educational attainment (primary school)						
Lower secondary	0.86	0.47	1.59	1.17	0.65	2.10
Prof. training	0.45*	0.28	0.71	0.43***	0.27	0.67
High school	0.32**	0.19	0.56	0.34***	0.20	0.57
University of appl. sciences	0.33**	0.19	0.60	0.49**	0.30	0.83
University	0.24***	0.13	0.45	0.25***	0.14	0.45
McFadden's R^2	*0.086*			*0.073*		

Key: *** < 0.001; ** < 0.01; * < 0.05

odds of worrying about not having enough money (OR 0.8). Respondents living in Ticino stand out as having by far the highest risk of financial worries (OR = 2.6), followed by Geneva (OR = 1.6), other background variables held constant.

The indicators of marital status recording a higher risk are the same for the Perceived as for the Self-Assessed Measure: compared to married couples, respondents whose current status is divorced/separated or widowed have a higher risk of worrying about not having enough money for current expenditures ($OR_{div./sep.} = 2.4$, $OR_{widow} = 1.5$).

Education has a clear protective effect against the propensity of worrying about finances: the odds of *not* perceiving oneself as economically vulnerable are by a factor of 3.9 higher for university graduates compared to the reference group who completed primary school. For respondents who have attended a university of applied sciences, high school or a professional training, the inversed odds are of a similar size: 2, 3 and 2.3, respectively. The overall effect sizes of the variable education on the Perceived Measure and the Self-Assessed Measure can thus be said to be very similar. Whether educational attainment level captures effects of economic status or whether it is more likely to reflect a sense of mastery of one's financial situation remains to be seen.

Comparing odds ratios and considering the overlap in confidence intervals of this initial model, the risk profile of the two subjectively vulnerable groups reveals striking similarities with regard to Background Characteristics. The most important nuance is that perceived economic vulnerability seems to have a rather younger and female 'face'. Another exception constitutes the large effect of living in the canton Ticino, though the confidence intervals for predicting the two outcomes do not overlap (sa_ev [1.37, 3.68], perc_ev [1.66, 4.05]). As will become clear in the next chapter, this difference is entirely accounted for by Economic Resources. Not surprisingly, Background Characteristics by themselves only explain a low share of the overall variance: McFadden's R^2 is 0.09 for predicting the Self-Assessed Measure and 0.07 for the Perceived Measure. Regarding our first research question on the difference between the two vulnerable groups in terms of their socio-demographic Background Characteristics we thus conclude that there are no statistically significant differences between the two population groups.

26.2 Regressing Psychosocial Consequences and Symptoms on the Perceived and the Self-Assessed Measure of Economic Vulnerability

In the extended model 2, variables operationalizing Psychosocial Consequences and Symptoms of economic vulnerability are added to the original model: the frequency of visiting friends, a self-worth scale, the worry about lack of recognition and a mastery scale. Recalling Pearlin's stress model, the presence of role strains and a diminished concept of self constitute secondary stressors that, together with the primary stressor (in this case, economic strain) provoke stress. In our model, they moreover reflect that individuals are not able to cope effectively with economic strain. Hence, we are expecting a greater effect size (odds ratios) of this set of covariates for the Perceived Measure compared to the Self-Assessed Measure.

Model 2 models of will be introduced simultaneously with two other models controlling for Economic Resources: Table 26.2 shows the odds ratios for each outcome measure of model 2 side by side with model 3, that includes income and model 4, to which wealth has been added.[3] In what follows we are going to jointly consider the effect of each covariate in models 2, 3 and 4 and compare these effects between the two outcome variables.

While there is no effect of sex on the Self-Assessed Measure in any of the models, the effect varies for the Perceived Measure: there is a statistically significant higher probability of worrying about not having enough money for women than for men, even after income (OR = 1.7, $p < 0.05$) and wealth (OR = 1.8, $p < 0.01$) are taken into account.

[3] The confidence intervals for each model can be found in the Appendix.

Table 26.2 Odds ratios and confidence intervals of Background Characteristics and Psychological Consequences and Symptoms on the Self-Assessed and the Perceived Measure

	Self-assessed measure			Perceived measure		
Variable (Ref.)	Model 2	Model 3	Model 4	Model 2	Model 3	Model 4
Gender (male)	0.84	0.97	1.06	1.45*	1.66**	1.81**
Age	0.95	0.94	0.92	0.83**	0.80**	0.79**
Canton (Basel)						
Genève	1.69*	1.58	1.33	1.59	1.53	1.21
Valais	1.62	1.19	1.55	1.38	1.15	1.28
Bern	0.91	0.86	1.07	1.00	0.95	1.00
Ticino	1.74*	1.48	1.26	1.86*	1.71*	1.49
Marital status (married)						
Single	0.68	0.50	0.34*	0.81	0.77	0.62
Divorced/sep.	3.21*	2.16**	1.32	2.09**	1.72*	1.18
Widowed	1.54***	1.35	1.16	1.40	1.42	1.23
Educational attainment (primary school)						
Lower secondary	0.84	1.07	1.29	1.35	1.53	1.86
Prof. training	0.52**	0.76	0.90	0.51**	0.65	0.74
Highschool	0.38**	0.71	0.91	0.42**	0.61	0.78
Univ. of appl. sciences	0.43**	0.82	1.09	0.64	0.97	1.25
University	0.29***	1.02	1.31	0.33***	0.69	0.87
Visit friends (at least once a month)						
Never	1.79*	1.37	1.07	1.59	1.34	1.21
At least once a year	0.79	0.64	0.67	0.96	0.91	0.97
At least once a week	1.17	1.00	0.86	1.55*	1.48	1.46
Every day	1.24	1.24	1.67	1.29	1.25	1.63
Selfworth (1–10)	1.01	1.00	0.97	0.97	0.97	0.95
Recognition (1–4)	0.71***	0.73**	0.76*	0.51***	0.52***	0.52***
Mastery (1–10)	0.86***	0.87***	0.88**	0.91**	0.92*	0.93*
Income (3′600–4′800)						
<2′400		8.34***	7.76***		2.51***	2.09**
2′400–3′600		5.07***	4.63***		1.75**	1.47
4′800–6′000		0.68	0.81		0.59*	0.67
>6000		0.13**	0.20*		0.38**	0.56
Wealth (>500′000)						
(almost) nothing			9.10***			5.56***
<60′000			2.89**			3.57***
60′000–150′000			1.40			1.23
150′000–500′000			0.86			0.88
McFadden's R2	0.133	0.267	0.341	0.14	0.176	0.236
Key: *** < 0.001; ** < 0.01; * < 0.05						

Age has no effect on the Self-Assessed Measure in any of the estimated models. In contrast, for the Perceived Measure, being 5 years older decreases the odds of worrying by approximately 25%. (OR = 0.8), holding all other variables constant.

Marital status has no effect on neither self-assessed nor perceived economic vulnerability after both income and wealth are taken into account. For respondents whose marital status is 'divorced' the odds of being vulnerable decrease when controlling for income and cease to be significant when adding wealth to the model. This confirms the previously stated observation that despite the formulation of these survey questions in such a way as to focus the response on monthly income, both measures are clearly influenced by the presence of financial assets.

As expected, education ceases to be significant for explaining either of the outcome measures once income and wealth are controlled for.

Turning to the variables that operationalize Consequences and Symptoms of economic vulnerability, the frequency of visiting or being visited by friends is only (and barely) significant in model 2. There seems to be an opposite effect for the Self-Assessed Measure compared to the Perceived Measure: while the difficulties in making ends meet is associated with less frequent contacts with friends, the contrary is true for respondents who worry about their finances: they tend to see their friends more frequently (at least once a week) than the average sample population (at least once a month). However, these differences in social behaviors are entirely explained by variations in income (model 3).

The self-worth scale turns out not to be significant in any of the models while the worry about lack of recognition is persistently significant even after controlling for income and wealth: in the case of the self-assessed outcome, income accounts for part of the effect of this covariate; but even after controlling for wealth, respondents who are less frequently worried[4] about not being recognized for what they do have a significantly lower risk of being vulnerable by the Self-Assesse Measure (OR = 0.8, $p < 0.01$). For the Perceived Measure, the effect size is greater and statistically highly significant (OR = 0.5, $p < 0.001$).

Mastery has a significant effect on both outcomes, but (contrary to our hypothesis) more so on the Self-Assessed Measure: while the effect on the Self-Assessed Measure remains highly robust and significant across the three models (OR = 0.9, $p < 0.001$), the significance of the effect on the Perceived Measure is reduced after controlling for wealth (OR = 0.9, $p < 0.05$).

Compared to the reference group[5] disposing of a monthly income of CHF 3600–4800, respondents who have to do with less than CHF 2400 per month have more than 8 times higher odds of self-assessing themselves as economically vulnerable, all other variables held constant. For the second to lowest income category (2400–3600) the odds are 5 times higher. Adding the variable wealth only results in a minor decrease of the effect of income. The effect of having a low income is much

[4]The items are ordered from 1 = strongly worried to 4 = not worried at all.

[5]The majority of the sample population (32%) is found in the monthly income category of ‚CHF 3600–4800'.

smaller for the perceived angle: with less than CHF 2400 the odds of worrying about not having enough money increase by a factor of 2.5 and for income category 2400–3600 the odds ratio is 1.8 ($p < 0.01$). This result lends support to our hypothesis that the Perceived Measure is less grounded in the material situation than the Self-Assessed Measure.

Analogous to income, the effect of wealth is much stronger on the Self-Assessed than on the Perceived Measure: the odds of having difficulties in making ends meet increase by a factor higher than 9 for respondents who have no wealth and by factor of 3 for those who have less than CHF 60,000 compared to the reference group[6] (more than CHF 500,000). For the Perceived Measure, the odds increase by a factor of 5.6 for those who own nothing and by a factor of 3.6 for those who have less than CHF 60,000. Neither of the two higher categories of wealth has any significant effect on the outcome variables. Still, the relative importance of wealth even for the Perceived Measure is worth noting.

Adding income to the variable set Consequences and Symptoms greatly improves the overall fit of the model for the Self-Assessed Measure (from $R^2 = 0.13$ to 0.27) and, to a lesser degree, for the Perceived Measure ($R^2 = 0.14–0.18$). The variable wealth adds considerable explanatory power to both models, causing the McFadden's R^2 to jump to 0.34 for the self-assessed outcome and to $R^2 = 0.24$ for the perceived outcome.

The answer to the question regarding the significance of the difference between the two vulnerable groups in terms of Economic Resources has to be nuanced: for income, the difference in effect is clearly significant as seen in the lack of overlap of the confidence intervals[7] (income category '<2400' in model 4 for the Self-Assessed Measure [4.59, 13.11] and for the Perceived Measure [1.34, 3.27]) whereas for wealth, the confidence interval for the Perceived Measure coincides to a large degree with the interval for the Self-Assessed Measure (wealth category 'almost nothing or nothing' in model 4 for Self-Assessed Measure [4.83, 17.16] and for Perceived Measure [3.25, 9.52]). Thus, the two subjective measures concordantly testify to the great importance of wealth for the subjective appraisal of economic vulnerability. The worry about not having enough money for current expenditures, however, tends to be much more detached from the actual level of income. The finding that the Perceived Measure is a less accurate indicator of the actual financial situation than the Self-Assessed Measure is in line with our theory-based predictions.

Regarding the significance of the relationship between the Perceived Measure and Psychosocial Symptoms and Consequences of economic vulnerability we cannot, based on the variables used for operationalizing Psychosocial Symptoms and Consequences, confirm this hypothesis: out of four explanatory variables, only the worry about not being recognized proved to have a significant effect on the Perceived Measure.

---

[6] The majority of the sample population (31%) is found in the wealth category of 'more than CHF 500′000'.

[7] Model 4 with confidence intervals can be found in Appendix Table 26.4.

Taking a closer look at the statistical significance of the difference of the effects we find that for worry about not being recognized, the difference between the two outcomes is statistically significant, as the lack of overlap of the confidence intervals indicates shows (in model 4 for the Self-Assessed Measure [0.61, 0.94] and for the Perceived Measure [0.44, 0.62]). This difference in effect can be interpreted as a further confirmation of the distinctiveness of the two measures of subjective economic vulnerability. For the variable mastery, the difference is also very obvious from the odds ratios, which are highly significant for the Self-Assessed Measure and insignificant for the Perceived Measure. Since there is some overlap in the confidence intervals, it is not possible to ascertain that the difference is statistically significant (for the Self-Assessed Measure [0.81, 0.95], for Perceived Measure [0.86, 0.99] in model 4). Even though the theoretical model predicted a stronger association with the Perceived Measure, the association of the Self-Assessed Measure with a low sense of mastery does not come as a total surprise; the experience of insufficient resources has actually been conceptualized as a primary source of stress (Pearlin et al. 1981). What is more unexpected is the absence of a statistically significant association with the Perceived Measure. Moreover, the observation that the variable mastery showed a stronger association with the Self-Assessed Measure than with the Perceived Measure – a trend that already showed up in the binary analysis – is contrary to the hypothesis stated in our theoretical model. Again, the statistically significant difference of these effects must be taken as further proof for the multidimensionality of subjective economic vulnerability.

In response to the guiding question of this part about the difference between the two subjective angles, the estimated regression models indicate that there is indeed a significant difference between the two measures, but less so in terms of psychosocial characteristics symptomatic of a stressful economic situation but rather in terms of Economic Resources. Among the variables operationalizing Psychosocial Symptoms and Consequences, only two out of four preserved a significant effect after adding Economic Resources: the worry about not being recognized and the mastery scale. These findings suggest that the relationship between the Self-Assessed and the Perceived Measure may be more complex and ambiguous than expected. A structural equation model will allow us to go further in our analysis by testing whether there are interaction effects with Economic Resources.

Appendix

Appendix Table 26.1 Model 1: Odds ratios and confidence intervals of regressing Background Characteristics on the Self-Assessed and the Perceived Measures

	Model 1: Background Characteristics					
	sa_ev			perc_ev		
Variable (Ref.)	OR	[95% CI]		OR	[95% CI]	
Sex (male)	ns	0.60	1.15	1.4*	1.03	1.90
Age	ns	0.98	2.60	0.83**	0.73	0.95
Canton (Basel)						
Geneva	ns	1.03	2.76	1.69*	1.08	2.67
Valais	ns	0.43	1.24	ns	0.95	2.41
Berne	ns	1.21	3.23	ns	0.61	1.55
Ticino	1.98**	0.35	1.50	2.58***	1.65	4.04
Marital status (married)						
Single	ns	2.31	5.24	ns	0.38	1.37
Divorced/sep.	3.48***	1.04	2.29	2.33***	1.56	3.49
Widowed	1.54*	0.54	1.87	1.48*	1.01	2.18
Educational attainment (primary school)						
Lower secondary	ns	0.31	0.76	ns	0.87	2.83
Prof. training	0.59*	0.20	0.58	0.46**	0.30	0.72
High school	0.40**	0.18	0.59	0.36***	0.21	0.61
University of appl. sciences	0.34**	0.13	0.47	0.51*	0.30	0.86
University	0.25***	0.13	0.40	0.28***	0.15	0.49
McFadden's R^2		*0.086*			*0.073*	
Key: *** < 0.001; ** < 0.01; * < 0.05; *ns* not statistically significant						

Appendix Table 26.2 Model 2: Odds ratios and confidence intervals of regressing Background Characteristics and Symptoms and Consequences of Economic Vulnerability on the Self-Assessed and the Perceived Measure

	Model 2: Background Characteristics + Consequences and Symptoms					
	sa_ev			perc_ev		
Variable (Ref.)	OR	[95% CI]		OR	[95% CI]	
Gender (male)	0.84	0.59	1.19	1.45*	1.04	2.02
Age	0.95	0.81	1.11	0.83**	0.71	0.94
Canton (Basel)						
Genève	1.69*	1.01	2.82	1.59	0.98	2.56
Valais	1.62	0.96	2.73	1.38	0.84	2.25
Bern	0.91	0.53	1.56	1.00	0.61	1.62
Ticino	1.74*	1.03	2.96	1.86*	1.14	3.02
Marital status (married)						
Single	0.68	0.32	1.45	0.81	0.42	1.57
Divorced/ sep.	3.21*	2.09	4.91	2.09**	1.37	3.19
Widowed	1.54***	1.00	2.38	1.40	0.92	2.12
Educational attainment (primary school)						
Lower secondary	0.84	0.43	1.63	1.35	0.71	2.58
Prof. training	0.52**	0.32	0.85	0.51**	0.31	0.84
High school	0.38**	0.21	0.68	0.42**	0.24	0.74
University of appl. sciences	0.43**	0.23	0.79	0.64	0.36	1.12
University	0.29***	0.15	0.57	0.33***	0.17	0.61
Feeling lonely (never)						
Always/often	1.42	0.84	2.40	1.09	0.65	1.83
Rarely	0.74	0.51	1.07	0.91	0.65	1.27
Visit friends (at least once a month)						
Never	1.79*	1.07	2.98	1.59	0.95	2.65
At least once a year	0.79	0.50	1.25	0.96	0.64	1.44
At least once a week	1.17	0.77	1.77	1.55*	1.06	2.26
Every day	1.24	0.60	2.55	1.29	0.64	2.60
Selfworth (1–10)	1.01	0.95	1.08	0.97	0.92	1.03
Recognition (1–4)	0.71***	0.59	0.86	0.51***	0.43	0.60
Mastery (1–10)	0.86***	0.80	0.93	0.91**	0.85	0.97
McFadden's R^2		0.133			0.14	
Key: *** < 0.001; ** < 0.01; * < 0.05						

Appendix Table 26.3 Model 3: Odds ratios and confidence intervals of regressing Background Characteristics and Symptoms and Consequences of Economic Vulnerability on the Self-Assessed and the Perceived Measure, controlling for income

	Model 3: Background Characteristics + Consequences and Symptoms + Income					
	sa_ev			perc_ev		
Variable (Ref.)	OR	[95% CI]		OR	[95% CI]	
Gender (male)	0.97	0.66	1.43	1.66**	1.18	2.34
Age	0.94	0.79	1.11	0.80**	0.69	0.92
Canton (Basel)						
Genève	1.58	0.90	2.75	1.53	0.93	2.49
Valais	1.19	0.68	2.09	1.15	0.69	1.90
Bern	0.86	0.48	1.52	0.95	0.57	1.56
Ticino	1.48	0.83	2.64	1.71*	1.04	2.83
Marital status (married)						
Single	0.50	0.22	1.11	0.77	0.39	1.52
Divorced/ sep.	2.16**	1.36	3.45	1.72*	1.11	2.66
Widowed	1.35	0.84	2.16	1.42	0.92	2.18
Educational attainment (primary school)						
Lower secondary	1.07	0.53	2.16	1.53	0.79	2.95
Prof. training	0.76	0.45	1.27	0.65	0.39	1.08
High school	0.71	0.38	1.33	0.61	0.34	1.11
University of appl. sciences	0.82	0.42	1.59	0.97	0.54	1.74
University	1.02	0.48	2.20	0.69	0.35	1.36
Visit friends (at least once a month)						
Never	1.37	0.80	2.35	1.34	0.80	2.26
At least once a year	0.64	0.40	1.03	0.91	0.60	1.37
At least once a week	1.00	0.64	1.56	1.48	1.00	2.18
Every day	1.24	0.56	2.75	1.25	0.61	2.57
Selfworth (1–10)	1.00	0.94	1.07	0.97	0.92	1.03
Recognition (1–4)	0.73**	0.60	0.89	0.52***	0.44	0.62
Mastery (1–10)	0.87***	0.81	0.94	0.92*	0.86	0.98
Income (3600–4800)						
< 2400	8.34***	5.11	13.60	2.51***	1.64	3.84
2400–3600	5.07***	3.15	8.15	1.75**	1.17	2.62
4800–6000	0.68	0.33	1.41	0.59*	0.35	0.99
> 6000	0.13**	0.03	0.56	0.38**	0.20	0.72
McFadden's R^2		0.267			0.176	

Key: *** < 0.001; ** < 0.01; * < 0.05

Appendix Table 26.4 Model 4: Odds ratios and confidence intervals of regressing Background Characteristics and Symptoms and Consequences of Economic Vulnerability on the Self-Assessed and the Perceived Measure, controlling for income and wealth

	Model 4: Background Characteristics + Consequences and Symptoms + Income + Wealth					
	Self-assessed measure			Perceived measure		
Variable (Ref.)	OR	[95% CI]		OR	[95% CI]	
Gender (male)	1.06	0.70	1.60	1.81**	1.26	2.59
Age	0.92	0.77	1.10	0.79**	0.68	0.92
Canton (Basel)						
Genève	1.33	0.73	2.43	1.21	0.72	2.02
Valais	1.55	0.84	2.87	1.28	0.76	2.17
Bern	1.07	0.58	1.98	1.00	0.60	1.66
Ticino	1.26	0.67	2.36	1.49	0.88	2.51
Marital status (married)						
Single	0.34*	0.15	0.81	0.62	0.30	1.25
Divorced/ sep.	1.32	0.79	2.20	1.18	0.74	1.88
Widowed	1.16	0.70	1.93	1.23	0.78	1.94
Educational attainment (primary school)						
Lower secondary	1.29	0.61	2.73	1.86	0.93	3.69
Prof. training	0.90	0.51	1.57	0.74	0.43	1.25
High school	0.91	0.46	1.80	0.78	0.42	1.47
University of appl. sciences	1.09	0.54	2.22	1.25	0.67	2.33
University	1.31	0.57	3.03	0.87	0.42	1.78
Visit friends (at least once a month)						
Never	1.07	0.59	1.92	1.21	0.70	2.10
At least once a year	0.67	0.40	1.11	0.97	0.63	1.49
At least once a week	0.86	0.53	1.40	1.46	0.97	2.19
Every day	1.67	0.73	3.85	1.63	0.78	3.40
Selfworth (1–10)	0.97	0.91	1.04	0.95	0.89	1.01
Recognition (1–4)	0.76*	0.61	0.94	0.52***	0.44	0.62
Mastery (1–10)	0.88**	0.81	0.95	0.93*	0.86	0.99
Income (3600–4800)						
< 2400	7.76***	4.59	13.11	2.09**	1.34	3.27
2400–3600	4.63***	2.80	7.66	1.47	0.96	2.25
4800–6000	0.81	0.38	1.72	0.67	0.39	1.15
> 6000	0.20*	0.04	0.89	0.56	0.28	1.12
Wealth (>500′000)						
(almost) nothing	9.10***	4.83	17.16	5.56***	3.25	9.52
< 60,000	2.89**	1.50	5.23	3.57***	2.12	6.01
60,000–150,000	1.40	0.72	2.70	1.23	0.71	2.13
150,000–500,000	0.86	0.44	1.68	0.88	0.52	1.48
McFadden's R^2		0.34			0.24	

Key: *** < 0.001; ** < 0.01; * < 0.05

References

Pearlin, L. I. (1989). The sociological study of stress. *Journal of Health and Social Behavior, 30*(3), 241–256.
Pearlin, L. I., Menaghan, E. G., Lieberman, M. A., & Mullan, J. T. (1981). The stress process. *Journal of Health and Social Behavior, 22*(4), 337–356.

Chapter 27
A Path Model of the Relationship Between the Self-Assessed and the Perceived Measure: Group Comparison by Wealth

The results found in the presented regression models confound the idea of a simple mediation effect of the model component Psychosocial Symptoms and Consequences that would explain the discrepancy between the Self-Assessed and the Perceived Measures. The issue may lie with the model design, which is of an additive type, and its inability to adequately capture the nature of the psychosocial relations between these model components. According to the results of our research so far it seems indeed likely that the effects of a diminished sense of self and role strains are not linear across levels of Economic Resources. In such a scenario, the relationship between the Self-Assessed and the Perceived Measure is no longer characterized by mechanisms of mediation but by a more complex type of interaction effect referred to as moderation.[1] We will test this moderation hypothesis by estimating a path model based on our theoretical model that allows the path estimates to vary by categories of wealth. The question of interest is which paths are significantly different for different levels of wealth and which paths can be treated as equal across wealth categories. Another question we are set to investigate is the relationship between income and the Perceived Measure via the Self-Assessed Measure. Here, we assume a mediation-type of relationship: the worry about not having enough money to pay current expenses is caused by difficulties in making ends meet with a monthly household income and indirectly influenced by the household income. In the initial model, income is an ordinal variable of five categories. Table 27.1 shows the correlation between the observed variables used in the model and Fig. 27.1 depicts the hypothesized path model.

The choice to use wealth as basis for the group comparison is an acknowledgement of the results of the previous chapter, which showed that the effect of wealth on the Perceived Measure is strong and indeed stronger than the effect of income. In order to test another one of our hypothesis concerning the mediation effect of the Self-Assessed Measure between Economic Resources and the Perceived Measure, income and the

[1] For an explanation of the concepts of moderation and mediation, see Footnote 122, p.158

J. Henke, *Revisiting Economic Vulnerability in Old Age*, Life Course Research and Social Policies 11, https://doi.org/10.1007/978-3-030-36323-9_27

Table 27.1 Correlation coefficients between observed variables used in the model

Observed variables	1	2	3	4	5	6
1. income	1.0	–	–	–	–	–
2. wealth	0.4	1.0	–	–	–	–
3. mastery	0.1	0.1	1.0	–	–	–
4. recognition	0.1	0.1	0.2	1.0	–	–
5. sa_ev	−0.4	−0.5	−0.2	−0.1	1.0	–
6. perc_ev	−0.2	−0.4	−0.1	−0.3	0.4	1.0

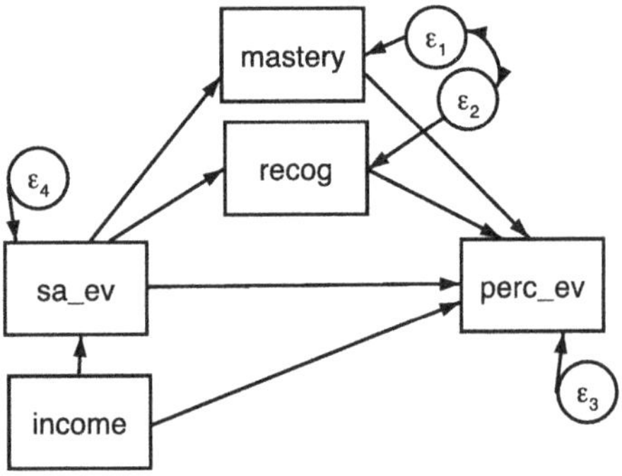

Fig. 27.1 Hypothesized path model for predicting the Perceived Measure (perc_ev)

Perceived Measure are linked through a direct path and an indirect path via the Self-Assessed Measure. The error terms of the psychosocial variables mastery and recognition are correlated, reflecting the assumption that there are other influences outside of the path model that may act on both of these variables. We used a maximum likelihood estimation method that allows the inclusion of missing values.[2]

Figure 27.2 shows the standardized coefficients and the model fit statistics of the initial path model, where wealth has not yet been added as a grouping variable. We see that the Perceived Measure is predicted most strongly by the Self-Assessed Measure ($\beta = 0.34$, $p < 0.001$), followed by the direct effect of worry about not being recognized ($\beta = -0.22$, $p < 0.001$). As expected, the path coefficient of mastery on the Perceived Measure is not significant. The small direct effect of income on the Perceived Measure ($\beta = -0.09$, $p < 0.001$) in comparison with the stronger effects via the Self-Assessed Measure confirms that the latter plays a mediating role. This effect will be examined in more detail in the next chapter.

[2]The Stata command for this SEM option is > *method(mlmv)*. This option assumes normality and missingness at random. It was our preferred method for the models based on psychosocial variables because these variables contained a considerable number of missing values and the default method (ml) would have excluded 96 observations. However, because some variables do not follow a normal distribution, the validity of the estimated coefficients was crosschecked by estimating the same model with an estimation method that does *not* assume normality (stata command is > *method(ml) vce(robust))*, which resulted in the exact same estimates.

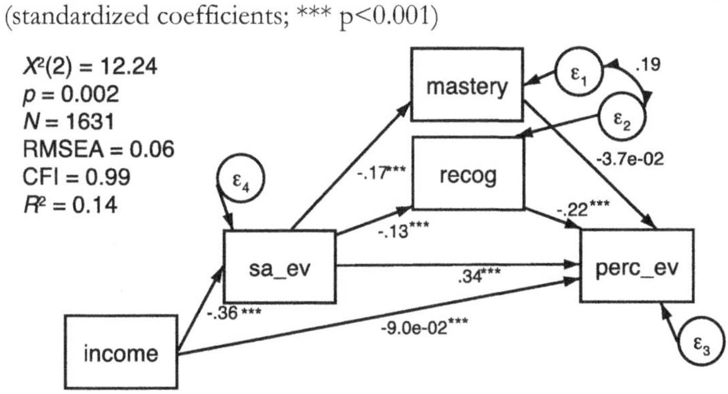

Fig. 27.2 The relationship between income, the Self-Assessed Measure and the Perceived Measure of economic vulnerability, mediated by mastery and recognition. (Standardized coefficients; ∗∗∗ p < 0.001)

The goodness-of-fit statistics at the model level indicate a good fit: though the chi-squared statistic of $\chi^2(2) = 12.24$ ($p = 0.002$) indicates that our model does not fully reproduce the original covariance matrix, the RMSEA of 0.06 is good and the CFI of 0.99 is excellent. The pseudo $R^2_{perc\_ev}$ shows that the model is able to explain 22% of the variance in the Perceived Measure.

Based on this initial model, the same structural coefficients are estimated in a multiple-group analysis, at first without any equality constraints. This model was estimated using the default method of maximum likelihood.[3] Now that wealth has been added as a grouping variable, the goodness-of-fit statistics improve noticeably: the chi-squared of $\chi^2(10) = 14.72$ ($p = 0.142$) and the RMSEA of 0.04 are now both indicating a very good fit while the CFI remains at the same high level. Table 27.2 shows standardized and unstandardized coefficients of the model for each wealth category and Fig. 27.3 visualizes the trends for standardized path coefficients as wealth levels increase from left to right. It was to be expected that the predictive power of the model is higher for the lowest wealth category ('nothing or almost nothing' results in $R^2_{perc\_ev} = 0.33$) compared to higher wealth categories ('>500,000' results in $R^2_{perc\_ev} = 0.10$). What is noteworthy, however, is the non-linear downward trend after the second to lowest wealth category, with a moderate increase in the middle categories, resulting in the same $R^2_{perc\_ev}$ for the second wealth categories '<60,000' as for the highest category '>500,000' (Table 27.2). This result is corroborated by the absence of any significant effect of income, which moreover changes algebraic sign from positive to negative. The direct effect of the Self-Assessed Measure on the Perceived Measure is strong only for the lowest wealth category ($\beta = 0.50$, $p < 0.001$) and drops by more than

[3] For this model, the method (ml) was used because some of the post estimation commands after group comparisons are not compatible with the method (mlmv). Thus, the sample size was reduced by 96 observations with missing values to $N = 1535$.

Table 27.2 Summary table for group comparison by wealth, unconstrained solution (*** p < 0.001, ** p < 0.01, * p < 0.05)

	Wealth categories									
	(almost) nothing		< 60,000		60,000–150,000		150,000–500,000		>500,000	
	(N = 187)		*(N = 208)*		*(N = 274)*		*(N = 380)*		*(N = 486)*	
Path	*β*	*B*	*β*	*B*	*β*	*B*	*β*	*B*	*β*	*B*
mastery										
<- sa_ev	−0.12	−0.6	−0.23***	−1.27***	−0.15*	−0.10*	−0.12*	−1.1*	−0.17***	−1.80***
perc_ev										
<- mastery	−0.07	−0.01	0.01	0.00	−0.12*	−0.02*	0.02	0.00	−0.06	−0.01
<- recog	−0.24***	−0.12***	−0.22***	−0.13***	−0.22***	−0.11***	−0.24***	−0.10***	−0.24***	−0.09***
<- sa_ev	0.50***	0.50***	0.18**	0.21**	0.23***	0.24***	0.27***	0.33***	0.14**	0.20**
<- income	0.04	0.02	−0.08	−0.04	−0.07	−0.02	−0.06	−0.02	−0.09*	−0.02*
recog										
<- sa_ev	−0.07	−0.14	−0.16*	−0.30*	−0.14*	−0.31*	−0.06	−0.18	−0.14**	−0.52**
sa_ev										
<- income	−0.28***	−0.13***	−0.38***	−0.17***	−0.31***	−0.09***	−0.24***	−0.05***	−0.26***	−0.04***
$R^2_{perc\_ev}$	0.33		0.10		0.17		0.14		0.10	
R^2_{group}	0.09		0.14		0.11		0.06		0.08	

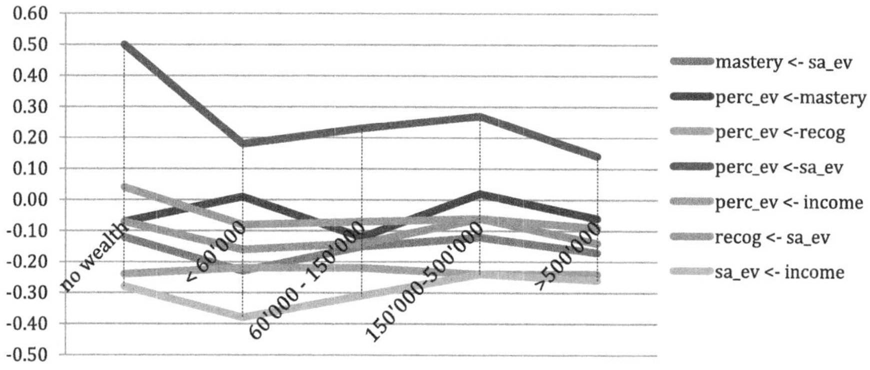

Fig. 27.3 Trends for standardized path coefficients, group by wealth, unconstrained solution

half of the effect size for the second wealth category before increasing again slightly for the middle categories. Again, the fact that the path coefficient for wealth categories '<60,000' and '>500,000' are virtually the same puts into serious question the anchorage of the Perceived Measure in a situation of real material deprivation.

Let us consider how the psychosocial effects vary as a function of different levels of financial capital. A close look at Table 27.2 reveals that for the Self-Assessed Measure, those individuals who are most affected by low levels of mastery are not those without any wealth but those who are found in the second to lowest wealth category '<60,000' ($\beta = -0.23$, $p < 0.001$). The effect of mastery on the Perceived Measure is only significant for wealth category '60,000–150,000' ($\beta = -0.12$, $p < 0.01$). Though, as we will see later, on the whole, this path does not differ across wealth categories in a statistically significant manner, this result is indicative for the by now well-known phenomena that it is not necessarily the most economically deprived who psychologically suffer the most from economic vulnerability.

The variable recognition follows a different trend in that it represents a very stable horizontal line across all wealth levels. This total lack of fluctuation may in fact be indicative of the overall dynamic that determines the worry about not having enough money for current expenditures: the Perceived Measure of economic vulnerability may reflect the propensity of a person to worry – about anything in life.[4] Among those who are vulnerable according to the Perceived Measure, 54% are in the lowest two wealth categories. However, the remaining 46% are still too many to be accounted for by a general anxiety[5] as will be discussed later. The variable mastery also does not show a clear trend, it even fluctuates between a positive and a negative association. The preliminary conclusion we draw from this is that the

[4] This being said, we have to remember that earlier in this volume we found that neuroticism, the personality trait that is more prone to worrying, was not associated with the Perceived Measure in any statistically significant way.

[5] The Berlin Aging Study found a prevalence of anxiety of 4.5% among the elderly community (aged 70 and older) (Schaub and Linden 2000).

Perceived Measure should only tentatively be interpreted as actual economic stress and ideally only in combination with another indicator of economic vulnerability.

The presented multiple-group model was examined in view of potential improvements using a post estimation command[6] for identifying those parameters that are significantly different across wealth levels. The result indicates that two invariant constraints are significant, sa_ev → perc_ev: $\chi^2(4) = 14.9$, $p = 0.005$; and income → sa_ev: $\chi^2(4) = 27.5$, $p < 0.001$. A second model, constraining all but these two paths was estimated, using the default maximum likelihood estimation.[7] Table 27.3 shows standardized and unstandardized coefficients of the constrained model for each wealth category.

As a result of constraining all paths to be equal except those from income to the Self-Assessed Measure and from the Self-Assessed to the Perceived Measure, the already very good statistics of model-fit attained the best possible values: the chi-squared of $\chi^2(30) = 29.0$, $p = 0.517$, a RMSEA of 0.00 and a CFI of 1 means that this model provides an excellent fit to the data. The stability of the model was moreover validated using a bootstrapping method.[8]

We will now return to the question whether the Self-Assessed Measure mediates the effect of income on the Perceived Measure by comparing the direct and indirect effects of income on the Perceived Measure. For this, we re-estimate the same model depicted in Fig. 27.1, this time using the standard maximum likelihood estimation method in order to be able to perform all post-estimation commands available in Stata. The standardized direct, indirect and total effects are shown in Table 27.4.

We find that the influence of monthly income on the Perceived Measure is substantially mediated by the Self-Assessed Measure: the total effect of income on the worry about not having enough money to pay current expenses (perc_ev) equals $\beta = -0.23$, $p < 0.001$, and is composed of 40% direct effect and 60% indirect effect via the Self-Assessed Measure. A detailed analysis of this effect, disaggregated levels of income, will be presented in the next chapter.

Based on a comparison of direct and indirect effects we now are able to quantify the small mediation effect of the psychosocial variables of which, as Fig. 27.2 already showed, only recognition is statistically significant: the indirect effect of the Self-Assessed Measure on the Perceived Measure via recognition represents only 10% ($\beta = 0.04$, $p < 0.001$) of the total effect. This result is not incongruent with our theoretical model positing that Psychosocial Symptoms and Consequences play a partially mediating role between the Self-Assessed and the Perceived Measure. However, based on the variables we used for operationalizing Psychosocial Symptoms and Consequences it would be more accurate to limit our conclusion to

---

[6] The stata command is > *estat ginvariant*

[7] For this model, the method (ml) was used because some of the post estimation commands after group comparisons are not compatible with the method (mlmv). Thus, the sample size was reduced by 96 missing observations to $N = 1535$.

[8] The stata option utilized is > vce(bootstrap, reps(1000) seed(111)). There was no difference in the estimated coefficients.

Table 27.3 Summary table for group comparison by wealth, constrained solution (∗∗∗ $p < 0.001$, ∗∗ $p < 0.01$, ∗ $p < 0.05$)

	Wealth Categories									
	Almost nothing or nothing		< 60,000		60,000–150,000		150,000–500,000		>500,000	
	(N = 187)		*(N = 208)*		*(N = 274)*		*(N = 380)*		*(N = 486)*	
Relationship	β	*B*	β	*B*	β	*B*	β	*B*	β	*B*
mastery										
<- sa_ev	−0.21∗∗∗	−1.09∗∗∗	−0.21∗∗∗	−1.09∗∗∗	−0.21∗∗∗	−1.09∗∗∗	−0.21∗∗∗	−1.09∗∗∗	−0.21∗∗∗	−1.09∗∗∗
perc_ev										
<- mastery	−0.03	−0.01	−0.03	−0.01	−0.03	−0.01	−0.03	−0.01	−0.03	−0.01
<- recognition	−0.20∗∗∗	−0.10∗∗∗	−0.20∗∗∗	−0.10∗∗∗	−0.20∗∗∗	−0.10∗∗∗	−0.20∗∗∗	−0.10∗∗∗	−0.20∗∗∗	−0.10∗∗∗
<- sa_ev	0.48∗∗∗	0.48∗∗∗	0.18∗∗	0.19∗∗	0.25∗∗∗	0.28∗∗∗	0.25∗∗∗	0.32∗∗∗	0.14∗∗	0.19∗∗
<- income	−0.04∗∗	−0.02∗∗	−0.04∗∗	−0.02∗∗	−0.04∗∗	−0.02∗∗	−0.04∗∗	−0.02∗∗	−0.04∗∗	−0.02∗∗
recog										
<- sa_ev	−0.14∗∗∗	−0.28∗∗∗	−0.14∗∗∗	−0.28∗∗∗	−0.14∗∗∗	−0.28∗∗∗	−0.14∗∗∗	−0.28∗∗∗	−0.14∗∗∗	−0.28∗∗∗
sa_ev										
<- income	−0.29∗∗∗	−0.14∗∗∗	−0.37∗∗∗	−0.17∗∗∗	−0.33∗∗∗	−0.09∗∗∗	−0.23∗∗∗	−0.05∗∗∗	−0.27∗∗∗	−0.05∗∗∗
$R^2_{\text{perc\_ev}}$	0.32		0.08		0.15		0.16		0.11	

Table 27.4 Standardized direct, indirect and total effects

Outcome	Direct effect	Indirect effect	Total effect
perc_ev			
mastery	−0.03	–	−0.03
recognition	−0.22***	–	−0.22***
sa_ev	0.33***	0.04***	0.37***
Income	−0.09***	−0.14***	−0.23***
mastery			
sa_ev	−0.17***	–	−0.17***
income	–	0.06***	0.06***
recog			
sa_ev	−0.13***	–	−0.13***
income	–	0.05***	0.05***
sa_ev			
income	−0.37***	–	−0.37***

The significance levels shown here are for the unstandardized solution. *** $p < 0.001$, ** $p < 0.01$, * $p < 0.05$

the model sub-component 'sense of diminishment' (variable recognition) since neither of the variables operationalizing social isolation, nor loneliness nor mastery had a significant mediation effect.

Reference

Schaub, R. T., & Linden, M. (2000). Anxiety and anxiety disorders in the old and very old – Results from the Berlin aging study (BASE). *Comprehensive Psychiatry, 41*(2, Suppl 1), 48–54.

Chapter 28
Integrating the Three Measures of Economic Vulnerability

28.1 Modeling the Relationship Among Objective, Self-Assessed and Perceived Measures

In what follows we are going to take a closer look at the relationship between all three measures of economic vulnerability and how they relate to the variable block Psychosocial Symptoms and Consequences. The decision to focus on the Objective Measure, after having previously examined the effects of Economic Resources in a more comprehensive way, is again motivated by the policy relevance of the income-based poverty indicator. A comparison of models based on different thresholds of the Objective Measure will allow us to observe trends in the strength of association between measures of economic vulnerability and psychosocial variables. Figure 28.1 shows the relationship among Objective, Self-Assessed, and Perceived Measures of economic vulnerability, mediated by the variables mastery and recognition. The model was improved as a result of the previous analysis: the variable mastery is now located as a mediator between the Objective and the Self-Assessed Measure. Like in the previous model, the error terms of the two psychosocial variables are correlated.

Though the objective of this model is not primarily to attain the best fit,[1] it is worth mentioning that all three post-estimation indicators (chi-squared statistic, RMSEA, CFI) have excellent values. The total variance explained leaves to be desired ($R^2 = 0.10$) and the R^2 for the Self-Assessed and the Perceived Measure are also relatively low ($R^2_{sa\_ev} = 0.11$, $R^2_{perc\_ev} = 0.21$).

[1] As laid out in the introduction of this chapter, the interest here is to model – as parsimoniously as possible – the relationship between the three proposed measures of economic vulnerability and their relationship with psychosocial variables. The variable wealth, which has previously been identified as a powerful predictor even for the Perceived Measure, was deliberately not integrated in this model.

J. Henke, *Revisiting Economic Vulnerability in Old Age*, Life Course Research and Social Policies 11, https://doi.org/10.1007/978-3-030-36323-9_28

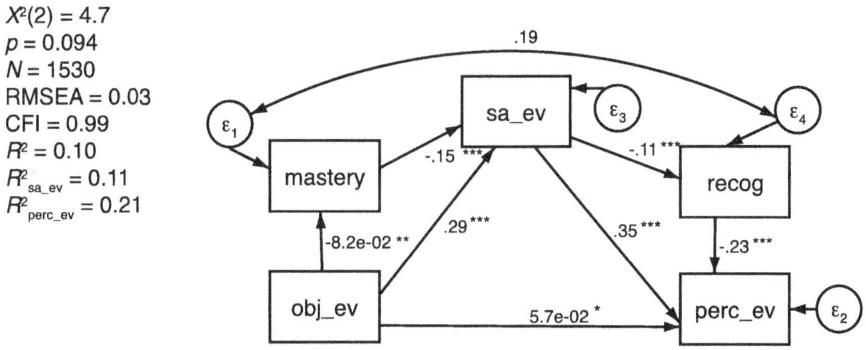

Fig. 28.1 The relationship among Objective, Self-Assessed, and Perceived Measures of economic vulnerability, mediated by Psychosocial Symptoms and Consequences. (Standardized coefficients, ***p < 0.001, **p < 0.01, *p < 0.05)

A second, almost identical model was estimated, the sole change being a higher threshold for the Objective Measure, which was set at CHF 3'600. We expect that this model provide a better fit because many of those suffering of Psychosocial Symptoms and Consequences are found in the income category just above the poverty line. This is confirmed (results not shown), though because of the already very good indices, the improvements appear only in the third decimal places and would therefore not show up in the reporting of the results. The path coefficients, too, are only modified very marginally (max. 0.02 difference) with the exception of the path leading from the Objective to the Self-Assessed Measure: identical to the effect in a previous model (Fig. 23.2) when the threshold of the income-based measure is raised from 2'400 to 3'600, the path coefficient to the sa_ev increases from $\beta = 0.29$ to $\beta = 0.38$. This result again testifies to the incongruence between the monetary poverty line and individuals' self-reports of economic vulnerability.

As could be expected after this analysis and the one based on a group-comparison by wealth, re-estimating the same model with yet a higher income threshold at CHF 4'800 only affects the path between the Objective to the Self-Assessed Measure: compared to the original model (obj_ev = 2'400) this path coefficient drops from $\beta = 0.29$ to $\beta = 0.22$, all other paths remaining close to constant (results not shown). The fit indices remain high but the explained variance drops for both the overall variance and the Self-Assessed Measure ($R^2 = 0.07$, $R^2_{sa\_ev} = 0.08$). Only the variance of the Perceived Measure increases minimally ($R^2_{perc\_ev} = 0.22$), confirming again the association of economic stress with middle to higher levels of economic resources.

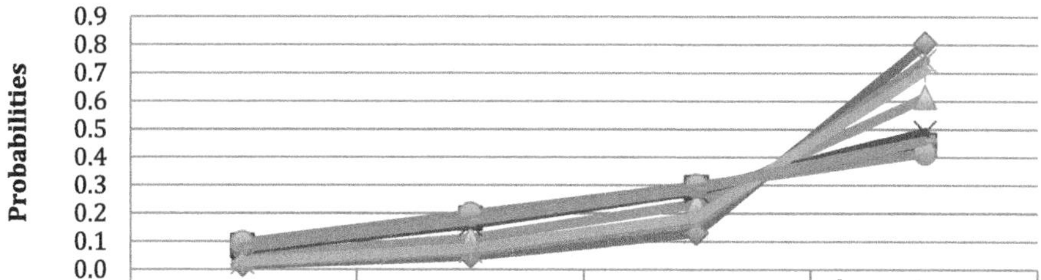

	strongly	somewhat	a little	does not concern me/no worry at all
AAA	0.0	0.0	0.1	0.8
AAB	0.1	0.2	0.3	0.4
ABA	0.0	0.1	0.2	0.6
ABB	0.1	0.2	0.3	0.5
BAA	0.0	0.1	0.2	0.7
BAB	0.1	0.2	0.3	0.4
BBA	0.0	0.1	0.2	0.7
BBB	0.1	0.2	0.3	0.5

Fig. 28.2 Outcome probabilities of vulnerability types for the dependent variable 'Worry about not getting sufficient recognition', controlling for Background Characteristics

28.2 A Typology of Economic Vulnerability Combining the Objective, the Self-Assessed, and the Perceived Measure

This chapter takes up the thread where we left it in Part IV. We will continue the assessment of the interaction of the three measures of economic vulnerability with regard to psychosocial variables. For this purpose, two ordinal regression models were estimated, successively regressing the variable recognition and an ordinal version of the variable mastery[2] on the Vulnerability Typology, yielding the probability outcomes of the vulnerability types for each item of the dependent variables.[3] Both models are based on complete sets of observations with no missing values (N = 1498); Background Characteristics (age, sex, canton, marital status, and education) are controlled for.

Figure 28.2 depicts outcome probabilities for the dependent variable recognition, connected by a line as to facilitate the interpretation of the response pattern for each vulnerability type. Two groups can easily be distinguished: the first one is composed

[2]The variable mastery was recoded into three categories that divide the sample population into three roughly equal-sized shares. The sample distribution of the recoded variable mastery is found in Appendix Table 28.1.

[3]For the interpretation of ordinal regression models, only outcome probabilities will be discussed in this chapter. A table reporting the odds ratios is shown in Appendix Table 28.2.

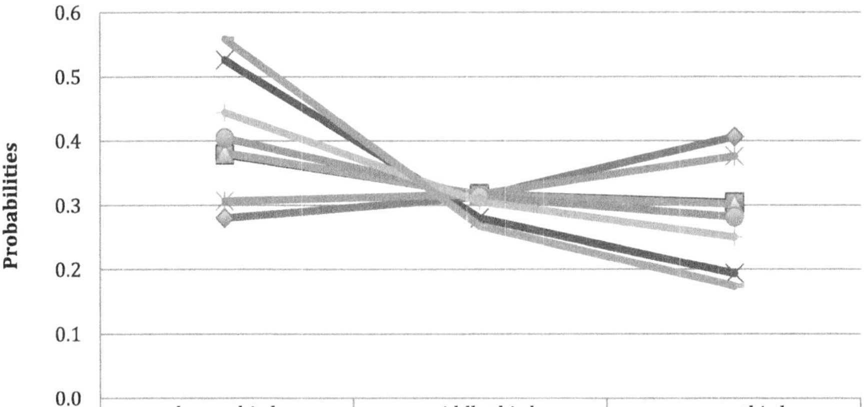

	lower third	middle third	upper third
AAA	0.3	0.3	0.4
AAB	0.4	0.3	0.3
ABA	0.4	0.3	0.3
ABB	0.5	0.3	0.2
BAA	0.3	0.3	0.4
BAB	0.4	0.3	0.3
BBA	0.4	0.3	0.3
BBB	0.6	0.3	0.2

Fig. 28.3 Outcome probabilities of vulnerability types for the dependent variable mastery, controlling for Background Characteristics

of types AAA, BAA, BBA. All of these vulnerability types have a high probability of responding that they either are 'not worried at all' about not being sufficiently recognized for what they do, or they feel that the question 'does not concern them' (0.7–0.8). For the second group, made up of types ABB, BBB, BAB, and AAB, the likelihood of not being affected by this concern is considerably lower, the probability outcomes being between 0.4 and 0.5. These four types are moreover the only ones for which the outcome of the variable recognition was predicted in a statistically significant manner. All four groups are of the generic type[4] XXB – they share in common that they are vulnerable in terms of their worry about not having enough money.

Figure 28.3 represents the outcome probabilities for the dependent variable mastery. For the probability outcomes of mastery, three subgroups can be distinguished: (1). types XBB, (2). types XAA, and lastly, the remaining four types. The 1st subgroup, types ABB and BBB, have a probability of 0.5 (0.6) to be found in the third of the population with the lowest sense of mastery. The 2nd subgroup, composed of types AAA and BAA, has a much lower probability of being in the lowest third of mastery (0.3). For the variable mastery, the combination of the Perceived and the Self-Assessed Measure of economic vulnerability clearly enhances the accuracy of predictions.

[4] X stands for either A or B.

Appendix

Appendix Table 28.1 Sample distribution of the variable mastery, recoded into three categories

Mastery scale	N	%	Cum. %
Lowest third	488	30.1	30.1
Middle third	491	30.2	60.3
Upper third	645	39.7	100.0

Appendix Table 28.2 Ordinal regression models with the Vulnerability Typology as independent variable predicting mastery and recognition, odds ratios are reversed

	Mastery (3 cat.)		Recognition	
Variable (Ref.)	Rev. OR	P > |z|	Rev. OR	P > |z|
Sex (male)	0.6832	< 0.001	1.3	< 0.001
Age	1.024	0.604	0.9	< 0.001
Canton (Basel)				
Geneva	0.9	0.67	0.8	0.236
Valais	0.8	0.108	0.9	0.784
Berne	1.0	0.808	1.0	0.855
Ticino	0.9	0.33	2.0	0
Marital status (married)				
Single	0.86	0.446	0.6	0.091
Divorced/sep.	0.88	0.429	1.1	0.477
Widowed	1.12	0.425	0.8	0.313
Educational attainment (primary school)				
Lower secondary	0.9	0.639	0.8	0.375
Prof. training	1.0	0.986	0.7	0.1
High school	1.1	0.798	0.8	0.444
University of appl. sciences	1.1	0.502	0.8	0.341
University	1.0	0.995	0.6	0.043
Vulnerability typology (ref. = AAA; obj = 0 / sa = 0 / perc = 0)				
AAB	1.5	0.026	3.7	< 0.001
ABA	1.5	0.136	1.7	0.069
ABB	2.7	< 0.001	3.0	< 0.001
BAA	1.0	0.822	0.9	0.757
BAB	1.6	0.191	3.9	< 0.001
BBA	1.9	0.031	1.1	0.779
BBB	3.1	< 0.001	3.3	< 0.001
McFadden's R^2	*0.019*		*0.058*	

Chapter 29
Conclusion Part V

With regard to demographic characteristics, whereas the effects on the Self-Assessed Measure are entirely explained by Economic Resources, the Perceived Measure is significantly predicted by age and sex, even after controlling for income and wealth: being female and younger are associated with an increased probability of worrying about not having enough money. Neither marital status nor educational attainment turn out to be predictors of financial worry once Economic Resources are controlled for. A cultural explanation, assuming the existence of regional differences in voicing complaints, is no longer plausible (it was considered at the bivariate stage because of the extremely high rate of 'frequent worriers' in Ticino): after controlling for wealth, the effect of the variable canton ceases to be significant for predicting economic stress.

Our results are only partly congruent with findings reported in a study on economic well-being of older adults in Germany. Weidekamp-Maicher (2008) examined the predictors of worry about the material situation among some 400 adults, aged 50–85. The dependant variable was a likert-type variable on the frequency of worry, ranging from 1 (never) to 7 (very frequently). Since the research design did not include income and wealth, we have to draw on the results reported in our regression model 1 (Table 19.1) for the comparison of results (the comparability of the results is, however, hampered by differences in the operationalization of worry). It is nonetheless noteworthy that, in in contrast to our results, the German research reported no significant differences according to age, sex, and marital status. Among Background Characteristics, the frequency of worry was influenced by low educational attainment, which is congruent with our findings.

The finding that the Perceived Measure is a less accurate indicator of the actual financial situation than the Self-Assessed Measure is in line with the predictions of our theoretical model: in the case of the Self-Assessed Measure, only the mode of assessment is subjective while the substance of assessment – adequacy of income – is an objective matter. By contrast, the Perceived Measure is based on the subjectivity of both mode and substance. The correlation between the Perceived Measure with income and wealth indeed turned out to be weak and highly non-linear: beyond

J. Henke, *Revisiting Economic Vulnerability in Old Age*, Life Course Research and Social Policies 11, https://doi.org/10.1007/978-3-030-36323-9_29

the very lowest categories of income and wealth, the Perceived Measure becomes so spurious that it has proven difficult to predict it within the limits of statistical significance.

The theory-based predictions for the Perceived Measure as revealing a greater distance from the objectively measurable financial situation while being more susceptible to influences of a psychological order is thus supported by our data though the accurate modeling of economic stress remained elusive, despite the use of a considerable array of variables for operationalizing the model component Psychosocial Symptoms and Consequences. This observation raises several questions with regard to our theoretical assumptions.

Firstly, we would like to come back to a question raised in the theoretical part about the distinctiveness of economic stress from potentially similar concepts captured by a sense of diminishment and mastery. Our results confirm that economic stress as operationalized by the worry about not having enough money to pay current expenditures is a distinct concept. It seems that it is more distinct still than what we expected: our hypothesis that economic stress is partly predicted by affective symptoms such as a diminished self-concept and a lower sense of mastery, has received very little confirmation. Consequently, little evidence was found for any significant mediation effect of psychosocial variables between economic strain and stress: only the worry about not getting enough recognition (one of the variables operationalizing the concept 'sense of diminishment') proved to mediate a small share of the effect of the Self-Assessed on the Perceived Measure. Given that among this sample population, the palette of psychosocial symptoms and consequences related to economic stress turned out to be so narrow, it may be promising to focus future research on the concept of stress proliferation, according to which primary stressors (economic strain) prompt secondary stressors, that are more related to role strains.

The second issue is related to the first one: in view of the unexpected finding that mastery is more strongly associated with the Self-Assessed than with the Perceived Measure, it is appropriate to return to our original assumptions concerning this variable. In Chap. 9 we introduced the variable mastery or 'sense of control' operationalized by Cohen's perceived stress scale (PSS), which we took as being very similar to Pearlin's scale of mastery. In hindsight, the close to identical wording of the items may have detracted from the fact that the PSS-items are introduced by the question 'in the last month, how often have you felt..' thus emphasizing the frequency of the subsequently stated experience of not being in control. It appears that the PSS comes closer to capturing the intensity of the *general* stress level rather than being an appropriate instrument for measuring the sense of mastery over different life situations. This reinterpretation of our variable mastery does not, however, diminish the interest of our findings: looking at the results from this angle, we could in fact conclude that among the Swiss retired population, there exists a significant difference between the experience of economic stress, operationalized by worry about not having enough money for current expenses, and stress in general. By way of accident, our models have thus provided support to the idea that economic stress is a distinct concept, distinguishing itself from both economic strain and general stress levels, and as such it merits to be analyzed in its own right. The perceived stress scale (PSS, our variably 'mastery') was developed in critical response to objective measures of stressful events and draws on Lazarus' theory of appraisal (Cohen et al. 1983). We recall that according to this theory, circumstances are only appraised as threatening if the person has high stakes in a situation and if he or she is afraid about not having the means to eliminate or modify the conditions that give rise to the problem (Pearlin and Schooler 1978). Coherently, we found that the absence of financial stocks, which represents a coping resource capable of changing the situation,[1] increases the probability of self-assessed economic strain. The Self-Assessed Measure and the PSS therefore seem to tap into a similar concept, one that may be more closely associated with the cognitive dimension than with the affective dimension that is more present in the Perceived Measure.

In what follows, we will round up the discussion on Psychosocial Symptoms and Consequences by looking at how the Perceived Measure is influenced by the two demographic variables sex and age, both of which seem themselves to be influenced by psychosocial factors.

Our finding that there exists a significant difference between the effect of the variable sex on the Perceived Measure (statistically significant) and the (absence of any) effect on the Self-Assessed Measure represents a contribution to the discussion of gender-differences in financial well-being. This debate has been a controversial one. Literature on psychological well-being has consistently reported that women have a higher tendency to experience negative emotions such as fear and worry (Lucas and Gohm 2000; Fischer 2000). On the other hand, specialized literature focusing on financial strain is more ambivalent: some report higher risks for financial strain among women (Antonucci u. a. 2002; Chan u. a. 2002) while others note that gender-differences are accounted for by differences in economic resources and other individual characteristics (Litwin and Sapir 2009), such as older women's lower probability of being married (Keith,1993).

Regarding the Self-Assessed Measure, our findings are in line with those studies that find factors such as income, education and marital status to fully account for the differences in economic *strain* experienced by women and men. However, based on our sample population, our conclusion diverges when it comes to the measure of economic *stress*: the effect size (OR 1.75) for the Perceived Measure reveals that older women in Switzerland have a considerably higher risk of worrying about their financial situation than men of a similar socioeconomic standing. This result merits further investigation, not least because previous research has shown that financial stress is likely to have negative spillover effects by affecting other life domains and in particular health, for example through sleep disturbances (Hall u. a. 2008; Dregan and Armstrong 2009). In view of the psychosocial variables that were controlled for in our models, among them the variable mastery that failed to be significant, another possible explanation for the observed gender gap could be unobserved gender dif-

[1] In contrast to this coping strategy, there exist strategies that aim at changing merely the *meaning* of a situation.

ferences in having a ‘sense of control’ that is more specific to the area of economics, such as financial literacy. A study on the relationship of financial literacy and retirement planning in Switzerland found that, even after controlling for income and education, women are on average more likely to score lower in financial literacy than men (Brown and Graf 2013). We conclude that, among the older population in Switzerland, the differentiation between the more cognitive and the more affective angles of subjective evaluations is clearly of practical relevance for capturing differences in the psychological dimension of economic quality of life between women and men.

The observation that older adults tend to worry less about money than equally endowed younger adults, or that they are more easily satisfied with less resources, has received such consistent empirical confirmation that the phenomenon has been labeled the ‘satisfaction paradox’ (Hansen et al. 2008). Again, the marked contrast between the Self-Assessed and the Perceived Measure requires that this result be discussed in more depth. In explaining older adults’ subjective appraisal of economic quality of life, the previously mentioned triad associated with biological and social age, the ‘age, cohort, and period’ effects must be considered. Their relative contribution to subjective economic quality of life (or low levels thereof) is object of an ongoing debate; in fact, the three types of effect are so intertwined that no single design of analysis, longitudinal or cross-sectional, suffices to disentangle them (Dregan and Armstrong 2009; Winship and Harding 2008). It could be argued that the age effect observed in our data is – at least partly – influenced by the period of observation: the VLV-survey having been conducted in 2011/2012, soon after the economic town-turn of 2008, it may very well be that younger pensioners who expect to live for several more decades are more easily stressed at the thought of the potential impact of the economic downturn on their old age provisions than more senior older adults. If this was true, it would represent an interaction-effect between age and period. The influence of period effects cannot be captured with the VLV survey, as it is cross-sectional data. It may be in order to cite a recent study based on longitudinal SHARE data on this very question, even though its results can only have an indirect bearing on our discussion here because it is focused exclusively on objective economic resources: Schulz and Heckhausen (1996) found that, in the case of Switzerland, the downturn after 2008 did not result in an increased probability of being in greater ‘financial distress’, defined as a lack of liquidity.[2] Another argument that could be raised against the influence of the economic downturn on the results observed in our data is the evidence that there exists a difference between micro and macro worries (Boehnke et al. 1998), the first one focusing on self and in-groups while the second one focusing on society and the world.

The second potential effect captured in the variable age, the cohort effect, is the result of long-term life course exposures to risk factors in early life or throughout the life course that were typical for an entire generation. Examples are generational

[2] The risk of financial distress in times of need was defined as a lack of liquidity if financial wealth, net of non-mortgage debt is less than 3 months of income, under the condition that income is not in the top 1/3 of the income distribution.

life style habits (Wolfe and Burney 1992), structural advantages for asset growth related to a flourishing economy or, to cite a negative example, increased job competition among birth cohorts that are larger in size (Maxim 1985). As discussed previously, it is true for the Swiss case that, as a result of the development of the pension system, younger cohorts of retired people are better endowed than pensioners of earlier birth years (Moser 2006). These contextual influences are accounted for in our models by the variables income and wealth. Context may also have a bearing on the perception of one's own financial situation through shared intra-biographical referencing because older cohorts were exposed to extended periods of economic adversity, shaping their money management skills (Mirowsky and Ross 1999) as well as patterns of consumption (Cook and Settersten 1995), habits which moreover have undergone profound changes over the twentieth century (Offer 2006). Thus, the particular circumstances a person has lived through have an crucial impact on referencing mechanisms or, to put it differently, 'the question ‚How well am I doing?' can be rewritten as the question ‚How well should I be doing, given my biography and historical context?' (Hazelrigg and Hardy 1997, p.4). It is very plausible, that the age effect reported in our models captures cohort differences in expectations about financial security in old age: net of income and wealth (and in contrast to self-assessed economic vulnerability where no age difference was found) older cohorts tend to worry less about not having enough money for current expenditures.

Increasing age has also been thought to be associated with less financial strain (in models that account for resources) because, as people grow older, the financial situation seems to loose relative importance compared to other concerns such as health and social relationships. Exploring the constituents of overall quality of life among the population group 65 and older in Britain, Ann Bowling found that financial circumstances and independence were among the top five things mentioned by older adults as 'giving quality of life' respectively as 'taking away quality of life'. However, in a subsequent regression analysis, none of the socioeconomic indicators were significant for self-rated quality of life; instead, psychosocial factors and health proved to be important (Bowling 2007). Our results can be seen as lending further support to this body of literature, indicating that older adults increasingly see finances as a means to an end rather than an end in itself and that their judgment of financial adequacy may be more strongly influenced by questions of adequacy and a fair and just treatment than by objective resources (George 1992; Liang et al. 1980).

The role of values is another potential explanatory factor located at the intersection of an age-cohort effect. There is evidence that the impact of financial strain on stress may depend on the value a person places on economic success: 'older adults may attempt to mitigate the effects of financial strain by devaluing the importance of economic achievement' (Krause and Baker 1992). Previous research has suggested that self-reported values have an influence on the probability of worrying (Schwartz et al. 2000). In our model, the influence of values in the form of life style expectations was theorized to be relevant for self-assessment of economic vulnerability and operationalized by the variable pre-retirement socio-professional category. It is recommended that future research on subjective economic vulnerability among pensioners in wealthy countries examine the influence of life style expectations by also taking into account nuances in values among persons of a similar socioeconomic standing.

Approaching the question of stress and anxiety from a wider psychological angle, important questions concerning its prevalence and nature remain subject to controversy. It is generally accepted that there is a relatively low prevalence of depression and anxiety disorders among older compared to younger adults (Carstensen et al. 1999; Jorm 2000). However, it has been argued that rates may be underestimated because symptoms are often not recognized due to comorbitity, including dementia syndromes, or as a result of the use of inappropriate diagnostic criteria for older persons (Jeste et al. 2005). The age effect found in our results seems to provide support to the more widely accepted view that old age has an inherently protective effect against stress. Given that the variable mastery appears to control for levels of general stress, and again, the insignificant effect of age on the Self-Assessed Measure, this finding could be attributed to an increased emotional control and psychological immunization to stressful experiences (Jorm 2000). Research based on the life-span theory of control behavior (Heckhausen and Schulz 1995) offers some promising insights. Wrosch and colleagues examined the relation between subjective well-being and three types of control strategies that individuals of different age might endorse when faced with stress: persistence in goal striving (primary control), positive reappraisals (secondary control), and lowering aspirations (secondary control) (Wrosch et al. 2000). While primary controls are directed at changing the situation objectively, secondary controls are self-protective in nature and aim at changing the emotional response towards a challenge (Schulz and Heckhausen 1996). As opportunities for goal attainment decrease with advancing age, individuals are thought to become more concerned with managing losses and maintaining levels of well-being (Baltes and Baltes 1993). Wrosch and colleagues found indeed that older adults tend to report higher levels of secondary control with respect to both indicators, lowering aspirations and positive reappraisals, when dealing with financial stress (Wrosch et al. 2000). Several other studies have confirmed that older adults more frequently use strategies that can be associated with secondary controls (Folkman et al. 1987; Heckhausen and Schulz 1995). These studies share in common that they are based on samples comprising a broader age-range, including working as well as retired individuals. In the case of our sample population, it cannot be assumed that there exist intrinsic age-graded differences in the opportunities for managing financial stress *within* the age group 65 and older. On the other hand, changes in control behavior have been observed with regard to various developmental goals and obstacles, not only in the area of finance, thus it appears legitimate to assume that secondary controls generally gain in significance with advancing age.

Nuancing this preliminary conclusion which tends towards psychosocial explanations for the effect of age on economic stress, a careful interpretation of the association between these two variables must not ignore the possibility of a selection effect: subjective economic hardship has been associated with a higher risk of mortality (Szanton u. a. 2008), meaning that the very old individuals in our sample population can be viewed as 'survivors': 'the increase in risk of mortality that is likely to follow increases in financial strain implies that age will have a positive relationship among individuals who exit the study, while those who remain in the study may experience little increase in strain with age or even a decrease' (Bierman 2014, p. 200). There has been evidence of this selection effect for individuals aged 80 and older in the region surrounding Lake Geneva in Switzerland: individuals with lower educational attainments died earlier, resulting in a reduced heterogeneity among the very old (Oris and Lerch 2009). Since lower educational attainment is associated with lower levels of economic resources, it can be deduced that the age group 80 and older in our sample has, on average, been less exposed to long periods of deprivation, contributing to explaining the observed negative relationship between the variable age and economic stress.

In summary, the discrepancy of the effect of the variable age on the two subjective measures of economic vulnerability shows the relevance of the distinction between the concepts of economic strain and economic stress for understanding economic quality of life among older adults in wealthy countries. Considering an array of explanatory factors that have emerged in previous research, we find that the effect observed in our models is best interpreted as a combination of psychosocial factors linked to biographical as well as cohort-history referencing, and to an age-related increase in the endorsement of secondary controls such as positive reappraisals and lowering aspirations.

References

Antonucci, T. C., Lansford, J. E., Akiyama, H., Smith, J., Baltes, M. M., Takahashi, K., Fuhrer, R., & Dartigues, J. (2002). Differences between men and women in social relations, resource deficits, and depressive symptomatology during later life in four nations. *Journal of Social Issues, 58*(4), 767–783.

Baltes, M. M., & Baltes, P. B. (1993). Psychological perspectives on successful aging: The model of selective optimization with compensation. In P. B. Baltes & M. M. Baltes (Eds.), *Successful aging: Perspectives from the behavioral sciences* (S. 1–34). Cambridge University Press.

Bierman, A. (2014). Reconsidering the relationship between age and financial strain among older adults. *Society and Mental Health*, 2156869314549675.

Boehnke, K., Stromberg, C., Regmi, M. P., Richmond, B. O., & Chandra, S. (1998, June 1). Reflecting the world 'out there': A cross-cultural perspective on worries, values and well-being. *Journal of Social and Clinical Psychology, 17*(2), 227–247. https://doi.org/10.1521/jscp.1998.17.2.227.

Bowling, A. (2007). Quality of life in older age. What older people say. In H. Mollenkopf & A. Walker (Hrsg.), *Quality of life in old age: International and multi-disciplinary perspectives* (Softcover reprint of hardcover 1st ed. 2007, S. 15–30). Springer.

Brown, M., & Graf, R. (2013). Financial literacy and retirement planning in Switzerland. *Numeracy, 6*(2).

Carstensen, L. L., Isaacowitz, D. M., & Charles, S. T. (1999). Taking time seriously: A theory of socioemotional selectivity. *American Psychologist, 54*, 165–181.

Cohen, S., Kamarck, T., & Mermelstein, R. (1983, December). A global measure of perceived stress. *Journal of Health and Social Behavior, 24*(4), 385–396.

Cook, F. L., & Settersten, R. A. (1995). Expenditure patterns by age and income among mature adults: Does age matter? *The Gerontologist, 35*(1), 10–23.

Dregan, A., & Armstrong, D. (2009, November 1). Age, cohort and period effects in the prevalence of sleep disturbances among older people: The impact of economic downturn. *Social Science & Medicine, 69*(10), 1432–1438. https://doi.org/10.1016/j.socscimed.2009.08.041.

Fischer, A. H. (2000). *Gender and emotion: Social psychological perspectives*. Cambridge University Press.

Folkman, S., Lazarus, R. S., Pimley, S., & Novacek, J. (1987). Age differences in stress and coping processes. *Psychology and Aging, 2*(2), 171–184.

George, L. K. (1992). Economic status and subjective Well-being: A review of the literature and an agenda for future research. In N. E. Cutler, D. W. Gregg, & M. P. Lawton (Eds.), *Aging, money, and life satisfaction: Aspects of financial gerontology* (pp. 69–99). New York: Springer Publishing Co.

Hall, M., Buysse, D. J., Nofzinger, E. A., Reynolds, C. F., III, Thompson, W., Mazumdar, S., & Monk, T. H. (2008). Financial strain is a significant correlate of sleep continuity disturbances in late-life. *Biological Psychology, 77*(2), 217–222.

Hansen, T., Slagsvold, B., & Moum, T. (2008). Financial satisfaction in old age: A satisfaction paradox or a result of accumulated wealth? *Social Indicators Research, 89*(2), 323–347.

Hazelrigg, L. E., & Hardy, M. A. (1997). Perceived income adequacy among older adults issues of conceptualization and measurement, with an analysis of data. *Research on Aging, 19*(1), 69–107.

Heckhausen, J., & Schulz, R. (1995). A life-span theory of control. *Psychology Review, 102*, 284–304.

Jeste, D. V., Blazer, D. G., & First, M. (2005). Aging-related diagnostic variations: Need for diagnostic criteria appropriate for elderly psychiatric patients. *Biological Psychiatry, 58*(4), 265–271.

Jorm, A. F. (2000). Does old age reduce the risk of anxiety and depression? A review of epidemiological studies across the adult life span. *Psychological Medicine, 30*(01), 11–22.

Keith, V. M. (1993). Gender, financial strain, and psychological distress among older adults. *Research on Aging, 15*(2), 123–147.

Krause, N., & Baker, E. (1992). Financial strain, economic values, and somatic symptoms in later life. *Psychology and Aging, 7*(1), 4–14.

Liang, J., Kahana, E., & Doherty, E. (1980, May 1). Financial Weil-being among the aged: A further elaboration. *Journal of Gerontology, 35*(3), 409–420. https://doi.org/10.1093/geronj/35.3.409.

Litwin, H., & Sapir, E. V. (2009). Perceived income adequacy among older adults in 12 countries: Findings from the survey of health, ageing, and retirement in Europe. *The Gerontologist, 49*(3), 397–406.

Lucas, R. E., & Gohm, C. L. (2000). Age and sex differences in subjective well-being across cultures. In E. Diener & E. M. Suh (Eds.), *Culture and subjective well-being* (pp. 291–317). Cambridge: The MIT Press.

Maxim, P. S. (1985). Cohort size and juvenile delinquency: A test of the Easterlin hypothesis. *Social Forces, 63*(3), 661–681.

Mirowsky, J., & Ross, C. E. (1999). Economic hardship declines with age: Reply to Hardy & Hazelrigg. *American Sociological Review, 64*(4), 577–584.

Moser, P. (2006). *Einkommen und Vermögen der Generationen im Lebenszyklus. Eine Querschnitts-Kohortenanalyse der Zürcher Stuerdaten 1992–2003*. Zürich: Statistisches Amt des Kantons.

Offer, A. (2006). *The challenge of affluence: Self-control and Well-being in the United States and Britain since 1950*. Oxford: OUP.

Oris, M., & Lerch, M. (2009). La transition ultime. Longévité et mortalité aux grands âges dans le bassin lémanique. In M. Oris, E. D. Widmer, A. de Ribaupierre, D. Joye, D. Spini, G. Labouvie-Vief, & J.-M. Falter (Hrsg.), *Transitions dans les parcours de vie et construction des inégalités* (S. 407–433). Lausanne: PPUR presses polytechniques.

Pearlin, L. I., & Schooler, C. (1978). The structure of coping. *Journal of Health and Social Behavior, 19*(1), 2–21.

Schulz, R., & Heckhausen, J. (1996). A life span model of successful aging. *American Psychologist, 51*(7), 702–714.

Schwartz, S. H., Sagiv, L., & Boehnke, K. (2000). Worries and values. *Journal of Personality, 68*(2), 309–346.

Szanton, S. L., Allen, J. K., Thorpe, R. J., Seeman, T., Bandeen-Roche, K., & Fried, L. P. (2008). Effect of financial strain on mortality in community-dwelling older women. *The Journals of Gerontology Series B: Psychological Sciences and Social Sciences, 63*(6), S369–S374.

Veenhoven, R. (2007). Subjective measures of well-being. In M. McGillivray (Ed.), *Human well-being: Concept and measurement* (S. 214–239). Basingstoke: Palgrave MacMillan.

Weidekamp-Maicher, M. (2008). *Materielles Wohlbefinden im späten Erwachsenenalter und Alter*. Berlin: Dissertationsverlag.

Winship, C., & Harding, D. J. (2008). A mechanism-based approach to the identification of age–period–cohort models. *Sociological Methods & Research, 36*(3), 362–401.

Wolfe, C. D. A., & Burney, P. G. J. (1992). Is stroke mortality on the decline in England? *American Journal of Epidemiology, 136*(5), 558–565.

Wrosch, C., Heckhausen, J., & Lachman, M. E. (2000). Primary and secondary control strategies for managing health and financial stress across adulthood. *Psychology and Aging, 15*(3), 387–399.

Chapter 30
Discussion and Final Conclusion

For any given construct, there exists a gap between the concept and the achievable measurement of its attributes in empirical research. The heterogeneity of both the needs- and the resource-component has been widely recognized as a complicating factor in measuring financial adequacy. This volume started out with questioning the traditional way of conceptualizing low levels of economic quality of life as income poverty and, more specifically, the operationalization of this concept based on an absolute monetary poverty line. The evidence reported in Part IV contributes to a stream of literature stating that income-based poverty measures fall short of identifying those individuals who are at the lowest end of the spectrum in terms of economic quality of life. Our findings lead us to caution against an isolated use of the absolute poverty line in the case of pensioners in Switzerland. In fact, we find that, in the Swiss context and applied among older adults, the absolute monetary poverty line is not measuring what most researchers assume they are measuring because income turns out not to be a reliable proxy for the general resource situation of pensioners in this country. In the case of the population aged 65 and older living in Switzerland, we are dealing with an extremely diverse group in terms of economic resources but also with regard to health-related financial needs and socially normed expectations. These population characteristics exacerbate the potential biases caused by conventional measures of economic vulnerability. Our analysis showed that the use of the monetary poverty line induces two main errors in identifying those groups who can legitimately be called 'economically vulnerable': it wrongly labels a considerable share of the population 'poor' even though they own a fortune, and it fails to identify another important share of the population who dispose of a limited budget just above CHF 2400 per month, who are struggling to make ends meet without having any financial stocks to draw on. Both of these misinterpretations have to do with the pivotal role of wealth in the household portfolio of older adults and the erroneous assumption that monthly income in retirement is an appropriate indicator of the general resource situation. Deconstructing the predominant view of objective measures as a 'view from nowhere' we argued that, from a quality of life perspective, the experience of the person labeled as 'vulnerable'

J. Henke, *Revisiting Economic Vulnerability in Old Age*, Life Course Research and Social Policies 11, https://doi.org/10.1007/978-3-030-36323-9_30

or 'not vulnerable' should take a central place in the definition of economic vulnerability. In search for an alternative or complementary measure, we proposed to evaluate whether a self-assessed indicator of economic vulnerability could be used as a global indicator of financial strain. Thus our goal was to empirically assess whether in the Swiss context, this measure is able to serve as a benchmark for identifying groups of pensioners who are experiencing low levels of economic quality of life.

Substantial evidence was found that the Self-Assessed Measure, operationalized by a binary indicator of difficulties in making ends meet, is strongly influenced by the availability of financial resources: among a wide array of economic variables tested, the Self-Assessed Measure is primarily predicted by income and wealth, more precisely, a combination of both. As far as its sensitivity to other indicators of low levels of economic quality of life is concerned, we found mixed evidence: among three variables measuring the frequency of social participation, two were more strongly associated with the Objective Measure than with the Self-Assessed Measure. It cannot be ascertained with our data whether differentials in levels of social participation are due to life style habits or the result of adaptations to a reduced income after retirement. As for the association with poor health, an important component of global quality of life in old age, neither the Objective nor the Self-Assessed Measure standing alone was able to get close to the predictive power of the two measures combined. The sum of these findings leads us to the paradoxical conclusion that the subjective assessment of the difficulties in making ends meet represents a relatively accurate indicator of an economic situation that is characterized by the lack of financial resources but that yields ambiguous results with regard to corresponding 'symptoms and consequences' of economic vulnerability.

At this point we must address the fact that the proposed Self-Assessed Measure is not without critique in literature. One of the main arguments raised against the use of subjective poverty indicators is that individuals who have experienced financial difficulties over an extended period of time may lower their expectations and adapt their aspirations and preferences, a situation that in our typology would correspond to many respondents of type BA. The phenomena of downward adaptation has indeed been observed in the Swiss context and, more specifically, using the same measure as the one used in this volume: based on the 2006–2010 waves of the Swiss Household Panel, Crettaz and Suter examined whether and to what degree indicators of material deprivation, subjective poverty and subjective well-being are affected by downward adaptations (Crettaz and Suter 2013). A longitudinal design allowed quantifying the degree of a downward adaptation as time progresses: the authors found that the odds of having more difficulties in making ends meet decreased by 10% after having spent 1 year in income poverty, and it decreased by 41% after 5 years of income poverty. These results cause them to be skeptical about the use of the Self-Assessed Measure: 'These findings suggest that subjective indicators are indeed often affected by the problem of adaptive preferences. In particular, downward adaptation due to (long-term) poverty, i.e., the fact that the poor get used to financial and material hardship, raises doubts about the suitability of subjective indicators for policy making' (Crettaz and Suter 2013, p.11).

The statistical effects described in the cited study are indeed substantial and their practical relevance cannot be questioned. Assuming that the psychosocial mechanisms of reference group comparison and intra-biographical referencing are comparable for older adults and that the effects would therefore be similar to those observed for the general Swiss population, we can assume the following: most respondents in our data set have seen their monthly income drop by an estimated 22–30% (Seifert and Pilgram 2009) and their financial situation is likely to have remained relatively stable since they left their professional activity (Moser 2006), implying that any adaptation process would already be well under way at the time of the interview.[1]

Moreover, while we appreciate the concern behind the authors' caution, we come to a different conclusion about the suitability of the Self-Assessed Measure for informing social policy. The criticism voiced by Crettaz and Suter resonates with an often-quoted statement by Sen: 'The most blatant forms of inequalities and exploitations survive in the world through making allies out of the deprived and exploited. The underdog learns to bear the burden so well that he or she overlooks the burden itself. (…) As people learn to adjust to the existing horrors by the sheer necessity of uneventful survival, the horrors look less terrible in the metric of utilities' (Sen 1984; p.308–309). In a situation where economic vulnerability is a question of survival, a subjective poverty assessment may in effect come closer to a test of how well people are informed about their lot (Kingdon and Knight 2006).

When talking about possible risks of a self-assessed evaluation of economic strain in the Swiss context, we cannot ignore that the notion of a minimal living standard in this country is based on a sociocultural concept, far above the subsistence level alluded to by Sen. As argued at different times throughout this volume, deprivation among Swiss pensioners is most likely to manifest in the area of social participation. The question we can attempt to respond to based on our data is therefore, whether respondents who are income poor but who do not self-assess themselves as economically vulnerable (type BA) are on average deprived of widespread social activities without realizing it. Our analysis has shown that it is rare among members of group BA to report that they never go out for a drink or a meal; in fact, they tend to go out more frequently than members of type AB. When it comes to more costly types of participation, such as taking a trip or a vacation, those whose economic vulnerability is based on self-reports tend to travel less frequently compared to those who are 'only' objectively vulnerable, thus confounding somewhat the suspicions evoked by Crettaz and Suter. These findings, based on sub-group comparisons, undermine the idea of an excessively low living standard of a population group who supposedly failed to self-assess themselves as economically vulnerable. Though the possibility of individual cases can never be ruled out (and later in

[1] It is obviously plausible that factors other than the loss of income at retirment may lead to financial strain.

this conclusion, we will mention examples thereof) we argue that the evidence for downward adaptation presented by Crettaz and Suter must be interpreted as an argument *in favor of* the reliability of the Self-Assessed Measure. That is, if it is used in a wealthy country and among a population group whose income can be expected to remain very stable because, as we have shown, there is little risk for adaptation to exceedingly low levels of quality of life,[2] and if the phenomenon of expensive taste – another point that is sometimes raised against subjective measures of economic well-being – is taken into account as far as possible. As we will expose later in this conclusion when addressing the Vulnerability Typology, these two undesirable scenarios related to referencing mechanisms could be identified if the Perceived Measure is drawn on.

The motivation to search for alternative indicators for measuring economic vulnerability emerged from the conceptual and methodological challenges of conventional measures, specifically the shortcomings of the absolute monetary poverty line in capturing the economically most vulnerable groups among the retired population. What is at stake is clearly the relevance of a given 'convention of equivalence of heterogeneous observations', admitting that no one way of dividing a group into 'vulnerable' and 'non-vulnerable' will be without failings. In a discussion about the usefulness of subjective measures, the risks and opportunities must therefore be assessed and weighed against each other, as well as against possible alternatives. From a quality of life perspective, the question of 'who decides the standards' is pivotal. We find that in the Swiss context, the Self-Assessed Measure offers a valid alternative for measuring low levels economic quality of life. By placing the individual at the core of the evaluation process, he or she in charge of determining the components and the threshold that are relevant in his or her own judgment. The economic situation of an individual, however challenging from a measurement perspective, represents a factual substance matter and the survey question 'difficulties in making ends meet' provides a clear criterion of evaluation by which to assess it. The presence of adaptation processes influencing the Self-Assessed Measure cannot be refuted; they are in fact an integral component of the appraisal process that must not be frowned upon but taken into consideration in the interpretation of the results. Granted that individual standards for judging economic vulnerability are socially and biographically conditioned (and thus relative, to some degree), the threshold below which a person begins experiencing difficulties in making ends meet must nonetheless be considered as 'absolute'[3] and therefore authoritative from a quality of life perspective: what matters is not so much having equal economic strain as others, but just not having economic strain, absolutely.[4] Let us not forget that eco-

[2] As exception, type BAB could be mentioned.

[3] This statement does not contradict the previous discussion about downward adaptation: here, we are referring to the experience at the time of measurement (the survey interview), which does not exclude the possibility of adaptation mechanisms that lead to a different assessment at a future point in time.

[4] The original quotation is 'Not so much having equal shame as others, but just not being ashamed, absolutely.' (Sen 1983, p.161).

nomic strain in old age has an especially bitter taste to it because it threatens to be a permanent lot.

Thus, weighing the advantages and disadvantages of the Objective Measure and the Self-Assessed Measure, we have come to the same conclusion as other researchers before us: 'an approach which examines the individual's own reported perception of well-being is less imperfect, or more quantifiable, or both, as a guide to forming that value judgment than are the other potential approaches' (Kingdon and Knight 2006, p.1220). Our results based on a systematic comparison of an absolute monetary line and the self-assessed difficulties in making ends meet make a strong case for the inclusion of self-assessed indicators among the standard monitoring tools that inform social policy in wealthy countries. Especially in the absence of information on financial stocks, it is highly recommended to use the Self-Assessed Measure as a complementary source of information on economic vulnerability. This being said, the Self-Assessed Measure based on the difficulties in making ends meet does not compete with the income-based poverty measure. In fact, we take objective measures of monetary resources to be play a pivotal role in the analysis of social well-fare and social change. As a proxy-measure they can do good service in assuring comparability of results over time and across different contexts. It is at the level of interpretation that caution is advised because the potential relevance of this measure from a societal point of view crucially depends on the interpretation of the results: 'While it is easy to calculate the same indices in different contexts, it is not a given that they reflect the same kind of information, even when calculated over equally defined variables – which entails the inclusion of the same components and the absence of verbal confusion on what is meant in similar questions posed in different surveys' (Laderchi 2000, p.14). Thus, the actual relevance of this measure can only be evaluated after taking into account the interpretation of the results, which must include a critical response to the questions of whether the defined minimum amount of Swiss francs is able to 'buy' the same level of the 'good life' for all groups assessed.

Moving on the Perceived Measure, the focus of the analysis presented in Part V was to empirically confirm the existence of two dimensions of subjective economic vulnerability. The social-indicator matrix by Veenhoven and Lazarus' theory of appraisal laid the theoretical foundation for our proposition that, from a quality of life perspective, it is relevant to distinguish between the Self-Assessed and the Perceived Measure of economic vulnerability. The empirical analysis revealed that the two groups that are vulnerable according to each of these subjective measures are dissimilar in many regards and that combining the two measures increases the likelihood of identifying those individuals who are affect by very low levels of economic quality of life.

The relatively lesser association of income and wealth with the Perceived Measure compared to the Self-Assessed Measure is in line with our hypothesis though the relationship between Economic Resources and the Perceived Measure was unexpectedly weak. The relation between the two subjective measures and a series of psychosocial variables, too, is less clear than expected. The nature and the strength of their correlations with the Perceived Measure are only partly congruent with our theoretical model: it would in fact be more accurate to exchange the model component Psychosocial Consequences and Symptoms with the more limited concept 'diminishment of self' as only the worry about not getting sufficient recognition had a mediating effect between economic strain and stress. The combination of the two subjective measures proved relevant for identifying the population group with the lowest sense of mastery: if the strain of not being able to make ends meet causes people to worry about not having enough money to pay their current expenditures, a possible consequence or symptom may naturally be the proliferation of the general experience of stress (Pearlin et al. 1997), as captured by our variable mastery/perceived stress scale.

The validity of the results presented in this volume needs to be seen in light of a few methodological limitations. Though our research question specifically focused on manifest vulnerability outcomes, it would have been ideal to additionally dispose of longitudinal data in order to be able to empirically test the dynamics of the risk chains leading to the observed vulnerability outcomes. Moreover, comparing data of same-aged respondents born at different points in time would have allowed disentangling the age effect observed for the Perceived Measure, that is, whether the lower susceptibility to economic stress of very old adults in our sample is due to a shared cohort history or characteristic of an age-related serenity.

There are certain limitations with regard to the variables used for operationalizing our theoretical concepts. Given the generally weak association of the Perceived Measure with any of the explanatory variables, the question of the cut-point may have to be revisited: the transformation of the likert-type variable into a binary indicator based on an arbitrary rupture of a continuum of perceived states may have hampered the predictive power of the variable worry about finances (perc_ev). Future research on the perceived measurement angle should rely on a variable that provides an inbuilt ranking that allows to more clearly identify a qualitative shift towards exceedingly low levels of perceived economic quality of life.

With regard to Economic Resources, more precise information on the type and amount of resources would have been an advantage, specifically being able to distinguish between assets that can easily be turned into liquidity from long-term investments such as real estate. Moreover, information on debts and mortgages would have provided a more complete picture of the individual's financial situation. With regard to social participation, it would have been recommended to discriminate the possibility that individuals choose not to undertake certain activities because of preference, as is state of the art for deprivation measures.[5] The operationalization of the model component Needs and Expectations, too, would have benefitted from custom-made items allowing for a more direct measurement of the factors that mediate the effect of Economic Resources on economic strain. Specifically, information on referencing mechanisms would have enhanced our research: examples thereof would be survey questions that elicit the specific life-style expectations of respondents,

[5] In the variables used here, were were not able to know whether individuals would have engaged in a social activity on a more frequent basis if they had more money at their disposition. For a discussion on this question with regard to deprivation measures, see Chap. 9, Sect. 9.2.3.3.

questions about reference groups and about comparing the current life style with pre-retirement standards and habits. Additional variables for the assessment of financial needs would have included items on living costs (housing costs, in particular) but also information on the financial implications of health problems.

For the testing of the hypothesis concerning Psychosocial Symptoms and Consequences, future research would benefit from variables that are more specifically tailored to measuring the concept of diminishment of self. To use the example of role strains, it would be expedient to have response items on a variety of ways by which scarcity of money may affect the person's fulfillment of important relationships. Instead of Cohen's perceived stress scale it would have been desirable to dispose of the mastery variable developed by Pearlin in order to capture the concept of sense of control with greater accuracy.

Another caveat that needs mentioning is the operationalization of some of the latent constructs presented in the structural equation models: the data available in the VLV-questionnaire being limited, several of the latent constructs were based on only two items instead of three or more, which would have increased robustness. Specifically, the above-mentioned Economic Resources, but also, the concept of social participation would have benefitted from additional items.

Lastly it needs to be stated that we do not wish to overgeneralize from the Swiss case: objective and subjective dimensions of economic vulnerability in old age may look quite different across countries. The theoretical and empirical work presented in this volume merely opens avenues for new research as the contextual factors that are unique to each nation and welfare system may lead to very distinct empirical results. For future research on subjective measures of economic vulnerability in wealthy countries, closer investigation about adaptive preferences would be desirable. Given the limitations of our data described above, our preliminary conclusion that this phenomenon does not undermine the validity of the Self-Assessed Measure should be substantiated or nuanced with more appropriate variables. Specifically, the question whether younger cohorts are on average more demanding with regard to minimal standards of living should be the object of closer investigation. In the context of biographical referencing, the role of values should be taken into account for a better understanding of the discrepancy between the Self-Assessed and the Perceived Measure. Instead of focusing only on the influence of failed coping resources in mediating the association between the two measurement angles, it would be promising to consider the role of values. What people value sheds light on the underlying motivation for aspiring certain states in life and 'worry is elicited when the attainment or maintenance of these desired states is threatened' (Boehnke u. a. 1998, p. 750). Lastly, we suggest that future reserach on the perceived measurement angle should include additional items on financial stress. Ideally, they should distinguish between intensity, duration and content of worries (Boehnke u. a. 1998). The marked gender-differences observed in our sample population call for further research on differences in coping resources and strategies between women and men. Similarly, the protective effect of age on economic stress deserves scientific attention. Preserving the distinction between the two subjective measurement angles in future analysis is pivotal for understanding the association of these two demographic variables with economic vulnerability.

We will now proceed to an assessment of the Vulnerability Typology by connecting our empirical findings with the pathways laid out in the vulnerability framework, taking up the discussion from the final chapters of Parts IV and V. The locus of interest here is to consider how risk factors and particular 'coping' interact for each vulnerability type, pointing to differentials in individual agency. Linking the dynamic vulnerability framework with the typology helps us reflect more systematically about the heterogeneity behind the observed outcome measures. Ultimately, the following is our concluding assessment of the validity of the typology to serve as an index that offers information on latent vulnerability derived from the underlying risk chain constellations. Beginning by the binary combination of the Objective with the Self-Assessed Measure and a discussion of the primary vulnerability types in terms of Economic Resources and Needs and Expectations, we then integrate the Perceived Measure and review the main findings on the psychosocial dimension of the vulnerability experience.

In Part IV, a look through the lenses of 'vulnerability types' revealed the contours of three[6] socioeconomic profiles of economically vulnerable older adults in Switzerland. The theoretical profiles, deduced from latent risk constellations, were indeed reflected in the manifest outcome constellations observed in our empirical analysis. In what follows, we will review the main findings concerning the three vulnerability types BB, AB and BA and in particular their association with wealth, social participation and health. For reasons of clarity, the labeling of the vulnerability types will include reference to the Perceived Measure, denoted as a generic X. The following discussion is based on outcome probabilities predicted by a series of ordinary regression models presented in Part IV.

Two theoretical risk constellations were identified as pathways to economic vulnerability type BBX,[7] the most negative of the four possible outcome combinations (Table 30.1). Group BBX is characterized by a high rate of individuals who have no wealth or less than CHF 60,000 in savings (55%). In both absolute and relative[8] terms, the most important socio-professional categories making up this group are former skilled manual workers and self-employed persons: over half of the respondents of group BBX belong to either of these two socio-professional categories. Among the three basic vulnerable types, BBX reports by far the lowest share of individuals aged 65–84 who benefit from good or very good health (43%). As far as costly social activities are concerned, this group reports the lowest frequency of participation: one third of respondents in this group never go out for a meal or a drink or only once a year, and one quarter reports to never take a day trip or a holi-

[6]The fourth type emerging at the intersection of the Objective and the Self-Assessed Measure is type AA, which is strictly speaking not a 'vulnerable' type.

[7]As mentioned previously, the distinction between risk constellation 3 and 4b) can only take place once the Perceived Measure (denoted by the X) is taken into account, which will be the case later in this chapter.

[8]This rate is higher than the one recorded for the overall sample or for any other vulnerable group.

Table 30.1 Overview of risk constellations and Vulnerability Typology

Group of respondents	*latent vulnerability – risks*		*manifest vulnerability – outcomes*			*Vulnerability Typology*
Risk constellation	**exposure + threat**	**coping**	**objective**	**self-assessed**	**perceived**	
1	↑	not applicable	non-vuln.	non-vuln.	non-vuln.	**Type AAA**
2 a)	↓	↑ success, sustainable	non-vuln.	vuln.	non-vuln.	**Type ABA**
2 b)	↓	↑ success, role-strain	non-vuln.	vuln.	vuln.	**Type ABB**
3	↓	↑ absent, insufficient	vuln.	vuln.	vuln.	**Type BBB**
4 a)	↓	↓ do not realize	vuln.	non-vuln.	non-vuln.	**Type BAA**
4 b)	↓	↓ do not want to mobilize coping resources	vuln.	vuln.	non-vuln.	**Type BBA**

day. As shown in the SEM analysis (Fig. 23.2), part of the explanation for the lower than average social participation may be that members of this group are affected by reduced mobility and poor functional health. Mostly, however, analysis of the influence of Financial Needs and Expectations suggest that more than any other vulnerable group, BBX is composed of individuals who have been suffering from low levels of economic quality of life for longer periods of time across the life course: on the one hand, the recorded low or inexistent financial stocks are tantamount to inexistent coping resources; on the other hand, a comparison with the population trends in social participation suggests an unusually modest life style, bordering on social isolation. Lastly, widespread health problems are another sign that long-term deprivation or financial hardship is taking its toll.[9]

Since our general hypothesis for the variable set Psychological Symptoms and Consequences found only limited empirical support, the distinction between risk constellation 3 and 4b that becomes possible with the Perceived Measure, has to be discussed with much caution. Outcome constellation BBB, the result of absent or insufficient coping resources, is expected to be associated with a diminished sense of self and low levels of mastery. Type BBA is thought to represent those cases, where the tight financial situation is recognized but deemed of little importance, or at least, not important enough to mobilize coping resources. For this reason, levels of mastery are expected to be unaffected and the intensity of worry about not getting enough recognition is likewise expected to be minimal. For the variable recognition, a small difference between the groups BBB and BBA was found, with BBA reporting levels close to type AAA, lending support to our hypothesis. For the variable mastery, members of type BBB clearly have a higher probability to be found among the population share with the lowest levels of mastery compared to type BBA.

From the perspective of our theoretical framework, risk constellation 2, leading to outcome ABX, stands out as a potential 'happy end' because, thanks to successful coping strategies, these respondents have been able to get above the objective vulnerability threshold. The subjective vulnerability report is expected to be negative because of a heightened awareness of scarcity of resources, though we have to keep

[9]This interpretation diverges from our original hypothesis that health would be a factor that increases financial needs. However, in light of previous findings, the presented conclusion seems more pertinent.

in mind that our theoretical model posits that whether individuals suffer psychologically from the situation or not will critically depend on (their assessment of) the sustainability of the solution or, on the flip side of things, whether the situation negatively affects their ability to fulfill social roles (Perceived Measure). In terms of the sustainability of financial coping resources, our analysis indeed showed that ABX features the most important share of respondents who do not have sustainable levels of capital stocks that can be drawn on: the outcome probability for 'owning nothing or almost nothing' is 33%; this is even higher than for group BBX.

As far as the impact of role strains is concerned, adding the Perceived Measure allows for further nuances within type ABX: risk constellation 2a (type ABA) concerns individuals who have to watch their budget while feeling confident in their ability to manage in the long run. In contrast, the almost identical type ABB (risk constellation 2b) is worried about not having enough money to pay current expenditures, possibly because for these people, the lack of finances feels to them like a limitation in the fulfillment of social obligations. Our analysis confirms, indeed, that members of group ABB are less satisfied with the way they are able to fulfill their social roles compared to group ABA, reflected in that they are more frequently worried about not being recognized for what they do. Moreover, type ABB has on average slightly lower levels of mastery than type ABA.

Our findings on the frequency of engaging in social activities must be interpreted against the backdrop of our assumptions concerning patterns of social participation in later life, specifically the idea of continuity and the likelihood of an increased frequency of participation during 'third life' – all other things remaining equal. Given that members of group ABX currently have a higher monthly income than members of groups BBX and BAX and that, on average, this relatively higher prosperity must already have been available in pre-retirement times, it is safe to expect that members of ABX will record a higher frequency in social participation relative to the other two groups. This has been confirmed for the frequency of going out to eat or drink something in a restaurant or a coffee shop: among the three vulnerable groups, type ABX is most similar to the non-vulnerable majority type AAX. This is a strong indicator that, compared to the other two groups, respondents of type ABX have not always been struggling financially but have previously enjoyed a day-to-day life style that is typical for the middle class. In this regard, it is interesting to remember that for the variable restaurant no variation by either level of wealth nor by pre-retirement socio-professional category was found. The absence of any effect of wealth suggests that this type of social activity is considered basic enough as not to justify consumption of financial stocks. The insignificant effect of the socio-professional category may mean that eating or drinking out does not generally represent a luxury in the Swiss context among those pensioners who have at least CHF 2400 per month to live on. When it comes to a one-time-off expenditure such as going on a day-trip or on holidays, this allegedly more socially active group scores lower than types BAX and BBX and substantially lower than AAX: the outcome probability of never travelling is 28%, compared to 11% for group AAX. Furthermore, group ABX has the highest outcome probability for never going to see a movie, a theater play or a concert.[10] This may be an indication that large shares of group ABX do have to make adjustments to a tighter financial situation by renouncing some more costly ways of social participation.

A review of the findings pertaining to type ABX seems to depict a group of people who have seen better times and who are expecting things to get worse, still. Their financial stocks are for the most part not sufficient to be considered a sustainable coping strategy. Everyday life style habits are similar to the non-vulnerable population (AAX), which of course, encompasses a wide range of patterns, leaving much room for personal adjustments that may be painful at the individual level. There is evidence of a life style change as a consequence of a shrunk budget that no longer allows those extra treats that used to be habitual. On the other hand it is important to reiterate that we did not find indications that the economic self-assessment of members of group ABX was generally affected by the phenomenon of expensive taste as could have been expected, given the binary nature of the monetary indicator that is open to the top (meaning that 'A' could stand for any level of income). If this bias had been prevalent among members of ABX, the group would have displayed a socio-cultural habitus – frequency of going to the movies, a theater play or a concert – similar to type AAX and not, as our data shows, quasi identical to the vulnerable groups BBX and BAX.

If poor health is interpreted as an indicator of the long-term effects of a life marked by economic hardship, the fact that health levels of type ABX are identical to those of the non-vulnerable type AAX is worth taking note of. This observation supports the idea that, among the three basic vulnerability types, ABX tends to represent individuals who have not always been faced with economic vulnerability but who are currently struggling despite attempts of adjusting to a lower budget. Having said this, we do not mean to imply that negative self-assessment is the direct consequence of the loss of income at retirement. In this context, a recent study about the transition to retirement and its influence on subjective economic vulnerability may be of interest. Using all waves[11] of the SHARE survey in 16 European countries, among them Switzerland, Fonseca and colleagues examined the effect of retirement on financial well-being, measured by indicators of relative poverty (60% of the median disposable equivalized income) and the self-assessed difficulties in making ends meet (Fonseca et al. 2013). Using an instrumental variable approach to account for the potential endogeneity of the effect of retirement[12] the authors found no evidence that the passage to retirement increased the risks of either relative poverty or of having greater difficulties in making ends meet in the case of Switzerland. This result based on longitudinal data helps contextualize our findings, indicating that

[10] As mentioned before, apart from this exception, the variable cinema barely showed any differences in the response pattern among the types BBX, ABX and BAX, which is why it will not be discussed further in this conclusion.

[11] 2004, 2006, 2010.

[12] The researchers used dummy variables to control for cross-country differences in eligibility, distinguishing between early and full pension ages.

factors other than the loss of income at retirement could have led to the observed self-assessed outcome for group ABX.

The risk chain constellation of the income-poor but self-assessed 'well-off' group BAX probably represents the least accurate fit of our theory to the real-life scenarios found in our empirical data. From a theoretical point of view, we saw that the activation of coping strategies requires that the individual recognize his or her situation as being adverse. Our analysis has shown, however, that the self-assessed angle is strongly influenced by the presence of wealth, prompting us to reframe some of our assumptions. The vulnerability experience represented by type BAX seems to be the least palpable because it combines two very distinct subgroups. On the one hand, it is made up of individuals who, despite being income poor, have enough savings that they are able to comfortably maintain their life style. The group of formerly self-employed exemplifies this atypical resource constellation: making up close to one fifth of group BAX, their outcome probability of being in the highest or second to highest wealth group (>500,000, 150,000–500,000) is 65%. For about half of the sub-group, those who are in the two highest wealth categories, the label 'vulnerable' does not seem appropriate. On the other hand, BAX is made up of individuals who have gotten used to a modest life style. This sub-group, representing roughly 50% of the group, suspicion is advised because of the optimistic outcome of the Self-Assessed Measure. Is it possible that these individuals, many of them skilled and unskilled workers, have given up on sharing in the commonly accepted level of social participation? Analogous to our observation for group ABX, no difference in life style appears with regard to the frequency of going to a restaurant or a coffee shop (see Appendix 4.5), which varies barely between socio-professional categories within type BAX. The only noticeable difference exists for the habit of travelling (see Appendix 4.6): unskilled workers and skilled manual workers (who together make up 40% of group BAX) are significantly more likely to 'never' take a trip or a holiday compared to former top-executives, academic or self-employed, whose activity rate is closer to the non-vulnerable type AAX. Again, we find that in the Swiss context, the threshold at which a social activity becomes object of budgetary considerations seems to settle around the amount necessary for taking a (however short) holiday – going out for a meal or drink remains affordable for most older adults, even among the economically vulnerable. This finding is congruent with previous research that established a very low degree of inequality in terms of quality of life (measured by CASP-19) among Swiss pensioners. The study, based on SHARE data (wave 2004) from ten European countries, found that the effect of income and net worth on levels of control, autonomy, self-realization and pleasure was insignificant or weak for older adults in Switzerland (Knesebeck et al. 2007).

The variable health positions group BAX as the healthiest group or, at least, the group that is most satisfied with their health, even more so than group AAX. The intra-group heterogeneity observed for this type was already object of discussion in Part II: based on the extended typology we found that for a series of examined health problems, type BAA was virtually identical to type AAA while type BAB fared significantly worse. This result points towards future avenues of research that may look at the distinct influence of economic stress (versus economic strain) on health outcomes. More central to our research question is the psychosocial difference between the two branches of type BAX: our theoretical model predicts a higher sense of mastery and low levels of worry about not being recognized for type BAA because these individuals 'do not realize' that they are economically vulnerable (risk constellation 4a); type BAB does not feature in the vulnerability framework because it makes no theoretical sense. The outcome probability of 'getting sufficient recognition: does not worry me/does not concern me' is indeed a high 0.7 for BAA, compared to 0.4 for type BAB. With regard to mastery, no significant differences are found between the two types. However, in view of our discussion of the two subgroups within type BAX, along the lines of socio-professional category, we find that the theoretical risk chain scenario is a poor fit to reality: type BAA is fundamentally the result of a discrepancy between income and wealth, especially among the formerly self-employed. The mystery surrounding BAB – the contradiction between the Self-Assessed and the Perceived Measure – cannot fully be explained but, once again, a look at the socio-professional category proves insightful: type BAB consists to 25% of unskilled workers; this is 6 times more than in the overall sample. Members of this group appear to be so accustomed to managing with very little money that they have strongly adjusted their life style habits. As a result, they are oblivious to the strain of making ends meet, yet on a psychological level, the worry about not having enough for bills, food and rent does not leave them in peace. The distilled findings on type BAX makes this group out to be the most complex and at the same time the most ambiguous among the three vulnerability types: on the one hand, BAX features individuals who ought to be classified as wealthy, while on the other hand, there is about the same proportion of individuals who seem to have succumbed to a life style that is below the Swiss average in terms of social participation and whose health testifies to a life course that was not conducive for their physical and psychological well-being.

In our theoretical discussion of the characteristics expected of an appropriate measurement instrument for economic vulnerability the idea of *relevance* – in terms of components to be considered as well as standards – was of central importance. Agreeing with Sen and Hawthorn (1987, p.20), we found that a good measure must be able to capture the *richness* of the idea of economic quality of life while being *practical* enough to be used in large-scale empirical research. Another requirement put forward in literature that we sought to respond to concerns the meaningful *interpretation of diverging measurement perspectives*: 'social policy makers need both objective and subjective indicators. The challenge of social reporting is to combine the strengths of these indicators and to make sense of the discrepancies they show' (Veenhoven 2002, p.43). The present Vulnerability Typology was developed in response to these claims. The Typology is based on integrating strands of information that should not be analyzed in isolation, providing a synthesis of the experiential spectrum of economic vulnerability. Allowing these different components and standards to coexist side-by-side is a way of preserving the richness of the idea of economic quality of life that spans material as well as psychological dimensions of economic well-being, pertaining to the collective interest as well as being sensitive to individual well-being. Given the complexity of factors that interact in shaping the individual experience in terms of resources, needs and expectations, as well as those factors related to social roles and identity, the combination of the Objective, the Self-Assessed and the Perceived Measure provides a way of distilling this wealth of information. Putting these measures side by side and allowing them to 'talk to each other' in an unmediated manner is a way of acknowledging that several perspectives may be equally valid. By simultaneously drawing on different vantage points, the typology thus enables the researcher to come closer to the ideal of 'reasoning trans-positionally' (Sen 1993, p.130). Moreover, we have empirically demonstrated that the observed incongruence in vulnerability outcomes has the potential of becoming an additional source of information if they are interpreted within a theoretical framework of probabilistic vulnerability: in juxtaposition with evidence-based risk chain constellations derived from an analysis of the local context, the joint analysis of three dimensions of economic vulnerability provides a glimpse into the 'black box' of human agency. In terms of its practical usefulness, it can be noted that the data used for operationalizing the proposed instrument can be collected by means of cross-sectional surveys and can be meaningfully interpreted with relatively basic statistical techniques; yet at the same time, the results encapsulated in the Vulnerability Typology also lend themselves to be used in a variety of more complex analyses. Of course, before this typology can be recommended as a useful instrument for analyzing economic vulnerability in other high-income countries, the theoretical assumptions underlying each of the vulnerability profiles would have to be reviewed in light of the most common risk factors in the local population.

It is our hope that the Vulnerability Typology may contribute to creating an open dialog between social sciences and the people who are the 'object' of social research: when researchers disclose the incongruence among different measurement perspectives, a need to discuss the lack of overlap arises and the chance that the normative character of methodological choices is made explicit increases. Since the existence of a neutral and value-free measurement is elusive, the systematic integration of the two steps of data analysis, measurement and interpretation, could contribute to moving toward a social science research that is more transparent with regard to the 'position' that it deems to be legitimate. In this sense, we believe that a triangular approach to measurement favors the kind of conceptual clarity that stands in contrast to what Gasper's referred to as the 'fallacy of misplaced concreteness', an analytical shortcut that is especially tempting when it comes to multidimensional constructs such as economic quality of life (or low levels thereof).

The observation that there is so little overlap between the three population groups identified as economically vulnerable according to objective and subjective measures calls for special attention not least from policy makers. According to the objectives laid down in the Swiss federal constitution the purpose of the Swiss pension system goes beyond granting a minimum level of existence; it is to allow the insured person to 'maintain his or her previous lifestyle in an appropriate manner.'[13] If we want to preserve the achievements of our welfare state, to allow everyone to grow old in dignity, we are well advised to listen to pensioners directly, trusting that they are the ones who are best positioned to make judgments about their own well-being. For the same reason, a triangular approach to measurement may also be of interest to other fields of research within social gerontology: at a life stage where cognitive and physical limitations tend to be on the increase, the reports of psychological well-being and objectively observable outcomes are likely to diverge consistently as people's priorities and internal frames of reference adjust to age-related challenges. Here, an adaptation of the Vulnerability Typology could do good service, for example in assessing health-related quality of life or aspects of social integration.

References

Boehnke, K., Schwartz, S., Stromberg, C., & Sagiv, L. (1998). The structure and dynamics of worry: Theory, measurement, and cross-national replications. *Journal of Personality, 66*(5), 745–782.

Crettaz, E., & Suter, C. (2013). The impact of adaptive preferences on subjective indicators: An analysis of poverty indicators. *Social Indicators Research, 114*(1), 139–152.

Fonseca, R., Kapteyn, A., Lee, J., Zamarro, G., & Feeney, K. (2013). Financial and subjective Well-being of older Europans. In A. Börsch-Supan, M. Brandt, H. Litwin, & G. Weber (Hrsg.), *Active ageing and solidarity between generations in Europe: First results from SHARE after the economic crisis* (S. 137–148). Berlin/Boston: Walter de Gruyter.

Kingdon, G. G., & Knight, J. (2006). Subjective Well-being poverty vs. income poverty and capabilities poverty? *The Journal of Development Studies, 42*(7), 1199–1224.

Knesebeck, O. V. D., Wahrendorf, M., Hyde, M., & Siegrist, J. (2007). Socio-economic position and quality of life among older people in 10 European countries: Results of the SHARE study. *Ageing and Society, 27*(02), 269–284.

Laderchi, C. R. (2000). *The monetary approach to poverty: A survey of concepts and methods.* Queen Elizabeth House: University of Oxford.

Moser, P. (2006). *Einkommen und Vermögen der Generationen im Lebenszyklus. Eine Querschnitts-Kohortenanalyse der Zürcher Stuerdaten 1992–2003.* Zürich: Statistisches Amt des Kantons.

Pearlin, L. I., Aneshensel, C. S., & Leblanc, A. J. (1997). The forms and mechanisms of stress proliferation: The case of AIDS caregivers. *Journal of Health and Social Behavior, 38*(3), 223–236.

[13] The original quote in French and in German:

'La prévoyance professionnelle conjuguée avec l'assurance-vieillesse, survivants et invalidité permet à l'assuré de maintenir de manière appropriée son niveau de vie antérieur. '*Constitution Fédérale, Article 117.*

'Die berufliche Vorsorge ermöglicht zusammen mit der Alters-, Hinterlassenen- und Invalidenversicherung die Fortsetzung der gewohnten Lebenshaltung in angemessener Weise.' *Schweizerische Bundesverfassung, Artikel 117.*

Seifert, K., & Pilgram, A. (2009). *Vivre avec peu de moyens. La pauvreté des personnes âgées en Suisse*. Zürich: Edition Pro Senectute.
Sen, A. (1983). Poor, relatively speaking. *Oxford Economic Papers, 35*(4), 153–169.
Sen, A. (1984). Rights and capabilities. In *Resources, values and development* (S. 307–324). Oxford: Blackwell.
Sen, A. (1993). Positional objectivity. *Philosophy & Public Affairs, 22*(2), 126–145.
Sen, A., & Hawthorn, G. (1987). The standard of living. In *The standard of living*. Cambridge: Cambridge University Press.
Veenhoven, R. (2002). Why social policy needs subjective indicators. *Social Indicators Research, 58*(1–3), 33–46.

The manufacturer's authorised representative in the EU is Springer Nature Customer Service Centre GmbH, Europaplatz 3, 69115 Heidelberg, Germany. If you have any concerns regarding our products, please contact ProductSafety@springernature.com

Printed and bound by CPI Group (UK) Ltd, Croydon, CR0 4YY
30/09/2026
02225447-0010